CLASSICAL
THERMODYNAMICS
OF FLUID SYSTEMS
Principles
and Applications

CLASSICAL THERMODYNAMICS
OF FLUID SYSTEMS
Principles
and Applications

Juan H. Vera

Grazyna Wilczek-Vera

CRC Press
Taylor & Francis Group
Boca Raton London New York

CRC Press is an imprint of the
Taylor & Francis Group, an **informa** business

CRC Press
Taylor & Francis Group
6000 Broken Sound Parkway NW, Suite 300
Boca Raton, FL 33487-2742

First issued in paperback 2020

© 2017 by Taylor & Francis Group, LLC
CRC Press is an imprint of Taylor & Francis Group, an Informa business

No claim to original U.S. Government works

ISBN-13: 978-1-4987-6727-9 (hbk)
ISBN-13: 978-0-367-78258-0 (pbk)

Visit the Taylor & Francis Web site at
http://www.taylorandfrancis.com

and the CRC Press Web site at
http://www.crcpress.com

Contents

SECTION II Mixtures

SECTION III   Applications

SECTION IV   Special Topics

SECTION III   Applications

SECTION IV   Special Topics

SECTION V   Appendices

Preface

"Thinking Straight in the Age of Information Overload"[†]

Classical thermodynamics is the human attempt to capture, with a few principles and a coherent mathematical framework, some of the natural facts that we observe in daily life. The mathematical superstructure, successful as it is, should not be confused with reality. The preface of the first edition of the classical text *Thermodynamics* by G. N. Lewis and M. Randall compared the study of this subject with a visit to an ancient cathedral. These authors said that, as with a cathedral, the perfection of thermodynamics inspires respect and awe. It is easy to forget that this perfect structure is the product of human efforts and that all the natural human imperfections, present during its construction, have been erased by the passing of time. They remarked that even the rude and loud language that could have been used by the workers at the time of construction is now silenced. One can hear only hushed voices respectfully admiring the imposing form of the final work. There are as many ways of visiting a cathedral as there are ways of studying thermodynamics. Different people search for different qualities from the same reality. Even the same person may see different details on different visits. This may explain why so many books have been written on the subject of thermodynamics. So many indeed that we feel compelled to justify why we have decided to write yet another one. The reason, as we see it, is the need to provide a unified view of seemingly independent topics of interest in the study of chemical processes. The author of a text makes choices and decides what to emphasize and for whom to write. This text is written mostly, but not exclusively, for instructors, researchers, and process engineers. It assumes knowledge of calculus and of the physical states of matter. Although not a prerequisite, an undergraduate course in chemical engineering thermodynamics or physical chemistry is desirable. This text is written for those who want to understand the connections between different thermodynamic subjects related to fluid systems. It may be used as a text for a graduate course and even by beginners seeking a strong start. In some sense, it is a look at the finished building with complete disregard for the historical steps involved in its construction. In doing so, an attempt has been made not to leave a single corner of the building unexplored, discussing the minute details that, if left unattended, may give a feeling of insecurity once at higher levels.

In a deeper sense, this book is a voice of praise to two American pioneers who, from Berkeley, California, changed the world of chemical thermodynamics. These pioneers were Gilbert N. Lewis and John M. Prausnitz. G. N. Lewis "translated" into practical terms the highly mathematical work of another American scientist, J. Willard Gibbs. J. M. Prausnitz put the equations to work and showed us all the road ahead.

With the ready availability of computer applications that can easily derive equations, find roots of polynomials, solve simultaneous equations, and even optimize

---

† Subtitle of Daniel Levitin's book, *The Organized Mind*, Penguin, Toronto, Canada, 2014.

processes, there seems to be little sense in dedicating effort to unnecessary algebra. Thus, the emphasis is placed on the clarification of concepts by returning to the conceptual foundation of thermodynamics. Special effort is directed to the use of a simple nomenclature and algebra. Section I presents, in a rational sequence unperturbed by arbitrary approaches or temporary fashions, the structural elements of classical thermodynamics of fluid systems. Section II covers the treatment of mixtures, with special emphasis given to the physical reality represented by equations and concepts. Section III shows both the usefulness and the limitations of classical thermodynamics for the treatment of practical problems related to fluid systems. Sections IV and V include selected topics and appendices. So, from this perspective, this is a collection of five books in one. The essential concepts are presented in Section I, "Fundamentals," and in Section II, "Mixtures." The 18 chapters that constitute these two parts are written in a terse style, without examples or references. Following the principle of Ockham's razor, these short chapters provide the essential elements needed to relate equations, written in a simple nomenclature, to experiences in daily life. In these chapters, as in the rest of the book, symbols are defined locally. An effort has been made not to give the same symbol different meanings in different chapters and to always give a word of warning when the situation was unavoidable. The last two chapters of Section II are specialized discussions of the behavior of organic mixtures and aqueous electrolyte solutions, respectively. Applications are given in Section III. There is a clear change in tempo between Sections II and III. Section III is written in a relaxed style, including examples and references, in an attempt to show where and how thermodynamics can help to solve problems in the real world.

In Section IV, we have included diverse topics that may be of interest to researchers and advanced students. Chapters 23 and 24 provide short introductions to the flow of compressible fluids and statistical thermodynamics, respectively. The application of the elements of statistical thermodynamics to the understanding of the basis for equations of state and excess Gibbs energy models is given in Chapters 25 and 26. Closing the section on selected topics, Chapter 27 discusses the activity coefficients of individual ions. This is a topic of current interest as, mostly due to the advent of biotechnology, the field of aqueous electrolyte solutions has started a new life after a stagnation of almost a hundred years. Finally, Section V contains four appendices of practical use, an introduction to material balances, and clear step-by-step procedures for using the virial equation of state, the Peng–Robinson–Stryjek–Vera equation of state, and the group method for activity coefficients' analytical solution of groups with the Kojima–Tochigi parameter. Parameters for the use of these two latter effective tools were previously published in the *Canadian Journal of Chemical Engineering* and the *Journal of Chemical Engineering of Japan*, respectively. As these two journals can be difficult to access, the parameters are reproduced here with permission from the journals. In addition, an extended table of PRSV parameters for more than 800 compounds is presented for the first time. These parameters were evaluated by F. L. Figueira, L. Lugo, and C. Olivera-Fuentes.

We are indebted to Professors Alberto Arce, Murray Douglas, Maen Husein, and Jaime Wisniak, in alphabetical order, for dedicating their time to reading selected chapters and offering invaluable suggestions. Special thanks are due to our colleague and friend, Professor Claudio Olivera-Fuentes, who read the complete text. His comments led to major improvements in content, form, and clarity. Any residual errors are solely ours. In addition to his contribution to the editing of the text, Professor Olivera-Fuentes enriched the content of this book by agreeing to incorporate in Appendix C the table prepared by his research group with PRSV parameters for more than 800 compounds. Thanks are also due to Professors J. Gmehling and K. Tochigi for their help in obtaining permission to reproduce the ASOG-KT parameters.

I (G.W.-V.) had my initial formation in thermodynamics in two places: in the Laboratory of Thermodynamics of Nonelectrolytes at the University of Warsaw, Poland, led by Dr. Bogusław Janaszewski, and in the Institute of Technical Chemistry II at the University of Erlangen-Nürnberg, West Germany (the political map of Europe was different then). I am grateful to my colleagues in both groups, particularly to Dr. Paweł Oracz, who contributed to my scientific growth. Special thanks are reserved for Professor Stanisław Malanowski, supervisor of my doctoral dissertation.

I (J.H.V.) am grateful to John M. Prausnitz, Alfonso Frick, Fernando Aguirre, Edward W. Funk, Gerald Ratcliff, and Donald Patterson, who, at one time or another, transferred to me their enthusiasm for the study of thermodynamics. I am also indebted to two former McGill students, Gillian Holcroft and Cecilia Latkowska, who gave me well-organized copies of the class notes they took in my graduate course, Thermodynamics. Their notes may be considered the first preedition of this text.

We thank McGill University for providing us with a stimulating environment and our many students and coworkers at McGill for enriching our intellects. Finally, we are thankful to our sons, Alonso, Felipe, and Martin, and their families for their continuous emotional support in this long journey.

We expect that by a rational understanding of the strength and limitations of classical thermodynamics, the reader will be empowered to apply its principles to problems of the real world. Welcome to the realm of classical thermodynamics.

Juan H. Vera and Grazyna Wilczek-Vera
Montreal, Canada

Authors

Juan H. Vera is Professor Emeritus, Department of Chemical Engineering, McGill University, Montreal, Canada. He received his doctorate (Ing. Quim.) from Universidad F. Santa Maria, Chile, and his master of science in chemical engineering from University of California, Berkeley. He has coauthored two textbooks, a manual on copper metallurgy (in Spanish), and authored more than 210 refereed publications in international journals. He has an international patent on extraction of proteins and a Canadian patent on extraction of heavy metals.

Grazyna Wilczek-Vera was Faculty Lecturer, Department of Chemistry, McGill University, Montreal, Canada. She co-authored 58 refereed publications. She received both her doctorate in chemical sciences (with distinction) and master of science in chemistry (with distinction) from the University of Warsaw, Poland. She received the Principal's Prize for Excellence in Teaching, 2008, at McGill University.

Section I

Fundamentals

The enormous usefulness of mathematics in the natural sciences is something bordering in the mysterious and there is no rational explanation for it.[†]

[†] Remark made by Eugene (Jano) P. Wigner (Nobel Prize in Physics, 1963) at the Richard Couvant Lecture in Mathematical Sciences, University of New York, 1959, as quoted in Wigner, E. P., *Symmetries and Reflections: Scientific Essays*, Woodbridge, CT: Ox Bow Press, 1979.

1 Basic Concepts and Definitions

Thermodynamics is an experimental science. It builds upon the results of experimental measurements, and its conclusions must be verified experimentally. On the other hand, well-established and experimentally confirmed thermodynamic models liberate us from the necessity of continuously performing experiments for each problem encountered.

This work presents a rational look into the structure of classical thermodynamics, with some insight into its molecular aspects based on simple elements of statistical thermodynamics. First, with the purpose of having a common language, it is necessary to agree on some elemental definitions and use of terms. In this work, *matter* and *energy* are considered to be primitive concepts that, similarly to space and time, do not need further definitions. In addition, we ignore the possible interconversion between matter and energy and consider the material and energy balances to hold independently of one another.

CONCEPTS OF SYSTEM AND SURROUNDINGS

Like most disciplines, thermodynamics has its own vocabulary. A *system* is the region of the space that we decide to study. A system can have real or imaginary *boundaries* or walls. The *surroundings* are the regions of the space in contact with and affected by the system. The *universe* includes the system and its surroundings. This term should not be confused be with the astronomical universe, of which we know little.

One way to introduce a topic is to consider first the most complex system that one can imagine and then, beginning from a general treatment, come down to specific cases. This work chooses instead to start building up from simple systems and add variables as the additional complexity of real problems may require.

The *simple system* chosen is fixed in space and not affected by any force field. It is homogeneous and isotropic. In addition, it is not electrically charged, it is large enough to make surface effects negligible, and it is chemically inert. As the properties of a simple system isolated from its surroundings will not change with time, such a system is said to be at a state of *equilibrium*. The *intensive state* of a simple system at equilibrium can be completely described by the values of its intensive parameters, *pressure* and *temperature*, and the values of its *composition variables*. The description of the *extensive state* of a simple system requires the specification of one additional *extensive variable* as, for example, its *total mass* or its *total volume*. The ratio of the values of two extensive properties of a simple system gives a value independent of the size of the system, such as *density* or its reciprocal value, the *specific* or *molar volume*. Temperature and pressure are sometimes called *field* variables to distinguish them from specific or molar quantities. As stated above, the study of more complex systems is possible by the addition of appropriate variables.

THERMODYNAMIC PROCESS AND STATE AND PATH FUNCTIONS

In a *thermodynamic process*, the system passes from one equilibrium state to another. In this process, energy and matter may be exchanged between the system and its surroundings. In the study of simple systems, we distinguish three forms of energy.

The *internal energy, U,* is the energy that a system has because it is formed by atoms and molecules in movement. *Heat* and *work* are forms of energy in transfer that can only be detected at the walls of a system during a thermodynamic process. Once the energy has been transferred to or from the system, it just shows as a change in internal energy, without distinction of the form in which it was transferred. Work, W, is defined as the energy transferred between a system and its surroundings in such a way that all net effects in the surroundings can be exactly reproduced by the effect of lifting or letting fall weights. Heat, Q, is the energy transferred in such a way that its effects on the surroundings cannot be reproduced by lifting or letting fall weights.

Energy is considered positive when it enters the system from the surroundings as heat or work.

The *internal energy of a system* is a *state function*, that is, a function that only depends on the state of the system. Thus, the internal energy change between two well-defined equilibrium states is independent of the path followed by the process.

On the other hand, heat and work are *path functions*, that is, functions whose value depends on the particular process by which energy was transferred between the system and the surroundings.

At this point, it is important to make the distinction between a system in *equilibrium* and a system in *steady state*. If heat, for example, enters and leaves the system at the same rate, the properties of the system will not change with time. This system is said to be at steady state.

However, if such a system is *isolated from its surroundings*, its properties will evolve toward their value at equilibrium, different from the value they had at steady state.

Real boundaries or walls enclosing a simple system can be of different types. They may be *permeable* or not. Sometimes, they can be semipermeable and allow some species to go through, while being nonpermeable to other species. A system enclosed by a nonpermeable wall is said to be a *closed system*, as no matter can be exchanged with the surroundings.

Real walls can be *movable* or *rigid*, thus allowing or not changes in the volume of the system. Movable walls allow energy transfer between the system and the surroundings in the form of work.

Similarly, real walls can be *diathermal* or *adiabatic*. Diathermal walls allow the transfer of heat, while adiabatic walls do not. A system enclosed by nonpermeable, rigid, and adiabatic walls is said to be an *isolated system*, as it cannot exchange matter or energy with its surroundings.

CONCEPT OF A REVERSIBLE PROCESS

One of the *basic problems of classical thermodynamics* is the evaluation of the change in value of the thermodynamic properties of a system after a particular process occurs between two equilibrium states.

A way of seeing this problem more clearly is by imagining two simple systems, each isolated under its own conditions, which are enclosed together inside a nonpermeable, rigid, and adiabatic reservoir. Thus, the two subsystems are part of a composite isolated system. Under these initial conditions, no thermodynamic process is possible. The internal energy of the composite system is the sum of the internal energies of the two subsystems. As the composite system is isolated, its overall internal energy would not change if a thermodynamic process occurs between the two subsystems.

In a thought experiment, we can remove one of the internal constraints separating both subsystems. By making the wall separating both subsystems diathermal, heat will be transferred from one subsystem to the other. By making it movable, energy in the form of work will be exchanged between the subsystems. Similarly, if it is made permeable, matter will be exchanged between the subsystems. One can do these changes in any order and each of these options will cause the subsystems to evolve to different but well-defined equilibrium states.

One important point to notice here is that the change considered is between two equilibrium states of the subsystems. The rate at which the change takes place is not important. Thus, it is possible to choose an *infinitely slow process* in which all conditions are kept under control at all times and use this path for calculating the changes in the state properties of each subsystem between its initial and final equilibrium states. This imaginary infinitely slow or *quasi-static process* is referred to as a *reversible process*, and it is extremely useful for thermodynamic calculations.

COMPRESSIBLE AND INCOMPRESSIBLE FLUIDS

In the study of the thermodynamics of fluid systems, we introduce a pragmatic division between *compressible* (gases) and *incompressible* (liquids) fluids. The volume of a pure compound compressible fluid is a function of pressure and temperature, while the volume of incompressible fluids, to a good approximation, is only a function of temperature. No real fluid is totally incompressible.

Both types of fluids in a closed system can exchange energy in the form of heat with the surrounding. As we have neglected the action of force fields and surface effects, incompressible fluids in a closed system exchange energy in the form of mechanical work only in the case of stirring. Compressible fluids can also exchange energy in the form of mechanical work by movement of the walls. In open flow systems, liquids and gases can exchange energy as work by the movement of a shaft. This is called shaft work.

The simplest compressible system is the so-called ideal gas. Its equation of state, that is, the equation that gives the volumetric properties as a function of field variables temperature (T) and pressure (P), is

$$V^* = \frac{nRT}{P} \qquad\qquad (1.1a)$$

or

$$v^* = \frac{V^*}{n} = \frac{RT}{P} \qquad\qquad (1.1b)$$

where V^* is the total volume of the system, v^* is the molar volume of the system, n is the number of moles of gas, and R is the universal gas constant:

$$R = 8.3144598(48) \left[\frac{J}{mol \cdot K} \right] = 1545.4 \left[\frac{ft \cdot lbf}{lbmol \cdot {}^{\circ}R} \right]$$

The ideal gas, in contrast with real gases, does not condense into liquid by lowering the temperature or increasing the pressure. Real gases at temperatures below a characteristic value called a critical temperature, T_c, condense into liquids by an increase in pressure above a certain value, called the saturation pressure, that is fixed for each temperature. The critical pressure, P_c, is the value of the saturation pressure of a compound at its critical temperature, above which no vapor-liquid equilibrium is possible. The ideal gas behavior is a good approximation for the behavior of real gases at low pressures and high temperatures.

2 The First and Second Laws of Thermodynamics

PRINCIPLE OF CONSERVATION OF ENERGY IN A CLOSED SYSTEM

Let us consider a simple system that is closed and fixed in space. As for most practical applications, the interconversion between mass and energy occurring in atomic or nuclear processes can be safely ignored; we consider here that the principles of conservation of mass and energy can be separately stated. In this case, a change in the internal energy, U, of this simple system is due only to an exchange of energy between the system and its surroundings, be this exchange in the form of heat, Q, or work, W. According to the International Union of Pure and Applied Chemistry (IUPAC) [1] convention of signs, energy is positive when it enters the system, thus increasing its internal energy. Using the symbol Δ to indicate the difference of the final value of a function minus its initial value, we write

$$\Delta U = Q + W \qquad (2.1a)$$

The mass of a closed system is constant, so this equation can be written per unit mass using lowercase symbols for the specific values,

$$\Delta u = q + w \qquad (2.1b)$$

For a process between two equilibrium states that are infinitely close one to the other, this equation takes the form

$$du = \delta q + \delta w \qquad (2.1c)$$

or

$$dU = \delta Q + \delta W \qquad (2.1d)$$

Equations 2.1a–d are forms of the *first law of thermodynamics*. It is important to observe that the symbol d is used for the exact differential of a state function, while the symbol δ is used to indicate small quantities of heat and work exchanged during the incremental process. This difference in symbols is required to distinguish a change in the values of a state function from transfers of energy that are path dependent.

REVERSIBLE OR QUASI-STATIC PROCESSES

Real processes are difficult to describe in mathematical terms. However, we are normally interested in the change of value of a state function after a process occurs so we can imagine an alternate process that would take the system from the same initial state to the same final state as in the actual process. Thus, we consider that the change between the final and the initial equilibrium states presented by the system in the real process occurs following an infinitely slow (*reversible*) process. For a reversible process in a closed system, we write

$$dU = \delta Q_{rev} + \delta W_{rev} \tag{2.2a}$$

or

$$du = \delta q_{rev} + \delta w_{rev} \tag{2.2b}$$

As the internal energy u is a state function, du has the same value for both the real and the reversible processes, while $\delta q_{rev} \neq \delta q$ and $\delta w_{rev} \neq \delta w$. By eliminating du from Equations 2.1c and 2.2b, we get

$$\delta q + \delta w = \delta q_{rev} + \delta w_{rev} \tag{2.3}$$

or

$$\delta w - \delta w_{rev} = \delta q_{rev} - \delta q \tag{2.3a}$$

It is instructive, now, to visualize alternative processes of expansion or compression of a closed simple system.

Figure 2.1 depicts in a *P-V* diagram two states of the system characterized by the values P_1, V_1 and P_2, V_2. For a closed system, the volume is a state function of the temperature and pressure. Although one normally associates a smaller volume of a closed system with a higher pressure, the $(P_1, V_1) \rightarrow (P_2, V_2)$ process presented in the figure has been purposely drawn to show that at the higher pressure the volume of the system is larger. As the temperature axis is not shown in the diagram, the heat transfer that occurred in this process is hidden.

As shown in Figure 2.1, the same initial and final states of the system can be joined by different lines, each representing a different path or process. Each of these well-defined paths represents a different infinitely slow, that is, reversible, process.

Although there is no need to restrict the nature of the system or the shape of the vessel, it is easier to visualize it as a gas contained in a cylindrical vessel closed by a weightless movable piston that can be displaced by the difference between the external and internal forces. The cylinder and the piston have diathermal walls, and the complete set is immersed in a thermostatic air bath at a temperature we can control.

Consider that at a temperature, T, the system has a volume, V, and exerts a pressure, P, over the inside wall of the piston. The system will be in a state of equilibrium with its surroundings when the internal pressure, P, is equal to the external pressure, P_E,

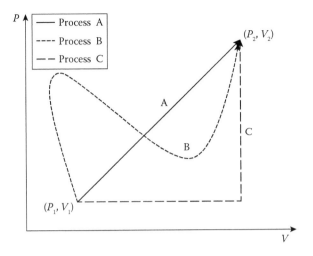

FIGURE 2.1   *P-V* diagram of $(P_1, V_1) \rightarrow (P_2, V_2)$ processes by three different paths, A, B, and C.

acting over the piston. If the piston has a weight, the pressure caused by this weight over the gas is added to the atmospheric pressure to get the value of P_E. To follow a reversible path, we may imagine that over the piston we have sand, the weight of which, acting over the surface area of the cylinder, plus the external atmospheric pressure, equals the internal pressure of the gas. By adding or removing sand from the top of the piston or changing slightly the temperature of the thermostatic bath, we can change the volume of the system following an arbitrarily chosen infinitely slow (reversible) path between the initial and final states that the system had in the real process.

The external force acting against the system is equal to the external pressure times the surface area of the piston, $F_E = P_E A$. An infinitesimal displacement, dL, of the piston times its surface area, A, gives the infinitesimal volume change $dV = A\, dL$ of the system. Thus, this infinitesimal volume change times the external pressure equals the work, done over the system by the external forces:

$$\delta W = -F_E\, dL = -P_E\, dV \qquad (2.4)$$

A negative sign is required for dV and dL to satisfy the convention of sign for work. For a compression, work is positive while the volume decreases. We can carry out a compression or an expansion process infinitely slowly by making the difference between the internal and the external pressure smaller and smaller for the small change in volume of the system. The reversible process is the imaginary limiting case when the infinitesimal expansion occurs against an external pressure, P_E, equal to the *internal pressure*, P, of the system. Hence, we identify the *reversible work* with

$$\delta W_{rev} = -P dV \qquad (2.5a)$$

or

$$W_{rev} = -\int_1^2 PdV \tag{2.5b}$$

The integral in Equation 2.5b is the surface area under the corresponding $P - V$ curve in Figure 2.1, and it differs for different reversible processes.

For the compression of a system under isothermal conditions, in a real process, the external pressure should be higher than the internal pressure. Thus, for a slight compression we write

$$P_E \geq P$$

The equal sign applies for the imaginary reversible compression. Multiplying both sides by $(-dV)$, which is a positive quantity, we obtain

$$\delta W \geq \delta W_{rev} \tag{2.6}$$

or, per unit mass,

$$\delta w \geq \delta w_{rev} \tag{2.7}$$

Thus, from Equation 2.3a we write

$$\delta w - \delta w_{rev} = \delta q_{rev} - \delta q \geq 0 \tag{2.8}$$

and

$$\delta q_{rev} = \delta q + (\delta w - \delta w_{rev}) \geq \delta q \tag{2.9}$$

CONCEPT OF ENTROPY, EXTREMUM PRINCIPLE, AND CLAUSIUS INEQUALITY

The difference between the actual and the reversible work can be interpreted as the *work lost* due to the irreversibility of the process, and it is given the symbol l_w:

$$w - w_{rev} = l_w \geq 0 \tag{2.10}$$

The difference between the reversible heat and the actual heat exchanged is called the *uncompensated heat*. As shown by Equation 2.8, both of these differences are the same and are always positive.

From Equations 2.2 and 2.5, we can now write the equation of the first law for a reversible process in a closed simple system as

$$dU = \delta Q_{rev} - PdV \tag{2.11}$$

We observe here that the reversible work is evaluated as the product of the *intensive parameter, P,* and change in the *extensive parameter, V*. A gradient in pressure is the potential for doing work, while the difference in volume is the effect shown in the system because work was done. The simple form of this equation suggests the introduction of a state function that for the reversible heat plays exactly the same role as the total volume for the reversible work. Thus, we introduce the concept of entropy through the following postulate:

> There exists an extensive state function S, called the entropy of the system, such that the heat exchanged in a differential reversible process can be expressed as the product of the intensive parameter T, the absolute temperature of the system, times the change in the value of this function in the process.

A gradient in temperature is the potential for heat exchange, while the difference in entropy is the effect of the reversible heat exchange.

Thus, we write

$$\delta Q_{rev} \equiv T dS \tag{2.12a}$$

or, per unit of mass, in specific or molar terms,

$$\delta q_{rev} \equiv T ds \tag{2.12b}$$

Consequently, for a closed system,

$$dU = T dS - P dV \tag{2.13a}$$

or

$$du = T ds - P dv \tag{2.13b}$$

Equation 2.13a, which relates entropy as a function of state to the other thermodynamic variables, is sometimes referred to as the *differential form of the fundamental equation of thermodynamics for a closed nonreacting system*. The purpose of rewriting the equations each time in terms of specific or molar quantities is to emphasize the difference between intensive parameters as P or T and extensive state functions as U, V, or S.

There is an important distinction between the state function entropy S and the state functions internal energy, U, and volume, V. While the values of the internal energy and the volume of the system change when the system exchanges heat or work with the surroundings in a reversible process, *the value of the entropy of the system is not affected by the exchange of reversible work*. By its definition, Equation 2.12, the entropy of the system only changes due to the exchange of reversible heat that would occur in an imaginary reversible process between the same initial and final states as in the real process. In a real process, the value of the entropy of the system also changes due to the action of (real) work, but this is only because of the work lost due to the irreversibility of the process, which, according to Equation 2.9, will have an effect on the magnitude of the reversible heat exchange involved.

There are other important conclusions to be drawn from the definition of entropy. From Equations 2.9 and 2.12a,

$$dS \geq \frac{\delta Q}{T} \qquad (2.14)$$

This relation, known as the *Clausius inequality*, can be converted into equality by adding to the right-hand side the term δS_{irr}, which is always positive for any irreversible process and equal to 0 for a reversible process.

$$dS = \frac{\delta Q}{T} + \delta S_{irr} \qquad (2.15)$$

Thus, a reversible adiabatic process is isentropic (S = constant), due to the definition of the entropy function by Equation 2.12. If the process is irreversible and adiabatic, Equation 2.15 shows that the entropy of the system will always increase.

The entropy of a system decreases in an irreversible process if the heat lost by the system is large enough to compensate for the increase in entropy due to the irreversibility of the process. Similarly, an irreversible process can be isentropic if the heat lost in the process exactly compensates for the increase in entropy due to irreversibility. Finally, a conclusion of great importance is that if an irreversible process occurs in an isolated system, the entropy of some of the subsystems may decrease, but the entropy of other subsystems must then increase in such an amount that the entropy of the overall isolated system will always increase. As the isolated system has constant internal energy, constant volume, and constant total mass, the above statement is expressed as

$$dS_{U,V,m} \geq 0 \qquad (2.16)$$

This *extremum principle*, generally identified as the *second law of thermodynamics*, is a natural consequence of the convention of the sign assigned to heat exchanged and the definition of an entropy change by Equation 2.12, more clearly seen in the form of Equation 2.15. When two bodies exchange heat, the entropy increase of the body at the lower temperature will always be larger than the entropy decrease of the body at the higher temperature. This is so because the same amount of heat exchanged is divided by a larger value of absolute temperature in the case of the hotter body. As all real processes are irreversible, if we accept that our physical universe is an isolated system, the conclusion is that the entropy of the universe increases with each and every process that occurs. From this general statement, it can be shown that natural processes occur in one and only one direction. Thus, for example, if a hot body is put in contact with a cold body, heat will always flow from the hot body to the cold body and never in the opposite direction unless there are other changes in the surroundings.

Confirmation of Agreement with the Directional Nature of Heat Flow

Consider a system, isolated from its surroundings, which is formed by two subsystems that can only exchange energy between each other in the form of heat. Designating the total system by t and the subsystems by α and β, we have

$$dS^t_{U,V,m} = (dS^\alpha + dS^\beta) \geq 0$$

If there is no work done, according to Equation 2.1c the heat exchanged in a real process is equal to the change in internal energy, and thus it is independent of the nature of the path followed. The same amount of energy can be exchanged between the two subsystems following a reversible path. Hence,

$$\frac{\delta Q^\alpha}{T^\alpha} + \frac{\delta Q^\beta}{T^\beta} \geq 0$$

As the total system is isolated, it follows that, $\delta Q^\alpha = -\delta Q^\beta$, or

$$\left[\frac{1}{T^\alpha} - \frac{1}{T^\beta}\right]\delta Q^\alpha \geq 0$$

If $\delta Q^\alpha \geq 0$, this means that heat enters the system α. Hence, to satisfy the above inequality, necessarily, $T^\beta \geq T^\alpha$ and vice versa. The process will continue until both temperatures are equal.

Heat can be transferred from a "cold" system to a "hot" system, but for this to occur, other changes in the surroundings, causing an overall increase in entropy of the universe, must take place. If a system decreases its entropy in an irreversible process, one or more parts of its surroundings increase their entropy in a larger amount, so the net change of entropy of the universe is always positive. This directional evolution of the universe may be referred to as the *arrow of time*.

Heat can be passed from a cold source to a hot sink by a cyclic process, as it happens in a refrigerator, in a home air conditioner, or in a heat pump. *In a cyclic process, after exchanging energy in the form of heat or work with the surroundings, the system returns to its initial state at the end of each cycle.* Thus, all net changes that mark that a process has occurred are to be found in the surroundings. We observe here that a compressor or pump transfers energy in the form of work to the system, and thus its work is assigned a positive sign, while a turbine extracts energy from the system, and thus its work, with respect to the system, is assigned a negative sign. Cyclic processes, employing a working fluid, are commonly used not only to transfer heat from a cold source to a hot sink, but also to produce useful work from heat energy. Both of these cases, which provide examples of the practical use of the concept of entropy, are considered below. For the analysis of cyclic devices, we are not concerned about the temperature of the circulating fluid. All we need to know is that for the fluid to receive heat from a source, its highest temperature should be below the temperature of the source, and to give heat to a sink, its lowest temperature should be above the temperature of the sink. These temperature differences required for heat transfer are causes of further irreversibility.

REFRIGERATORS, AIR CONDITIONERS, AND HEAT PUMPS

Most refrigerators, air conditioners (including dehumidifiers), and heat pumps use the *vapor–compression refrigeration cycle* shown in Figure 2.2.

In each cycle, the working fluid passes through four units. It enters the refrigeration unit as a liquid, it receives an amount of heat q_L from the cold source at a constant low temperature T_L, and it evaporates. The vapor is then compressed in an adiabatic compressor, raising its pressure and temperature. The hot compressed vapor is sent to a heat exchanger where it condenses giving an amount of heat q_H to a hot sink at a constant high temperature T_H. The liquid is then passed through an adiabatic valve or through an adiabatic turbine to produce some work, and closes the cycle reentering the refrigeration unit. To visualize this, in a home refrigerator the cold source is the interior of the fridge, while the air exterior to the refrigeration unit serves as a hot sink.

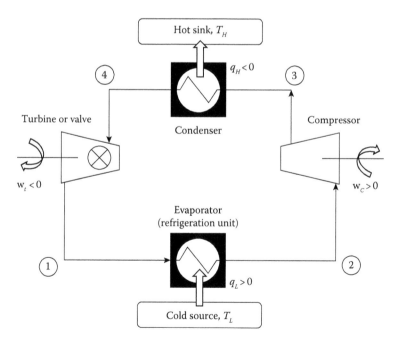

FIGURE 2.2 Refrigeration device (signs for heat and work refer to the circulating fluid). Note that T_H and T_L refer to the temperatures of the hot sink and cold source, respectively, and not to the temperatures of the circulating fluid.

According to the first law, Equation 2.2,

$$\Delta u_{cycle} = q_{net} + w_{net} = 0$$

After each cycle, there is no change in internal energy, or in entropy, for the circulating fluid and the energy received by the fluid in the form of work and as heat from the cold source is passed as heat to the hot sink.

$$|q_H| = |q_L| + |w_{net}| \tag{2.17}$$

Equation 2.17 shows that the absolute value of the heat released by a refrigerator to the surroundings is larger than the absolute value of heat received from the cold source. This is why air conditioners and heat pumps have their high temperature heat exchanger outside the space that is being cooled. Otherwise, the unit would act as a heater. A refrigerator or a dehumidifier heats the room where it operates. For the surroundings, the signs for heat and work are reversed with respect to the signs of these quantities for the working fluid; consequently, by using their absolute values the above relation is valid for both the circulating fluid and the surroundings. The performances of the devices used for transferring heat from a cold source to a hot sink are measured by the magnitude of ratio of the value of the desired effect to the value of the required duty. This ratio is known as the coefficient of performance (*COP*), and with the exception of heat pumps, it takes the form

$$(COP) = \frac{|q_L|}{|w_{net}|} \tag{2.18}$$

The comparison of performance of different devices of the same kind is better done by referring their (*COP*) to a common basis, namely, the (*COP*) of a reversible device operating between the same temperatures of the source and the sink of heat:

$$(COP)_{rev} = \frac{|q_L|_{rev}}{|w_{net}|_{rev}}$$

This value is obtained recalling that there is no entropy change due to the action of reversible work. Thus, the only places where an entropy change can be found at the end of a cycle are the cold source and the hot sink. According to the second law of thermodynamics, for the reversible process to be possible, the following condition must hold:

$$\Delta s_{U,V,m} = \Delta s_H + \Delta s_L = \frac{|q_H|_{rev}}{T_H} - \frac{|q_L|_{rev}}{T_L} \geq 0 \tag{2.19a}$$

Combining this expression with Equation 2.17 for the reversible process, we obtain

$$\Delta s_{U,V,m} = |q_L|_{rev}\left[\frac{1}{T_H} - \frac{1}{T_L}\right] + \frac{|w_{net}|_{rev}}{T_H} \geq 0 \tag{2.19b}$$

If no net work is put into the cycle, that is, $|w_{net}|_{rev} = 0$, as $T_L < T_H$, the entropy change of the universe in one cycle would be negative, indicating the impossibility of carrying out this process. Although the reversible work, as such, does not produce a change in entropy, its conversion into heat released, shown in Equation 2.17, increases the entropy of the universe in each cycle.

From Equation 2.19, the minimum work required for a reversible cycle to pass heat from a cold source to a hot sink is given by

$$|w_{net}|_{rev} = |q_L|_{rev} \left[\frac{T_H - T_L}{T_L} \right]$$

Rearranging, we obtain

$$(COP)_{rev} = \frac{|q_L|_{rev}}{|w_{net}|_{rev}} = \left[\frac{T_L}{T_H - T_L} \right] \tag{2.20}$$

Therefore, the second law efficiency of a refrigerating device is given by

$$Refrigerator\ Second\ Law\ Efficiency = \left[\frac{|q_L|}{|w_{net}|} \right]_{actual} \frac{T_H - T_L}{T_L} \tag{2.21}$$

For a heat pump, on the other hand, the desired effect is the heat delivered to the hot sink, and the coefficient of performance is defined as

$$(COP) = \frac{|q_H|}{|w_{net}|} \tag{2.22}$$

Hence, repeating the same algebra, we obtain

$$(COP)_{rev} = \frac{|q_H|_{rev}}{|w_{net}|_{rev}} = \left[\frac{T_H}{T_H - T_L} \right] \tag{2.23}$$

and

$$Heat\ Pump\ Second\ Law\ Efficiency = \left[\frac{|q_H|}{|w_{net}|} \right]_{actual} \frac{T_H - T_L}{T_H} \tag{2.24}$$

HEAT ENGINES

One of the major steps made by humanity toward modern society was the conversion of heat into useful work using cyclic processes. Heat coming from wood, coal, oil, natural or liquefied gas, or even atomic energy can be used to replace human effort in doing work. The second law, in this case, says that it is impossible to convert *all* the heat into work in a cyclic process. This conclusion comes directly from the considerations given above.

Let us discuss a reversible process to see this effect. As the reversible work does not contribute to entropy changes, the single effect observed when converting all the heat into work in a cycle is that the hot source gives up an amount of heat, q_H, at a constant high temperature T_H, thus decreasing its entropy. This decrease in entropy would be the change in entropy of the universe caused by the process, thus making it impossible. On the other hand, if part of the heat received by the working fluid from the hot source would be given away to a cold sink, increasing the entropy of the cold sink by a larger amount, the process would become possible.

A *cyclic heat engine*, as shown in Figure 2.3, uses a working fluid operating in a cycle that is effectively the reverse of the one described for the vapor–compression refrigeration devices.

In this case, a liquid is vaporized in a high-temperature, high-pressure evaporator and passed through a turbine to produce work. The low-pressure vapor coming from the turbine is sent to a lower-temperature condenser. The liquid from the condenser is then pumped back to the evaporator.

We consider that the turbine and the pump operate reversibly and adiabatically, and thus do not change the entropy of the operating fluid or the entropy of its surroundings. Similarly, we consider that the hot source gives an amount of heat, q_H, at a constant high temperature, T_H, and the cold sink receives an amount of heat, q_L, at a constant lower temperature, T_L.

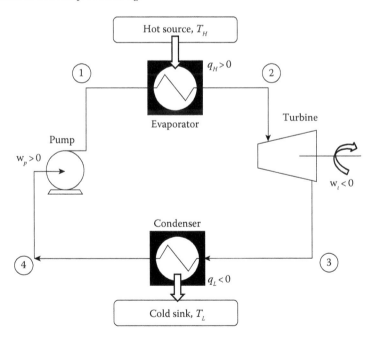

FIGURE 2.3 Heat engine (signs for heat and work refer to the circulating fluid). Note that T_H and T_L refer to the temperatures of the hot source and cold sink, respectively, and not to the temperatures of the circulating fluid.

Equation 2.17 expresses the first law for the working fluid in one cycle. As necessarily some heat must be lost to a cold sink to make the process feasible, it is of interest to know what would be the maximum possible efficiency of such a cyclic engine. As the heat from the hot source is the one that has a cost and the work obtained is the desired effect, the *thermal efficiency* of a cycle, η_{th}, is defined as the ratio of the net work obtained to the amount of heat received:

$$\eta_{th} = \frac{|w|}{|q_H|} \tag{2.25}$$

Combining with Equation 2.17, we obtain

$$\eta_{th} = 1 - \frac{|q_L|}{|q_H|}$$

Hence, the greater the amount of heat passed to the cold sink, the lower is the efficiency. As the second law indicates that the value of q_L cannot be zero, the changes in entropy of the hot source and cold sink are used to determine the limiting value for the ratio $|q_L/q_H|$.

For this cycle, in a reversible engine it is necessary to have

$$\Delta s_{U,V,m} = \Delta s_H + \Delta s_L = -\frac{|q_H|_{rev}}{T_H} + \frac{|q_L|_{rev}}{T_L} \geq 0$$

or

$$\frac{|q_L|_{rev}}{|q_H|_{rev}} \geq \frac{T_L}{T_H}$$

Thus, the maximum possible efficiency of a cyclic thermal engine is

$$\eta_{th} \leq \left(1 - \frac{T_L}{T_H}\right) \tag{2.26}$$

As an illustration, we consider that the temperature of the hot source is about 700 K and heat is released to the atmosphere, which is at about 300 K. The maximum possible efficiency of such a reversible (imaginary) thermal engine will not exceed 57%. As the temperature of the atmosphere is not a controllable variable, one way to increase the efficiency of the cyclic engine is to increase the temperature of the hot source. Therefore, in practice, the circulating fluid is evaporated and the vapor is then superheated in contact with a high-temperature hot source.

An interesting aspect to ponder is whether it would be possible to have a device that converts all the heat into work. The answer is that a noncyclic reversible device may seem to approach this total conversion but will never reach this limit.

Consider a gas contained in a cylinder–piston device that has a heater at its base. The complete system, including the movable piston, is thermally insulated from the surroundings. As heat is added to the system, the gas expands and moves the piston against the atmospheric pressure. At first, it would seem that all heat added was converted into work. However, as the volume of the gas has changed in the process, while the pressure has been kept constant, its temperature has also changed (did it increase or decrease?). So, some of the energy given by the heater is accumulated by the gas. The state of the gas at the end of the process is not the same as at the beginning. Any attempt to return the gas to its initial state would require to perform work on the system or release heat.

REFERENCE

1. International Union of Pure and Applied Chemistry, Physical and Biophysical Chemistry Division. 2007. *Quantities, Units and Symbols in Physical Chemistry*, 3rd ed. Cambridge, UK: RSC Publishing, section 2.11.

3 Conservation of Energy in an Open Flow System. Definition of Enthalpy

In a completely rigorous line of thought, it can be argued that principles of classical thermodynamics are only applicable to systems at equilibrium. In fact, there is a separate branch of thermodynamics dedicated to nonequilibrium cases. However, being aware of the approximation involved, there is one simple application of the energy balance that is worth discussing here.

CLOSED SYSTEM MOVING IN SPACE

First, consider the case of a closed system of mass, m, moving in space. In this case, we distinguish three components of its total energy, E: its internal energy, U, its kinetic energy, E_K, with respect to an external point of reference, and its potential energy, E_P, with respect to a fixed plane of reference. Thus,

$$E = U + E_K + E_P \tag{3.1}$$

or, per unit mass,

$$e = u + e_k + e_p \tag{3.2}$$

OPEN FLOW SYSTEM AT UNSTEADY STATE

Now, let us consider a control volume in which fluid enters by several streams, i, and fluid leaves by several streams, j. Per unit time, indicated by a dot over the symbol, in each stream i enters a mass $\delta \dot{m}_i$ of fluid and in each stream j exits a mass $\delta \dot{m}_j$. In addition, per unit time, the system exchanges an amount of heat, $\delta \dot{Q}$, and an amount of work, $\delta \dot{W}$, with the surroundings. Thus, the change of energy in the control volume, per unit time, is

$$d\dot{E} = \sum_{in} (u + e_K + e_P)_i \, \delta \dot{m}_i - \sum_{out} (u + e_K + e_P)_j \, \delta \dot{m}_j + \delta \dot{Q} + \delta \dot{W} \tag{3.3}$$

For the work, we can distinguish two components: the work associated with a moving shaft, or shaft power $\delta \dot{W}_s$, and the work associated with the fluid entering it in each stream against a pressure P_i and with the fluid going out in each stream against a pressure P_j. Per unit mass, the work associated with the movement of the fluid is equal to the product of its specific volume and the prevailing pressure at the point of interest. This can be visualized considering that the force is the pressure times the surface area over which it acts, and when multiplying by the displacement necessary to move a unit of mass, the product of the surface area times the displacement gives the specific volume of the fluid. Hence,

$$\delta \dot{W} = \delta \dot{W}_s + \left(\sum_{in} (Pv)_i \delta \dot{m}_i - \sum_{out} (Pv)_j \delta \dot{m}_j \right) \tag{3.4}$$

Combining both equations, we write

$$d\dot{E} = \sum_{in} (h + e_K + e_P)_i \delta \dot{m}_i - \sum_{out} (h + e_K + e_P)_j \delta \dot{m}_j + \delta \dot{Q} + \delta \dot{W}_s \tag{3.5}$$

DEFINITION OF ENTHALPY

In Equation 3.5, we have used the symbol h for the sum of the internal energy u and the product Pv:

$$h = u + Pv \tag{3.6a}$$

or

$$H = U + PV \tag{3.6b}$$

All terms of the right-hand side of Equation 3.6 are a function of the state of the system. Thus, the function h *is also a state function and it is called enthalpy. Obviously, the definition of enthalpy is independent of the movement of a fluid, so it is equally valid for a closed system.*

OPEN FLOW SYSTEM AT STEADY STATE

One important application of Equation 3.5 is the case of a steady state in which the mass and energy of the system, as well as all flows of mass and energy, are constant with time. In this case, we remove the symbol δ that was introduced to account for possible variations of the flows with time. Thus, setting $d\dot{E} = 0$ in Equation 3.5, for a system at steady state we write

$$\sum_{out} (h + e_K + e_P)_j \dot{m}_j - \sum_{in} (h + e_K + e_P)_i \dot{m}_i = \dot{Q} + \dot{W}_s \tag{3.7}$$

In this case, the total mass entering the system is equal to the total mass leaving the system:

$$\sum_{in} \dot{m}_i = \sum_{out} \dot{m}_j \tag{3.8}$$

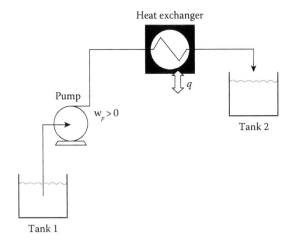

FIGURE 3.1 Open flow system.

A common situation encountered in engineering is to have a system with only one stream entering and one stream leaving, as the one shown in Figure 3.1. For this case, at steady-state operation, we write

$$\dot{m}_i = \dot{m}_j = \dot{m} \tag{3.9}$$

Hence, the energy balance equation takes the form

$$(\Delta h + \Delta e_K + \Delta e_P)\dot{m} = \dot{Q} + \dot{W}_s \tag{3.10}$$

where Δ indicates the difference between the value of the property at the exiting stream minus the value of the property at the intake stream. Dividing both sides by $\dot{m}$,

$$\Delta h + \Delta e_K + \Delta e_P = q + w_S \tag{3.11}$$

All quantities in this equation are expressed per unit mass of fluid. The kinetic energy term, written in terms of the linear velocity υ of the fluid, that is, the volumetric flow rate over the surface area of the duct in which it flows, takes the form

$$\Delta e_K = \frac{(\upsilon_{out})^2 - (\upsilon_{in})^2}{2} \tag{3.12}^\dagger$$

Similarly, the potential energy term, written in terms of the elevation of the point over an arbitrarily set reference plane, Z, takes the form

$$\Delta e_P = g(Z_{out} - Z_{in}) \tag{3.13}$$

---

† From equations for momentum transport, it has been shown [1] that when there is a well-developed velocity profile in a cross section area, the kinetic energy term has the form $\Delta e_k = \frac{1}{2}\Delta\frac{\langle \upsilon^3 \rangle_{ave}}{\langle \upsilon \rangle_{ave}}$. For most cases, the kinetic energy term represents a small contribution; thus, Equation 3.12 is commonly used.

where g is the acceleration of gravity. These two terms are usually negligible in comparison with the enthalpy change, and they are normally omitted. Equation 3.7 without these terms is sometimes referred to as the *enthalpy balance*. For the case of negligible contribution of the kinetic and potential energy terms, Equation 3.11 reduces to

$$\Delta h = q + w_S \tag{3.14}$$

Equation 3.14 has a direct similitude with Equation 2.1, the first law for closed systems. It is particularly useful for the analysis of compressors and turbines dealing with gases or steam, and also in the study of distillation columns, evaporators, and concentration tank operations.

MECHANICAL ENERGY BALANCE

Finally, a commonly used equation for the treatment of flow of incompressible fluids in pipes is the so-called mechanical energy balance (MEB). This equation can be obtained from fluid mechanics by integration of the equation of motion without any need of passing through the energy balance. To obtain this equation from the energy balance, we observe that the term Δu is the change in internal energy of a unit of mass of the fluid passing, as a closed system, from the state at the inlet to the state at the outlet. In this process, the unit mass of fluid exchanged a net amount of heat q. As the fluid considered here is incompressible, the molar volume is constant and there is no reversible work of expansion or compression, so the net energy due to work received by the system, according to Equation 2.10, is equal to the lost work, $w = l_w$, appearing from the irreversibility effects caused by friction.

$$\Delta h = \Delta u + \Delta(Pv) = (q + l_w) + v\Delta P \tag{3.15}$$

Replacing Equations 3.12, 3.13, and 3.15 into Equation 3.11, we write the MEB as

$$v(P_{out} - P_{in}) + \frac{\dot{v}_{out}^2 - \dot{v}_{in}^2}{2} + g(Z_{out} - Z_{in}) - w_s + l_w = 0 \tag{3.16a}$$

or

$$v\Delta P + \frac{\Delta \dot{v}^2}{2} + g\Delta Z - w_s + l_w = 0 \tag{3.16b}$$

As stated above, the linear velocity of the fluid is equal to the volumetric flow rate of the fluid divided by the surface area of the duct. Thus, with the exception of l_w, all terms in this equation are measurable and the value of l_w can be back-calculated. This has been done for many typical flow configurations of practical interest, and engineering correlations are available to calculate the value of l_w for different designs.

While the energy balance for an open flow system, Equation 3.11, is valid for both compressible and incompressible fluids, the MEB as written, Equation 3.16, is only valid for incompressible fluids. The fluid mechanics of compressible fluids is complex, and it is better treated through equations pertaining to transport phenomena. The fact that the molar volume of compressible fluids is a function of not only temperature but also pressure creates a completely new reality when the fluid flows.

At ordinary (room) temperatures and pressures, real gases, with the exception of hydrogen, helium, and neon, cool down upon expansion. The three latter gases experience the same effect, but only at much lower temperatures. The steady flow characteristics of a compressible fluid in a horizontal duct of a constant cross surface area depend on whether the velocity at the entrance is subsonic or supersonic. When the gas enters at subsonic velocity, the fluid expands as the pressure drops and the temperature decreases. The entropy and velocity increase until reaching the velocity of sound at which the entropy is at its maximum. If this maximum length is exceeded, chocking occurs and the entrance conditions cannot be maintained. Chocking is a discontinuity; there is an abrupt increase in pressure, temperature, and density that reduces the downstream velocity to the subsonic range.

Supersonic flow at the entrance of the duct can be accomplished with a special nozzle. In this case, there is a compression of the gas as it flows. Thus, there is an increase in pressure, density, temperature, and entropy down the duct. The velocity decreases. Stable flow conditions can only be maintained if the length of the duct is less than or equal to a critical length at which a sonic discharge velocity is produced while the entropy is at its maximum value. If the duct is longer, a shock wave will be produced with an abrupt increase in pressure, temperature, and density that reduces the downstream velocity to subsonic values. A further increase of the duct length will cause the shock wave to move back toward the entrance, traveling at a velocity larger than sonic. When it reaches the entrance, only subsonic velocities will exist in the duct. Beyond this, the shock wave will move into the nozzle feeding the duct.

The thermodynamics of compressible fluid flow requires elements of equations of state, which is covered in Chapter 5. A more detailed thermodynamic treatment of adiabatic flow of gases is presented in Chapter 23.

USE OF THE ENTHALPY FUNCTION IN CLOSED SYSTEMS

Although the enthalpy function defined by Equation 3.6 is particularly useful for the treatment of open flow systems, it is also applicable for the study of closed systems undergoing changes at constant pressure. If the pressure of the closed system is constant, according to Equation 2.4 the expansion or compression work exchanged with the surroundings is directly given by the product of the pressure times the change in volume of the system. Thus, in this case, we can write Equation 2.2 as

$$du_P = \delta q + \delta w_P = \delta q - P\,dv$$
$$d(u + Pv)_P = \delta q$$
$$dh_P = \delta q$$

Thus, for a constant pressure process in a closed system, we write

$$\Delta h_P = q \tag{3.17}$$

As the heat exchanged in a process is a measurable quantity, Equation 3.17 provides values for enthalpy changes in mixing processes.

REFERENCE

1. Bird, R.B. 1957. The equations of change and the macroscopic mass, momentum, and energy balances. *Chem. Eng. Sci.* 6: 123–181.

4 The Algebra of State Functions. The Helmholtz and Gibbs Functions

Until now, we have encountered four state functions of a system characterized by the measurable variables: absolute pressure, P; absolute temperature, T; and the composition of the system expressed by the number of moles of each compound present, n_i. These state functions are the total volume, V; the internal energy, U; the entropy, S; and the enthalpy, H. Two more state functions are used in thermodynamics. These are the *Helmholtz function*,

$$A = U - TS \tag{4.1}$$

and the *Gibbs function*,

$$G = H - TS \tag{4.2}$$

THERMODYNAMIC FUNCTIONS IN OPEN SYSTEMS. THE CHEMICAL POTENTIAL

The systems studied in the previous chapters were closed, so changes in composition were not considered. Thus, the dependence of state functions on the number of moles of each compound was left implicit. State functions have exact differentials, meaning that the value obtained by integration of their differentials between two states is independent of the path followed. We observe that Equation 2.13 suggests that the natural variables for the function U, internal energy, in a differential expansion should be S and V. If we write the total differential of the internal energy, considering its dependence on composition, we have

$$dU = \left(\frac{\partial U}{\partial S}\right)_{V,n} dS + \left(\frac{\partial U}{\partial V}\right)_{S,n} dV + \sum_i^c \left(\frac{\partial U}{\partial n_i}\right)_{S,V,n_{j \neq i}} dn_i \tag{4.3}$$

The number of moles of each compound may change because the system is open to the surroundings and mass enters or leaves the system, or because there are chemical reactions transforming moles of some compounds into moles of some other compounds. For a closed nonreacting system, the last term of the right-hand side cancels out; thus, comparing with Equation 2.13, we conclude that

$$T = \left(\frac{\partial U}{\partial S}\right)_{V,n} \tag{4.4}$$

and

$$-P = \left(\frac{\partial U}{\partial V}\right)_{S,n} \tag{4.5}$$

In addition, for simplicity, we introduce the symbol μ_i for the partial derivative of the total internal energy with respect to n_i at constant $n_{j\neq i}$, S, and V, and call it the *chemical potential*:

$$\mu_i \equiv \left(\frac{\partial U}{\partial n_i}\right)_{S,V,n_{j\neq i}} \tag{4.6}$$

Hence, we rewrite Equation 4.3 as

$$dU = T\,dS - P\,dV + \sum_i^c \mu_i dn_i \tag{4.7}$$

Equation 4.7 is sometimes referred to as the *differential form of the fundamental equation of thermodynamics*. For a system of constant composition, Equation 4.7 reduces to Equation 2.13. The composition of the system can change because it is open or because chemical reactions occur.

For the enthalpy of a system with changes in composition during a process, from Equation 3.6,

$$dH = dU + P\,dV + V\,dP \tag{4.8}$$

Thus, using Equation 4.7 we can write

$$dH = T\,dS + V\,dP + \sum_i^c \mu_i\,dn_i \tag{4.9}$$

Again, from the expression of the exact differential of H, in terms of S, P, and n, we write

$$dH = \left(\frac{\partial H}{\partial S}\right)_{P,n} dS + \left(\frac{\partial H}{\partial P}\right)_{S,n} dP + \sum_i^c \left(\frac{\partial H}{\partial n_i}\right)_{S,P,n_{j\neq i}} dn_i \tag{4.10}$$

So, we obtain

$$T = \left(\frac{\partial H}{\partial S} \right)_{P,n} \tag{4.11}$$

$$V = \left(\frac{\partial H}{\partial P} \right)_{S,n} \tag{4.12}$$

$$\mu_i = \left(\frac{\partial H}{\partial n_i} \right)_{S,P,n_{j \neq i}} \tag{4.13}$$

A similar exercise with Equation 4.1 for A and Equation 4.2 for G gives

$$-S = \left(\frac{\partial A}{\partial T} \right)_{V,n} \tag{4.14}$$

$$-P = \left(\frac{\partial A}{\partial V} \right)_{T,n} \tag{4.15}$$

$$\mu_i = \left(\frac{\partial A}{\partial n_i} \right)_{T,V,n_{j \neq i}} \tag{4.16}$$

$$-S = \left(\frac{\partial G}{\partial T} \right)_{P,n} \tag{4.17}$$

$$V = \left(\frac{\partial G}{\partial P} \right)_{T,n} \tag{4.18}$$

$$\mu_i = \left(\frac{\partial G}{\partial n_i} \right)_{P,T,n_{j \neq i}} \tag{4.19}$$

Thus, to gain perspective, we group the expressions involving the same property as

$$T = \left(\frac{\partial U}{\partial S} \right)_{V,n} = \left(\frac{\partial H}{\partial S} \right)_{P,n} ; \qquad \text{(4.4) and (4.11)}$$

$$-P = \left(\frac{\partial U}{\partial V} \right)_{S,n} = \left(\frac{\partial A}{\partial V} \right)_{T,n} ; \qquad \text{(4.5) and (4.15)}$$

$$V = \left(\frac{\partial H}{\partial P}\right)_{S,n} = \left(\frac{\partial G}{\partial P}\right)_{T,n} ; \qquad \text{(4.12) and (4.18)}$$

$$-S = \left(\frac{\partial A}{\partial T}\right)_{V,n} = \left(\frac{\partial G}{\partial T}\right)_{P,n} ; \qquad \text{(4.14) and (4.17)}$$

In the above differentiations, the numbers of moles of all the compounds are held constant, so the same relations can be written in terms of the molar properties of the system.

For the chemical potential, Equations 4.6, 4.13, 4.16, and 4.19 give

$$\mu_i \equiv \left(\frac{\partial U}{\partial n_i}\right)_{S,V,n_{j \neq i}} = \left(\frac{\partial H}{\partial n_i}\right)_{S,P,n_{j \neq i}} = \left(\frac{\partial A}{\partial n_i}\right)_{T,V,n_{j \neq i}} = \left(\frac{\partial G}{\partial n_i}\right)_{P,T,n_{j \neq i}}$$

We observe that in all cases, the natural or "canonical" variables for U are S and V, for H are S and P, for A are T and V, and for G are P and T. The main use of the above relations is for the derivation of the so-called *Maxwell's relations*, which are the stepping-stones for the evaluation of the thermodynamic functions from measurable variables.

MAXWELL'S RELATIONS

For a state function of several variables, the order of partial differentiation with respect to two of the variables is immaterial. This is, in fact, a property of all functions having an exact differential. Hence, in general, for a function X of variables x_1 and x_2, among other variables that are held constant, we write

$$\frac{\partial}{\partial x_1}\left[\left(\frac{\partial X}{\partial x_2}\right)_{x_1}\right]_{x_2} = \frac{\partial}{\partial x_2}\left[\left(\frac{\partial X}{\partial x_1}\right)_{x_2}\right]_{x_1}$$

Thus, taking the variables V and S for the function U, we have

$$\frac{\partial}{\partial V}\left[\left(\frac{\partial U}{\partial S}\right)_{V,n}\right]_{S,n} = \frac{\partial}{\partial S}\left[\left(\frac{\partial U}{\partial V}\right)_{S,n}\right]_{V,n}$$

or from Equations 4.4 and 4.5,

$$\left[\frac{\partial T}{\partial V}\right]_{S,n} = -\left[\frac{\partial P}{\partial S}\right]_{V,n} \qquad \text{(4.20)}$$

Similarly, from the state function H, with variables P and S, from Equations 4.11 and 4.12, we write

$$\left[\frac{\partial T}{\partial P}\right]_{S,n} = \left[\frac{\partial V}{\partial S}\right]_{P,n} \tag{4.21}$$

From A, with variables T and V, from Equations 4.14 and 4.15, we get

$$\left[\frac{\partial P}{\partial T}\right]_{V,n} = \left[\frac{\partial S}{\partial V}\right]_{T,n} \tag{4.22}$$

and from G, with variables P and T, from Equations 4.17 and 4.18,

$$-\left[\frac{\partial S}{\partial P}\right]_{T,n} = \left[\frac{\partial V}{\partial T}\right]_{P,n} \tag{4.23}$$

In addition to these "classic" Maxwell's relations, which are particularly useful to relate the thermodynamic functions with the measurable variables P, T, and V, there are another two that are important for μ_i, the chemical potential. From Equations 4.17 and 4.19,

$$\frac{\partial}{\partial T}\left[\left(\frac{\partial G}{\partial n_i}\right)_{T,P,n_{j\neq i}}\right]_{P,n} = \frac{\partial}{\partial n_i}\left[\left(\frac{\partial G}{\partial T}\right)_{P,n}\right]_{T,P,n_{j\neq i}}$$

$$\left[\frac{\partial \mu_i}{\partial T}\right]_{P,n} = -\left[\frac{\partial S}{\partial n_i}\right]_{T,P,n_{j\neq i}} \tag{4.24}$$

and from Equations 4.18 and 4.19,

$$\left[\frac{\partial \mu_i}{\partial P}\right]_{T,n} = \left[\frac{\partial V}{\partial n_i}\right]_{T,P,n_{j\neq i}} \tag{4.25}$$

We observe here that the evaluation of the chemical potential function, μ_i, from the Gibbs function G by Equation 4.19, and its temperature and pressure derivatives at constant composition, from Equations 4.24 and 4.25, respectively, involves the differentiation of an extensive property with respect to the number of moles of compound i, keeping constant the temperature, pressure, and number of moles of all other compounds different from i. This particular form of partial derivative is discussed in detail in Chapter 5.

OTHER USEFUL MATHEMATICAL RELATIONS

Additional equations can be obtained using the following three mathematical relations:

$$\left(\frac{\partial X}{\partial Y}\right)_Z = \frac{1}{\left(\dfrac{\partial Y}{\partial X}\right)_Z} \qquad \text{Inversion} \qquad (4.26)$$

$$\left(\frac{\partial X}{\partial Y}\right)_Z = \left(\frac{\partial X}{\partial W}\right)_Z \left(\frac{\partial W}{\partial Y}\right)_Z \qquad \text{Chain rule} \qquad (4.27)$$

$$\left(\frac{\partial X}{\partial Y}\right)_Z = -\left(\frac{\partial X}{\partial Z}\right)_Y \left(\frac{\partial Z}{\partial Y}\right)_X \qquad \text{Change of variable} \qquad (4.28)$$

Finally, a particularly useful transformation for any property of a closed system considered a function of V and T is obtained from the exact differential:

$$dX = \left(\frac{\partial X}{\partial V}\right)_{T,n} dV + \left(\frac{\partial X}{\partial T}\right)_{V,n} dT$$

For a change of temperature at constant pressure, we obtain

$$\left(\frac{\partial X}{\partial T}\right)_{P,n} = \left(\frac{\partial X}{\partial V}\right)_{T,n} \left(\frac{\partial V}{\partial T}\right)_{P,n} + \left(\frac{\partial X}{\partial T}\right)_{V,n} \qquad (4.29)$$

In order to fix ideas with respect to the difference between state and path functions, we consider here a simple example. It is strongly suggested that you try to solve the example before reading the solution.

EXAMPLE

Consider the following two differentials:

$$df = (9x^2y^2 - 2y + 7)dx + (6x^3y - 2x)dy$$

and

$$\delta w = (9x^2y^2 - 2y + 7)dx + (6x^3y + 2x)dy$$

1. Test whether these differentials are exact or not.
2. For both cases, evaluate the line integral from the point $(x = 1, y = 3)$ to the point $(x = 2, y = 6)$ along the following two paths:
   Path 1: $y = 3x$;   Path 2: $y = x^2 + 2$

3. Classify the functions f and w in thermodynamic terms as either a path or a state function. Decide whether the value of the line integral of the first of these functions should be designated by f or Δf; equally, decide whether the line integral of the second function should be designated by w or Δw.

Solution

1. In order to decide whether the differentials of the functions are exact, we verify whether the functions preceding the differentials of the variables x and y meet Maxwell's conditions. For clarity, we consider the following general form:

$$dF = \left(\frac{\partial F}{\partial x}\right)_y dx + \left(\frac{\partial F}{\partial y}\right)_x dy$$

An exact differential should meet Maxwell's relation of the form

$$\frac{\partial}{\partial y}\left[\left(\frac{\partial F}{\partial x}\right)_y\right]_x = \frac{\partial}{\partial x}\left[\left(\frac{\partial F}{\partial y}\right)_x\right]_y$$

For the function f,

$$\frac{\partial}{\partial y}\left[9x^2y^2 - 2y + 7\right]_x = 18x^2y - 2 \text{ and } \frac{\partial}{\partial x}\left[6x^3y - 2x\right]_y = 18x^2y - 2$$

Thus, df is an exact differential. For the function w, however, the partial differential with respect to y of the coefficient of dx is the same as above, while the partial differential with respect to x of the coefficient of dy is

$$\frac{\partial}{\partial x}\left[6x^3y + 2x\right]_y = 18x^2y + 2$$

indicating that the expression for δw is not an exact differential.

2. First, we verify that both paths of integration pass through the two extreme points of the integration. After this verification, we proceed with the integration of both functions following path 1. For this, we replace $y = 3x$ and $dy = 3dx$ in the expressions for df and δw. After reducing terms, we obtain

$$df = (135x^4 - 12x + 7)dx \text{ and } \delta w = (135x^4 + 7)dx$$

Thus, integrating between $x = 1$ and $x = 2$,

$$\int_1^2 df = 826 \text{ and } \int_1^2 \delta w = 844$$

For the integration following path 2, we replace $y = x^2 + 2$ and $dy = 2x\ dx$ in the expressions for df and δw. After reducing terms, we obtain

$$df = 3(7x^6 + 20x^4 + 10x^2 + 1)dx \quad \text{and} \quad \delta w = (21x^6 + 60x^4 + 38x^2 + 3)dx$$

Thus, integrating between $x = 1$ and $x = 2$,

$$\int_1^2 df = 826 \quad \text{and} \quad \int_1^2 \delta w = 844\frac{2}{3}$$

As expected from the test of the validity of Maxwell's equation, the value resulting from integration of the function f is independent of the path, while the value of the integration of function w is path dependent.

3. In thermodynamic terms, f is a state function like U, H, S, A, or G, and its integral should be designated by Δf, while w is a path function and its integral should be designated by w.

5 Calculation of Changes in the Value of Thermodynamic Properties

MEASURABLE PROPERTIES

In this chapter, we consider the calculation of changes in internal energy, enthalpy, and entropy for a closed system undergoing a process between an initial equilibrium state at a pressure P_1 and temperature T_1 and a final equilibrium state at P_2 and T_2. The internal energy, enthalpy, and entropy are nonmeasurable properties. A preliminary consideration of Equations 4.7 and 4.9 shows that in order to make any progress, we must first eliminate the entropy from these equations.

The pressure, temperature, composition, and volume of a closed system are *measurable properties*. Two additional measurable properties are the *coefficient of thermal expansion*,

$$\alpha = \frac{1}{V}\left(\frac{\partial V}{\partial T}\right)_{P,n} = \frac{1}{v}\left(\frac{\partial v}{\partial T}\right)_{P,n} \tag{5.1}$$

and the *coefficient of isothermal compressibility*,

$$\kappa_T = -\frac{1}{V}\left(\frac{\partial V}{\partial P}\right)_{T,n} = -\frac{1}{v}\left(\frac{\partial v}{\partial P}\right)_{T,n} \tag{5.2}$$

As the volume of a closed system decreases with an increase in pressure at constant temperature, in order to have positive values for the coefficient of isothermal compressibility, a negative sign is used in the right-hand side of Equation 5.2. The heat exchanged between a closed system and its surroundings, during a process carried out under controlled conditions, is also measurable. Considering that the heat required to change the temperature of a closed system by $1°$ depends on its initial temperature, it is necessary to refer to an infinitesimal change in temperature and write, per unit mass,

$$c \equiv \lim_{\Delta T \to 0}\left[\frac{\delta q}{\Delta T}\right] \tag{5.3}$$

The quantity c, so defined, is called the *specific heat* and is a function not only of temperature but also of the path followed by the process. Two paths are commonly used. For a constant pressure (isobaric) process, the external pressure, P_E, and the internal pressure, P, of the system are the same and constant. Thus, from Equations 2.11 and 3.6,

$$\delta q_P = du + P_E dv = du + d(Pv) = dh_P$$

Once the path of the process is defined as isobaric, the heat exchanged with the surroundings becomes equal to the change in the state function enthalpy and the relation for a reversible process between the same two states is applicable. From Equation 4.9 at constant pressure and number of moles,

$$\delta q_P = \delta q_{rev,P} = dh_P = T \, ds_P \tag{5.4}$$

An important conclusion from this relationship is that for a closed system undergoing an isobaric process, the *specific heat at constant pressure*, at a particular temperature T, can be written as

$$c_P = \left(\frac{\partial h}{\partial T}\right)_{P,n} = T\left(\frac{\partial s}{\partial T}\right)_{P,n} \tag{5.5}$$

The subscript n is kept as a reminder that this equation is valid for a closed system. Thus, as the heat exchanged to change by a small increment the temperature of a closed homogeneous system is measurable, the value of the specific heat can be experimentally obtained.

Similarly, by defining a constant volume path for the process, the heat exchanged between the system and its surroundings from Equations 2.2 and 2.3 takes the form

$$\delta q_v = \delta q_{rev,v} = du_v = T ds_v \tag{5.6}$$

Thus, the *specific heat at constant volume* can then be written as

$$c_v = \left(\frac{\partial u}{\partial T}\right)_{v,n} = T\left(\frac{\partial s}{\partial T}\right)_{v,n} \tag{5.7}$$

Usually, the specific heat at constant volume is not measured independently, but if needed, it is obtained from the exact relation

$$c_v = c_P - \frac{Tv\alpha^2}{\kappa_T} \tag{5.8}$$

The derivation of this equation is given at the end of this chapter, but it is strongly recommended that the reader attempts to obtain it before seeing how it is done.

CALCULATION OF PROPERTY CHANGES FOR SOLID AND LIQUID PHASES

Once experimental values for the specific heat at constant pressure have been measured and tabulated, for most practical applications solids and liquids can be safely considered to be incompressible and the small effects caused by volume changes with pressure are ignored. Enthalpy changes can be calculated by integration of rearranged Equation 5.5, where superscript c indicates a "condensed phase."

$$\Delta h^c = \int_{T_1}^{T_2} c_P \, dT \tag{5.9}$$

Using the arithmetic average of the C_P values at T_1 and T_2, we write

$$\Delta h^c \approx \overline{c_P} \, \Delta T \tag{5.10}$$

For the change in internal energy, if ever needed, although the correction for pressure is usually negligible, we may use

$$\Delta u^c = \Delta h^c - v \Delta P \tag{5.11}$$

For the calculation of entropy changes, from Equation 5.5,

$$\Delta s^c = \int_{T_1}^{T_2} \frac{c_P}{T} \, dT \tag{5.12}$$

or, considering an average value for specific heat at constant pressure,

$$\Delta s^c \approx \overline{c_P} \ln \frac{T_2}{T_1} \tag{5.13}$$

The superscript c has been used in these equations as a reminder that these expressions are only applicable to condensed, incompressible phases.

CALCULATION OF PROPERTY CHANGES FOR PHASE CHANGE OF A PURE COMPOUND

If the process is isobaric and isothermic, from Equation 5.4 we write

$$q_{P,T} = \Delta h_{P,T} = T \Delta s_{P,T}$$

A particular process that occurs at constant pressure and temperature is the *change of phase for a pure compound*, from either solid to liquid, liquid to gas, or solid to gas. As the *phases in equilibrium are said to be "saturated,"* instead of indicating by subscripts that the pressure and temperature are kept constant, a superscript s is used to indicate the change in the property between saturated phases. Thus, we write

$$\Delta s^s = \frac{\Delta h^s}{T_s} \tag{5.14}$$

As shown by Equation 5.4, the enthalpy change in a phase change is equal to the heat required, which is also a measurable quantity. For this reason, the enthalpy change of vaporization, for example, is simply called the *heat of vaporization*. Similarly, we use the terms *heat of condensation, heat of sublimation, heat of melting*, and so forth. Once the enthalpy of phase change is measured at a particular temperature, the entropy change can be calculated with Equation 5.14 and the internal energy change, if ever required, can be obtained from Equation 3.6. However, once the enthalpy change is measured at one temperature, it is desirable to have some way of calculating it at different temperatures using properties that are more easily measured. The treatment that follows considers the phase change from saturated liquid (sL) to saturated vapor (sV), but the equations obtained are applicable to any two phases in equilibrium. Rearranging Equation 5.14, we write

$$h^{sV} - T_s s^{sV} = h^{sL} - T_s s^{sL} \tag{5.14a}$$

or, from Equation 4.2,

$$g^{sV} = g^{sL} \tag{5.14b}$$

For an incremental temperature change from T_s to $(T_s + dT_s)$,

$$g^{sV} + dg^{sV} = g^{sL} + dg^{sL}$$

By comparison between the last two equations we get

$$dg^{sV} = dg^{sL}$$

From the general expression for the exact differential of the Gibbs energy, for a closed system we have

$$dg = \left(\frac{\partial g}{\partial T}\right)_{P,n} dT + \left(\frac{\partial g}{\partial P}\right)_{T,n} dP$$

Using Equations 4.17 and 4.18, we obtain

$$dg = -sdT + vdP$$

As we know from the physical phenomenon of phase change, once the temperature of equilibrium changes, the saturation pressure of the system changes accordingly. Thus, considering the saturation values for pressure as a function of temperature as independent variable, and eliminating the subscript s for the independent variable T, for the change in the saturation equilibrium condition we write

$$-s^{sV} dT + v^{sV} dP^s = -s^{sL} dT + v^{sL} dP^s$$

Rearranging we get

$$\frac{dP^s}{dT} = \frac{\Delta s^s}{\Delta v^s} \qquad (5.15a)$$

Combining this expression with Equation 5.14,

$$\frac{dP^s}{dT} = \frac{\Delta h^s}{T \Delta v^s} \qquad (5.15b)^\dagger$$

For work with volumetric properties, it is usual to introduce the definition of the *compressibility factor, z*:

$$z \equiv \frac{Pv}{RT} \qquad (5.16)$$

In this definition, which is valid for any physical state of matter, R is the so-called gas constant and v is the molar volume. Thus, introducing the compressibility factor we write

$$\frac{dP^s}{dT} = \frac{\Delta h^s}{RT^2 \Delta z^s} P^s$$

or

$$\frac{d \ln P^s}{dT} = \frac{\Delta h^s}{RT^2 \Delta z^s} \qquad (5.15c)$$

This exact relation is known as the *Clausius–Clapeyron equation*, and it is valid for any phase change. For the particular case of vapor–liquid equilibrium, from experimental data it has been found that the ratio between heat of vaporization and the change due to vaporization of the compressibility factor is a very weak function of temperature. Hence, we can write

$$\frac{\Delta h^s}{R \Delta z^s} = B + DT + ET^2 + \ldots$$

---

† For liquid–solid equilibrium at a temperature T, as for most compounds $v^{sL} > v^{sS}$, an increase in pressure increases the melting temperature, so at the temperature T, liquid freezes to solid. For a few compounds, including water, $v^{sL} < v^{sS}$. Thus, for these compounds an increase in pressure lowers the melting temperature, and at the temperature T, solid melts. This effect makes ice skating easier, as a liquid phase is formed under the pressure of a skate.

where B, D, E, ..., are empirical constants that can be evaluated from experimental data for each pure compound. Thus,

$$d \ln P^s = \left[\frac{B}{T^2} + \frac{D}{T} + E + ... \right] dT$$

The indefinite integral of this equation gives

$$\ln P^s = A - \frac{B}{T} + D \ln T + ET + ... \tag{5.17a}$$

where A is the integration constant. This semiempirical correlation for the vapor pressure of pure compounds is known as the *Riedel equation*. A simple empirical equation for the vapor pressure applicable over a limited temperature range, closely related to the Riedel equation, is the *Antoine equation*. It has the form

$$\log_{10} P^s = A_A - \frac{B_A}{t + C_A} \tag{5.17b}$$

where the temperature t is in degrees Celsius and the constants A_A, B_A, C_A are also empirically evaluated.

Rewriting Equation 5.15c, we get

$$\Delta h^s = RT^2 \Delta z^s \frac{d \ln P^s}{dT} \tag{5.18}$$

The fact that this exact equation is verified using experimental data can be considered among the best evidence of the usefulness of the definition of entropy. The derivation of this equation was possible thanks to the introduction of the concept of entropy by a postulate. After some algebra, the entropy was eliminated, and the final result, Equation 5.18, confirmed by experiment supports the theory. In addition, this exact equation coupled with a semiempirical correlation like the Riedel equation provides a valuable tool to evaluate enthalpy and entropy changes of vaporization. For approximate calculations for the case of vapor–liquid equilibrium, Equation 5.15b is often simplified. Two parameters used to characterize the volumetric behavior of fluids are the reduced temperature, T_r, and reduced pressure, P_r, defined by

$$T_r = \frac{T}{T_c} \tag{5.19}$$

and

$$P_r = \frac{P}{P_c} \tag{5.20}$$

where T_c and P_c are the critical temperature and critical pressure of the fluid, respectively.

For $T_r \geq 0.8$ the empirical relation of Haggenmacher gives a reasonable approximation for the difference in the compressibility factors of the saturated phases. This equation has the form

$$\Delta z^s \approx \left[1 - \frac{P_r}{T_r^3} \right]^{0.5}$$

For $T_r \leq 0.8$, the molar volume of the liquid is orders of magnitude smaller than the molar volume of the vapor. Thus, the value of the liquid compressibility factor is negligibly small compared with its value for the vapor. Hence, considering the vapor phase as an ideal gas,

$$\Delta z^s \approx 1$$

USE OF EQUATIONS OF STATE FOR A PURE COMPOUND OR FOR A MIXTURE OF CONSTANT COMPOSITION

Due to their compressible nature, gases are the real problem for the calculation of the value of property changes. In this case, we need to use equations of state (EOSs). Any thermodynamic function in terms of its canonical variables can be used to define the state of the system. For example, an expression for H in terms of P, S, and the set of number of moles n, as suggested by Equation 4.10, is an EOS. For practical purposes, however, unless otherwise specified, *the term* EOS *refers to the relation of the volume of the system as a function of temperature, pressure, and composition.* These are all measurable properties. Moreover, for our purposes, we only consider either a pure compound or a mixture of fixed composition.

Depending on the variable that it expressed explicitly, there are two forms of EOS commonly used. These are the *volume explicit form*:

$$v = v(P, T)$$

and the *pressure explicit form:*

$$P = P(v, T)$$

The symbol used in the right-hand side of the above expressions indicates that the property shown outside the brackets is a function of the variables in brackets.

CASE OF INDEPENDENT VARIABLES V AND T, $P = P(v, T)$

For a closed system, the exact differential of the internal energy, u, in terms of temperature and volume as variables takes the form

$$du = \left(\frac{\partial u}{\partial T} \right)_{v,n} dT + \left(\frac{\partial u}{\partial v} \right)_{T,n} dv$$

The first partial derivative of the right-hand side is equal to the measurable specific heat at constant volume, Equation 5.7. For the second partial derivative, from the fundamental equation (4.7),

$$du = Tds - Pdv$$

From here we write

$$\left(\frac{\partial u}{\partial v}\right)_{T,n} = T\left(\frac{\partial s}{\partial v}\right)_{T,n} - P$$

In addition, from the Maxwell relation, Equation 4.22,

$$\left(\frac{\partial s}{\partial v}\right)_{T,n} = \left(\frac{\partial P}{\partial T}\right)_{v,n}$$

Thus, we finally can write for du,

$$du = c_v\, dT + \left[T\left(\frac{\partial P}{\partial T}\right)_{v,n} - P\right]dv \tag{5.21}$$

For the calculation of enthalpy changes, in this case we use

$$dh = du + d(Pv) \tag{5.22}$$

To calculate changes in entropy, we follow a similar path and write

$$ds = \left(\frac{\partial s}{\partial T}\right)_{v,n} dT + \left(\frac{\partial s}{\partial v}\right)_{T,n} dv$$

For the first partial derivative we use Equation 5.7, and for the second one we use the Maxwell relation, Equation 4.22, as used above; thus,

$$ds = \frac{c_v}{T} dT + \left(\frac{\partial P}{\partial T}\right)_{v,n} dv \tag{5.23}$$

Integration of these equations between two equilibrium states will give the value of the property change as a function of T_1, T_2, v_1, and v_2. However, the variables normally measured are T_1, P_1, T_2, and P_2. Thus, before doing any calculation of a property change, it is necessary to use the EOS and evaluate v_1 using T_1, P_1, and also evaluate v_2 using T_2, P_2. The direct path between the two equilibrium states presents an additional problem. The specific heat of the gas is a function of temperature and pressure, and the latter dependence is usually unknown. However, as the value of a property change is independent of the path followed between the two equilibrium states considered, the calculations are done using a three-step path. First, the gas is taken at constant temperature from T_1, P_1 to T_1, P^*, where P^* is a pressure low enough that the gas behaves as an ideal gas. In the second step, the temperature is changed from T_1 to T_2, at constant pressure P^*. Finally, in the third step, the gas is taken from T_2, P^* to T_2, P_2. This path has the advantage that the specific heat capacities of different gases as an ideal gas are known as a function of temperature.

CASE OF INDEPENDENT VARIABLES P AND T, $v = v(P, T)$

For a closed system,

$$dh = \left(\frac{\partial h}{\partial T}\right)_{P,n} dT + \left(\frac{\partial h}{\partial P}\right)_{T,n} dP$$

From Equation 4.9, in molar terms for a closed system,

$$dh = Tds + vdP$$

Thus, for the second differential of the right-hand side we write

$$\left(\frac{\partial h}{\partial P}\right)_{T,n} = T\left(\frac{\partial s}{\partial P}\right)_{T,n} + v$$

and using the Maxwell relation, Equation 4.23, we eliminate the term containing the entropy and write

$$\left(\frac{\partial h}{\partial P}\right)_{T,n} = v - T\left(\frac{\partial v}{\partial T}\right)_{P,n}$$

In terms of measurable properties, the first differential of the right-hand side is given by Equation 5.5. Hence, finally,

$$dh = c_P \, dT + \left[v - T\left(\frac{\partial v}{\partial T}\right)_{P,n}\right] dP \tag{5.24}$$

As before, for du we use directly

$$du = dh - d(Pv) \tag{5.25}$$

For entropy change with independent variables P and T,

$$ds = \left(\frac{\partial s}{\partial T}\right)_{P,n} dT + \left(\frac{\partial s}{\partial P}\right)_{T,n} dP$$

The first differential of the right-hand side is given by Equation 5.5, and the second one is given by the same Maxwell relation, Equation 4.23, used above to obtain Equation 5.24; thus,

$$ds = \frac{c_P}{T} dT - \left(\frac{\partial v}{\partial T}\right)_{P,n} dP \tag{5.26}$$

Again in this case, when calculating the change in the value of a thermodynamic property between a state at P_1, T_1 and a state at P_2, T_2, it is necessary to do a three-step calculation. First, the gas is taken at constant temperature from T_1, P_1 to T_1, P^*, where P^* is a pressure low enough that the gas behaves as an ideal gas. Then the temperature is changed from T_1 to T_2, at constant pressure P^*, and finally, the gas is taken from T_2, P^* to T_2, P_2.

CHANGES IN THE VALUES OF THE THERMODYNAMIC PROPERTIES OF THE IDEAL GAS

The ideal gas behavior gives a good approximation to the behavior of nonpolar gases, such as nitrogen, oxygen, or air, at room or higher temperature and at around atmospheric pressure. The compressibility factor of an ideal gas is unity, $z^* = 1$, and Equation 5.16 takes the form

$$P^* = \frac{RT}{v^*} \tag{5.27}$$

The asterisk is used here to indicate ideal gas behavior. We observe that there is a subtle difference between Equation 1.1 and Equation 5.27. In Equation 1.1, the molar volume of the ideal gas is evaluated at a pressure, P, and a temperature, T. As the pressure in this case is an independent variable, there is no need to mark it with an asterisk. In Equation 5.27, the volume of the ideal gas and the temperature are the independent variables, so the value of the pressure obtained is only valid for the case of an ideal gas, and it is marked with an asterisk. For the change in internal energy of the ideal gas from Equation 5.21,

$$du^* = c_v^* dT + \left[T\left(\frac{\partial P^*}{\partial T}\right)_{v,n} - P^* \right] dv^*$$

But, from Equation 5.27,

$$\left[T\left(\frac{\partial P^*}{\partial T}\right)_{v,n} - P^* \right] = 0$$

so

$$du^* = c_v^* \, dT \tag{5.28a}$$

Thus, the internal energy of the ideal gas is only a function of temperature. From Equations 3.6a and 5.27, the enthalpy of the ideal gas can be written as

$$h^* = u^* + P^*v^* = u^* + RT$$

or

$$dh^* = du^* + RdT = \left(c_v^* + R\right)dT$$

Thus, the enthalpy of the ideal gas is only a function of temperature. In addition, from Equation 5.5,

$$dh^* = c_P^* \, dT \tag{5.28b}$$

Hence,

$$c_P^* = c_v^* + R \tag{5.29}$$

This result can also be obtained applying Equation 5.8 for the ideal gas. For the entropy change of the ideal gas, from Equation 5.26 we write

$$ds^* = \frac{c_P^*}{T} dT - \frac{R}{P} dP \tag{5.30}$$

Thus, the entropy of the ideal gas is a function of both temperature and pressure.

ADIABATIC REVERSIBLE COMPRESSION OR EXPANSION OF THE IDEAL GAS

As an approximation, we consider here constant values of the specific heats of the ideal gas, $\overline{c}_P^*$ and $\overline{c}_v^*$. An adiabatic reversible thermodynamic process is isentropic. Thus, from Equation 5.30 we write

$$\overline{c}_P^* \, d\ln T = Rd \ln P$$

Integrating between a state at P_1, T_1 and a state at P_2, T_2, and rearranging, we obtain

$$\frac{T_2}{T_1} = \left(\frac{P_2}{P_1}\right)^{\frac{R}{c_P^*}} = \left(\frac{P_2}{P_1}\right)^{\frac{\gamma-1}{\gamma}} \tag{5.31}$$

In Equation 5.31, we have defined

$$\gamma \equiv \frac{\overline{c}_P^*}{\overline{c}_v^*} \tag{5.32}$$

Combining Equation 5.31 with the ideal gas equation, Equation 1.1,

$$\frac{T_2}{T_1} = \left(\frac{v_1}{v_2}\right)^{\gamma-1} \tag{5.33}$$

Finally, combining Equations 5.31 and 5.33,

$$P_1 v_1^\gamma = P_2 v_2^\gamma \tag{5.34}$$

DEVIATIONS FROM THE IDEAL GAS BEHAVIOR

For nonpolar gases, such as oxygen, nitrogen, carbon dioxide, or air, at normal ambient temperature or higher, and at pressures below 5 atm, the following truncated forms of the virial equation give a satisfactory approximation of their volumetric behavior:

$$z = 1 + \frac{B}{v} + \frac{C}{v^2} + \cdots \tag{5.35}$$

and

$$z = 1 + \frac{B}{RT}P + \frac{C - B^2}{(RT)^2}P^2 + \cdots \tag{5.36}$$

In Equations 5.35 and 5.36, B and C are the second and third virial coefficients, which are usually given as a function of the reduced temperature, defined by Equation 5.19, by expressions of the form

$$B = B^{(0)}(T_r) + \omega B^{(1)}(T_r) \tag{5.37}$$

where ω is the so-called Pitzer's acentric factor, defined by

$$\omega = -\log_{10}(P^s \text{ at } T_r = 0.7) + \log_{10}P_c - 1.000 \tag{5.38}$$

Further discussion of the virial EOS is presented in Chapter 25 and Appendix B. Larger deviations from the ideal behavior are treated with cubic EOSs. One such equation is discussed in Chapter 25 and Appendix C.

PREPARATION OF TABLES AND PLOTS OF THERMODYNAMIC PROPERTIES

Tables and plots of thermodynamic properties are constructed on the basis of extensive collections of values of the measurable properties. At constant temperature, molar volumes are measured as a function of pressure. Values of the specific heats, vapor pressure, and heats of vaporization are independently measured and tested for consistency. Once the data are available, it is necessary to assign a zero value to the nonmeasurable properties, like enthalpy and entropy. Normally, the property for the saturated liquid at a well-defined temperature is assigned the zero value. When combining information from differences sources, it is imperative to control that they are all using the same reference point. Otherwise, a displacement of the relative scale is required. Once the zero is fixed, the enthalpy of the saturated vapor at that equilibrium temperature is evaluated using the exact Equation 5.15b, and the entropy using Equation 5.14. From these points, the (relative) values of the properties at any state in the vapor or liquid phases can be obtained using the equations for a single phase discussed above. For precise work, once volumes of the fluid have been measured as a function of pressure at constant temperature, the data are correlated using extended empirical virial EOSs of the form

$$P = \frac{RT}{v}\left[1 + \sum_{k=1}^{n} \frac{C_k}{v^k}\right] \tag{5.39}$$

For a pure compound, the adjustable parameters C_k are only a function of temperature.

Derivation of Equation 5.8

We start from the definition (Equation 5.7):

$$c_v = T\left(\frac{\partial s}{\partial T}\right)_{v,n}$$

From the transformation (Equation 4.29),

$$c_v = T\left[\left(\frac{\partial s}{\partial T}\right)_{P,n} - \left(\frac{\partial s}{\partial v}\right)_{T,n}\left(\frac{\partial v}{\partial T}\right)_{P,n}\right]$$

But from Equation 5.5,

$$c_P = T\left(\frac{\partial s}{\partial T}\right)_{P,n}$$

Thus,

$$c_v = c_P - T\left[\left(\frac{\partial s}{\partial v}\right)_{T,n}\left(\frac{\partial v}{\partial T}\right)_{P,n}\right]$$

From Equations 4.22 and 4.28, we get

$$\left(\frac{\partial s}{\partial v}\right)_{T,n} = \left(\frac{\partial P}{\partial T}\right)_{v,n} = -\frac{\left(\frac{\partial v}{\partial T}\right)_{P,n}}{\left(\frac{\partial v}{\partial P}\right)_{T,n}}$$

so

$$c_v = c_P + \frac{T\left[\left(\frac{\partial v}{\partial T}\right)_{P,n}\right]^2}{\left(\frac{\partial v}{\partial P}\right)_{T,n}}$$

Using the definitions of α (Equation 5.1) and κ_T (Equation 5.2),

$$c_v = c_P - \frac{Tv\alpha^2}{\kappa_T}$$

For the ideal gas,

$$\frac{Tv^*\alpha^{*2}}{\kappa_T^*} = R$$

Thus,

$$c_P^* - c_v^* = R$$

Section II

Mixtures

$$\phi_i = \gamma_i \, \phi_i^{0 \; \dagger}$$

† Equation from Wilczek-Vera, G., and Vera, J. H. 1989. A consistent method to combine PRSV EOS with excess Gibbs functions, *Fluid Phase Equilib.* 51: 197–208.

6 Partial Molar Properties and Property Changes by Mixing

PARTIAL MOLAR PROPERTIES AND THE GIBBS DUHEM EQUATION

For a single-phase system, the thermodynamic state functions V, U, S, H, A, and G are all *extensive* and can be expressed in terms of the state variables P, T and composition. Volume is a measurable property and provides a more visual sense of the algebra. However, the treatment is general and expressions similar to Equation 4.3, but with independent variables P and T, can be written for all other extensive functions U, S, H, A, or G. Thus, for the total differential of any of the state functions, taking the volume as an example, we write

$$dV = \left(\frac{\partial V}{\partial T}\right)_{P,n} dT + \left(\frac{\partial V}{\partial P}\right)_{T,n} dP + \sum_{k}^{c} \left(\frac{\partial V}{\partial n_k}\right)_{T,P,n_{j\neq k}} dn_k \tag{6.1}$$

For the particular case of the total volume, the partial derivatives with respect to temperature and pressure are *measurable properties* and are given a symbol and a name. In both of these derivatives, the numbers of moles of all compounds are kept constant, so they refer to a closed system. Therefore, we can consider these two terms per unit mass. For an open system, the third term on the right-hand side represents the effect on volume due to a change in the numbers of moles of the components of the system. For simplicity, we use the symbol $\bar{v}_i$ for the partial derivative of the total property V with respect to the number of moles of component i, at constant temperature, constant pressure, and constant number of moles of all other compounds, different from i.

$$\bar{v}_i \equiv \left(\frac{\partial V}{\partial n_i}\right)_{T,P,n_{j\neq i}} \tag{6.2}$$

Similar relations can be written for the partial molar properties of U, S, H, and G. The derivatives have dimensions. The derivative of an extensive variable with respect to the change in another extensive variable, as in Equation 6.2, gives a value that is independent of the size of the homogeneous system. If the number of moles of each component of the mixture is multiplied by a factor λ, at constant temperature and pressure, the volume V_λ of the system formed by λ times the moles of each species will be λ times the volume V of the original system:

$$\lambda V(T, P, n_1, n_2, \ldots, n_c) = V_\lambda(T, P, \lambda n_1, \lambda n_2, \ldots, \lambda n_c) \qquad (6.3)$$

For clarity, in Equation 6.3 we have shown in parentheses the variables for each function. From here on, however, this detail will be omitted. To demonstrate the intensive character of $\overline{v}_i$, we differentiate Equation 6.3 with respect to the number of moles of compound i, n_i at constant temperature, pressure, and moles of all other compounds different from i,

$$\lambda\left(\frac{\partial V}{\partial n_i}\right)_{T,P,n_{j\neq i}} = \left(\frac{\partial V_\lambda}{\partial \lambda n_i}\right)_{T,P,\lambda n_{j\neq i}}\left(\frac{\partial \lambda n_i}{\partial n_i}\right) = \left(\frac{\partial V_\lambda}{\partial \lambda n_i}\right)_{T,P,n_{j\neq i}}\lambda$$

or simplifying by dividing by λ:

$$\left(\frac{\partial V}{\partial n_i}\right)_{T,P,n_{j\neq i}} = \left(\frac{\partial V_\lambda}{\partial \lambda n_i}\right)_{T,P,\lambda n_{j\neq i}} \qquad (6.4)$$

We conclude that the value of $\overline{v}_i$ as defined by Equation 6.2 remains unchanged by a change in the size of the system at constant composition, temperature, and pressure. If now we differentiate Equation 6.3 with respect to a change of the arbitrary size factor λ,

$$\left(\frac{\partial \lambda V}{\partial \lambda}\right)_{T,P,n} = \sum_k^c \left(\frac{\partial V_\lambda}{\partial \lambda n_k}\right)_{T,P,\lambda n_{j\neq k}}\left(\frac{\partial \lambda n_k}{\partial \lambda}\right) = \sum_k^c \left(\frac{\partial V}{\partial n_k}\right)_{T,P,\lambda n_{j\neq k}} n_k$$

Thus, we write

$$V = \sum_k^c n_k \overline{v}_k \qquad (6.5)$$

This equation can also be written in molar terms as

$$v = \sum_k^c x_k \overline{v}_k \qquad (6.6)$$

In Equation 6.6, we have used the definition of the mole fraction of the component i:

$$x_i = \frac{n_i}{\sum\limits_k n_k} \qquad (6.7)$$

Equation 6.6 clearly shows that $\bar{v}_i$, as defined by Equation 6.2, is the contribution of compound i to the molar property of the mixture, in this case the molar volume. Thus, $\bar{v}_i$ is known as the *partial molar volume* of compound i in the mixture, and it is a function of temperature, pressure, and composition. Equations 6.2, 6.5, and 6.6 are general forms, not limited to the volume of the system, and with the appropriate change of symbols, they directly apply to U, S, H, A, or G. From Equation 6.5, we then write the general relation

$$dV = \sum_k^c n_k\,d\bar{v}_k + \sum_k^c \bar{v}_k\,dn_k$$

On the other hand, from Equations 6.1 and 6.2, we write

$$dV = \left(\frac{\partial V}{\partial T}\right)_{P,n} dT + \left(\frac{\partial V}{\partial P}\right)_{T,n} dP + \sum_k^c \bar{v}_k\,dn_k$$

Combining the two equations above, we write

$$\sum_k^c n_k\,d\bar{v}_k = \left(\frac{\partial V}{\partial T}\right)_{P,n} dT + \left(\frac{\partial V}{\partial P}\right)_{T,n} dP \tag{6.8}$$

or, in molar terms,

$$\sum_k^c x_k\,d\bar{v}_k = \left(\frac{\partial v}{\partial T}\right)_{P,n} dT + \left(\frac{\partial v}{\partial P}\right)_{T,n} dP \tag{6.9}$$

In Equations 6.8 and 6.9, we have not introduced Equations 5.1 and 5.2, which, being particular for volume, do not have a counterpart for the nonmeasurable thermodynamic properties U, S, H, and G. The all-important Equation 6.9 is known as the *Gibbs–Duhem equation. It shows that the partial molar properties of the compounds in a mixture are not all independent.*

Observing that the two derivatives of the right-hand side are taken at constant composition, we combine Equations 6.6 and 6.9 and write

$$\sum_k^c x_k\,d\bar{v}_k = \sum_k^c x_k\left(\frac{\partial \bar{v}_k}{\partial T}\right)_{P,n} dT + \sum_i^c x_k\left(\frac{\partial \bar{v}_k}{\partial P}\right)_{T,n} dP$$

Rearranging,

$$\sum_k^c x_k\left[d\bar{v}_k - \left(\frac{\partial \bar{v}_k}{\partial T}\right)_{P,n} dT - \left(\frac{\partial \bar{v}_k}{\partial P}\right)_{T,n} dP\right] = 0$$

As this equation is valid for any fixed composition, *independent of the particular set of mole fractions considered, for a mixture of fixed composition* the term in square brackets must be identically zero; thus, we write

$$d\,\overline{v}_i = \left(\frac{\partial \overline{v}_i}{\partial T}\right)_{P,n} dT + \left(\frac{\partial \overline{v}_i}{\partial P}\right)_{T,n} dP \qquad (6.10)$$

Equation 6.10 is the general Gibbs–Duhem equation in terms of partial molar properties *for a mixture of fixed composition*. As an example, it is written here for the partial molar volume, but it also applies to $\overline{u}$, $\overline{h}$, $\overline{s}$, and $\overline{g}$.

MAXWELL RELATION APPLIED TO PARTIAL MOLAR PROPERTIES

As discussed in Chapter 4, for a state function of several variables, the order of partial differentiation with respect to two variables is immaterial. Thus, this property of all functions having an exact differential can be combined with the definition of partial molar property, Equation 6.2, to obtain

$$\frac{\partial}{\partial n_i}\left[\left(\frac{\partial V}{\partial n_j}\right)_{T,P,n_{k\neq j}}\right]_{T,P,n_{k\neq i}} = \frac{\partial}{\partial n_j}\left[\left(\frac{\partial V}{\partial n_i}\right)_{T,P,n_{k\neq i}}\right]_{T,P,n_{k\neq j}}$$

or

$$\left(\frac{\partial \overline{v}_j}{\partial n_i}\right)_{T,P,n_{k\neq i}} = \left(\frac{\partial \overline{v}_i}{\partial n_j}\right)_{T,P,n_{k\neq j}} \qquad (6.11)$$

Equation 6.11, not often found in literature, shows again that the partial molar properties of different compounds in a mixture are not independent of each other. One important point to keep in mind is that when the expressions for the partial molar properties of all compounds in a mixture are obtained from a single expression for the total property, these expressions always satisfy the Gibbs–Duhem equation and also the Maxwell's relations. This is an important conclusion not properly emphasized in the literature.

GENERALITY OF THE RELATIONS BETWEEN THERMODYNAMIC PROPERTIES

The derivation of Equation 6.10 from Equation 6.9 shows that any equation in terms of molar properties can be directly written in terms of partial molar properties if the algebra does not include a change in composition and the relation between the functions is linear in composition. As a simple example, consider the definition of enthalpy by Equation 3.6a:

$$h = u + Pv \qquad (3.6a)$$

$$\sum_{k}^{c} x_k \overline{h}_k = \sum_{k}^{c} x_k \overline{u}_k + P \sum_{k}^{c} x_k \overline{v}_k \quad \text{or} \quad \sum_{k}^{c} x_k \left(\overline{h}_k - \overline{u}_k - P\overline{v}_k \right) = 0$$

For a mixture of fixed composition, with the same argument used to derive Equation 6.10, we obtain

$$\overline{h}_i = \overline{u}_i + P\overline{v}_i \tag{6.12}$$

A similar algebra can be followed to derive other relations involving partial molar properties from the equation relating the molar properties of a simple system.

LIMITING VALUES OF THE PARTIAL MOLAR PROPERTIES

Partial molar properties at particular values of P and T are functions of the composition of a homogeneous mixture. For liquid mixtures, but not for gaseous mixtures, they can be considered to be independent of pressure. Considering the general form of Equation 6.6, it is evident that as the mixture becomes more and more concentrated in one of the compounds, the value of the partial molar property of that particular component will get closer and closer to the value of the property for that pure compound. Thus, taking the volume as an example, we write

$$\lim_{x_i \to 1} \overline{v}_i = v_i^0 \tag{6.13}$$

In Equation 6.13, we have introduced the superscript 0 to denote the value of a property for a pure compound. Equations similar to Equation 6.13 can be written for all other partial molar properties. However, on the other extreme of composition, that is, when a compound i gets more and more dilute in a mixture, the value of its partial molar property will get further and further away from the value of the property for the pure compound. The limiting value at infinite dilution is certainly characteristic of the composition of the solvent mixture. The partial molar volume of methanol at infinite dilution in water is not the same as the partial volume of methanol at infinite dilution in ethanol, and different from its value in a mixture of ethanol and water. Thus, in this case, it is necessary to specify the composition of the solvent, and we write

$$\lim_{x_i \to 0} \overline{v}_i = \overline{v}_i^{\infty,r} \tag{6.14}$$

In Equation 6.14, the superscript ∞ indicates infinite dilution, and the superscript r is a reminder of the need to specify the composition of the solvent mixture. In practice, when the composition of the solvent is understood from the physical situation, the latter subscript is omitted.

PARTIAL MOLAR PROPERTIES AS A FUNCTION OF MOLE FRACTIONS

The use of the volume as an example to derive expressions for partial molar properties is particularly useful in obtaining a physical sense of the meaning of a partial molar property at infinite dilution. Consider, for example, the case that the molar volume v of a binary mixture is measured as a function of composition, expressed by the mole fraction of component 1, x_1. In a plot of v versus x_1, at $x_1 = 0$, $v = v_2^0$, and at $x_1 = 1$, $v = v_1^0$. However, it is not immediate where to find $\bar{v}_1$ or $\bar{v}_2$. For generality, consider a multicomponent mixture containing c compounds for which the molar volume at a pressure P and a temperature T is known as a function of the mole fraction of $(c - 1)$ components, which are independent variables. As the mole fractions add to unity, one of these values, say x_i, is not an independent variable. Hence, in terms of independent variables we write

$$v = v(x_1, x_2, \ldots, x_{i-1}, x_{i+1}, \ldots, x_c)$$

In the variables for the function above, we have omitted P and T, which are kept constant in the derivation of the partial molar property, and also x_i, which is selected to be the dependent mole fraction. From Equation 6.2,

$$\bar{v}_i \equiv \left(\frac{\partial V}{\partial n_i}\right)_{T,P,n_{j\neq i}} = \left(\frac{\partial nv}{\partial n_i}\right)_{T,P,n_{j\neq i}} = v\left(\frac{\partial n}{\partial n_i}\right)_{n_{j\neq i}} + n\left(\frac{\partial v}{\partial n_i}\right)_{T,P,n_{j\neq i}}$$

However, $\left(\dfrac{\partial n}{\partial n_i}\right)_{n_{j\neq i}} = 1$, and applying the chain rule,

$$\left(\frac{\partial v}{\partial n_i}\right)_{T,P,n_{j\neq i}} = \sum_{k\neq i}\left(\frac{\partial v}{\partial x_k}\right)_{\substack{T,P,x_r \\ r\neq i, r\neq k}}\left(\frac{\partial x_k}{\partial n_i}\right)_{k\neq i}.$$

Moreover, in the latter expression,

$$\left(\frac{\partial x_k}{\partial n_i}\right)_{k\neq i} = -\frac{x_k}{n}$$

Thus, finally,

$$\bar{v}_i = v - \sum_{k\neq i} x_k\left(\frac{\partial v}{\partial x_k}\right)_{\substack{T,P,x_r \\ r\neq i, r\neq k}} \tag{6.15}$$

Equation 6.15 is useful for the derivation of partial molar properties from an expression of the molar property as a function of mole fractions.

EQUATIONS AND PLOTS FOR BINARY MIXTURES

For a binary mixture, from Equation 6.15 we can write for component 2

$$\bar{v}_2 = v - x_1 \left(\frac{\partial v}{\partial x_1} \right)_{T,P} \tag{6.16a}$$

where no mole fraction is kept constant for the differentiation. If x_1 changes, x_2 must change accordingly. For component 1, we write

$$\bar{v}_1 = v - x_2 \left(\frac{\partial v}{\partial x_2} \right)_{T,P}$$

or, considering $x_1 = 1 - x_2$ and $\left(\frac{\partial v}{\partial x_2} \right)_{T,P} = \left(\frac{\partial v}{\partial x_1} \right)_{T,P} \left(\frac{\partial x_1}{\partial x_2} \right) = -\left(\frac{\partial v}{\partial x_1} \right)_{T,P},$

$$\bar{v}_1 = v + (1 - x_1) \left(\frac{\partial v}{\partial x_1} \right)_{T,P} \tag{6.16b}$$

A simple geometrical exercise helps to give a physical sense for the partial molar properties and their values at infinite dilution. From Equations 6.16a and 6.16b,

$$\left(\frac{\partial v}{\partial x_1} \right)_{T,P} = \frac{(\bar{v}_1 - v)}{(1 - x_1)} = \frac{(v - \bar{v}_2)}{x_1} \tag{6.17}$$

Thus, drawing a tangent line to the curve representing v versus x_1, the partial molar volumes are located at the points where the tangent line cuts the corresponding vertical axis at $x_1 = 0$ and $x_1 = 1$. A tangent line at one of the extremes of the curve representing v versus x_1 will give the corresponding pure compound molar volume at that extreme of the line and the value of the partial molar volume of the compound at infinite dilution at the other extreme of the line (Figure 6.1).

Once the partial molar volumes of the components of a binary mixture are known, they can be plotted as a function of composition. A typical example is shown in Figure 6.2.

The curve for $\bar{v}_1$ versus x_1 runs from $\bar{v}_1^\infty$ at $x_1 = 0$ to v_1^0 at $x_1 = 1$, and we observe that it reaches zero slope at the latter end. This necessary condition is a result of the restrictions imposed by the Gibbs–Duhem equation. From Equation 6.9, for a binary mixture at constant pressure and temperature,

$$\left[x_1 d\bar{v}_1 + x_2 d\bar{v}_2 = 0 \right]_{T,P}$$

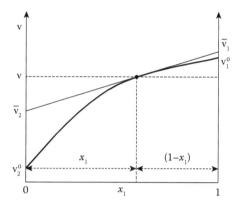

FIGURE 6.1   Determination of partial molar properties.

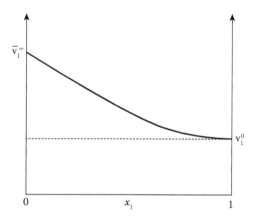

FIGURE 6.2   Partial molar volume as a function of concentration.

Thus, for a change in composition, taking x_1 as independent variable,

$$\left[x_1 \frac{d\bar{v}_1}{dx_1} + x_2 \frac{d\bar{v}_2}{dx_1} = 0 \right]_{T,P}$$

For the limit $x_1 \to 1$, $x_2 \to 0$, necessarily

$$\lim_{x_i \to 1} \left[\frac{d\bar{v}_1}{dx_1} \right]_{T,P} = 0$$

This limiting behavior is valid for volume or enthalpy but not for entropy or entropy-related functions, like the Gibbs or Helmholtz functions or the chemical potential. In Chapter 7, attention is focused on what is perhaps the most important of all partial molar properties: the chemical potential.

PROPERTY CHANGES BY MIXING

In general, when pure compounds at a pressure P and temperature T are mixed to form a mixture at P and T, the molar property of the mixture is different from the weighed addition of the molar properties of the compounds. When mixed, the compounds contribute to the molar properties of the mixture by their partial molar property values. Thus, for example, the volume change of mixing is

$$\Delta v_m = \sum_1^c x_k \left(\overline{v}_k - v_k^0 \right) = v - \sum_1^c x_k v_k^0 \tag{6.18}$$

As discussed in the presentation of Equation 5.4, for a process happening in a closed system, the heat exchanged with the surroundings is equal to the enthalpy change. Thus, the enthalpy change of mixing of compounds under these conditions is properly called *heat of mixing*.

$$\Delta h_m = \sum_1^c x_k \left(\overline{h}_k - h_k^0 \right) = h - \sum_1^c x_k h_k^0 \tag{6.19}$$

Similar expressions can be written for the change of other thermodynamic properties upon mixing.

IDEAL MIXTURES

For the treatment of fluid mixtures, in gaseous or liquid phases, it is convenient to distinguish between ideal and nonideal behavior. A mixture that forms without heat effect and without volume change at constant pressure and temperature is called an *ideal mixture*. This measurable physical evidence indicates that the molecules mix without interfering with each other; that is, a mole of component k occupies the same volume in the mixture as it would as a pure substance. We may say that the molecules do not "see" each other as being different species.

$$\Delta v_m^{id} = \sum_1^c x_k \left(\overline{v}_k^{id} - v_k^0 \right) = v_m^{id} - \sum_1^c x_k v_k^0 = 0 \tag{6.20}$$

and

$$\Delta h_m^{id} = \sum_1^c x_k \left(\overline{h}_k^{id} - h_k^0 \right) = h_m^{id} - \sum_1^c x_k h_k^0 = 0 \tag{6.21}$$

With the same argument used to derive Equation 6.10, in the composition range that Equations 6.20 and 6.21 are valid, we can write

$$\overline{v}_k^{id} = v_k^0 \tag{6.22}$$

and

$$\overline{h}_k^{id} = h_k^0 \tag{6.23}$$

Mixtures formed from compounds that are close to each other in a chemical homologous series approximate closely this behavior. For example, mixtures of benzene and toluene, methanol and ethanol, or acetic acid and propionic acid form with negligible volume change or heat effect. Chapter 15 discusses a generalization of the concept of ideal mixture. As shown there, the entropy change and the Gibbs energy change, when forming an ideal mixture, are different from zero.

7 The Chemical Potential and the Gibbs–Helmholtz Equation

CHEMICAL POTENTIAL AS A PARTIAL MOLAR PROPERTY

Due to the importance of the chemical potential in thermodynamics, we recapitulate here some of the equations related to it. Comparison of Equations 4.19 and 6.2 written for the Gibbs function G shows that what was defined as the chemical potential by Equation 4.19 is shown by Equation 6.2 to be the partial molar Gibbs energy.

$$\mu_i \equiv \bar{g}_i = \left(\frac{\partial G}{\partial n_i} \right)_{P,T,n_{j \neq i}} \tag{7.1}$$

Thus, using the general form of Equation 6.6, we write

$$g = \sum_k^c x_k \bar{g}_k = \sum_k^c x_k \mu_k \tag{7.2a}$$

or, alternatively,

$$G = \sum_k^c n_k \bar{g}_k = \sum_k^c n_k \mu_k \tag{7.2b}$$

Notice that *the chemical potential is not the partial molar property for the internal energy, the enthalpy, or the Helmholtz energy.* Equations 4.6, 4.13, and 4.16 show that for these functions, the differentiation with respect to the number of moles of compound i is not taken at constant pressure and temperature, as required by the definition of a partial molar property.

CHEMICAL POTENTIAL DEPENDENCE ON TEMPERATURE AND PRESSURE

The change of the chemical potential with temperature and pressure is given by Equations 4.24 and 4.25, which can be rewritten in terms of partial molar properties as

$$\left[\frac{\partial \mu_i}{\partial T} \right]_{P,n} = -\left[\frac{\partial S}{\partial n_i} \right]_{T,P,n_{j \neq i}} = -\bar{s}_i \tag{7.3}$$

61

and

$$\left[\frac{\partial \mu_i}{\partial P}\right]_{T,n} = \left[\frac{\partial V}{\partial n_i}\right]_{T,P,n_{j \neq i}} = \bar{v}_i \qquad (7.4)$$

GIBBS–HELMHOLTZ EQUATION

With the same procedure used for the derivation of Equation 6.12, from Equation 4.2 in molar terms we write

$$\mu_i \equiv \bar{g}_i = \bar{h}_i - T\bar{s}_i \qquad (7.5)$$

Equation 7.5 is a key relation to derive the all-important *Gibbs–Helmholtz equation*. For this purpose, we consider the following mathematical identity:

$$\left(\frac{\partial \left(\mu_i/RT\right)}{\partial T}\right)_{P,n} \equiv -\frac{\mu_i}{RT^2} + \frac{1}{RT}\left(\frac{\partial \mu_i}{\partial T}\right)_{P,n}$$

Thus, using Equation 7.3 and combining with Equation 7.5, we write

$$\left(\frac{\partial \left(\mu_i/RT\right)}{\partial T}\right)_{P,n} = -\frac{1}{RT^2}\left[\mu_i + T\bar{s}_i\right] \equiv -\frac{\bar{h}_i}{RT^2} \qquad (7.6a)$$

Equation 7.6a is the Gibbs–Helmholtz equation in terms of partial molar properties. As the partial derivative is taken at constant composition, this expression can be directly written in terms of molar or total properties. In the latter case, it takes the form

$$\left(\frac{\partial \left(G/RT\right)}{\partial T}\right)_{P,n} \equiv -\frac{H}{RT^2} \qquad (7.6b)$$

GIBBS–DUHEM EQUATION IN TERMS OF CHEMICAL POTENTIALS

Although all equations valid for partial molar properties can be directly written in terms of the chemical potential, due to its importance we rewrite Equations 6.9 and 6.10

in this form here. From the general form of Equation 6.9, *mutatis mutandis*, we write the all-important *Gibbs–Duhem equation in terms of chemical potentials*,

$$\sum_k^c x_k \, d\mu_k = \left(\frac{\partial g}{\partial T}\right)_{P,n} dT + \left(\frac{\partial g}{\partial P}\right)_{T,n} dP$$

or, using Equations 4.17 and 4.18, in molar terms,

$$\sum_k^c x_k \, d\mu_k = -s \, dT + v \, dP \tag{7.7a}$$

This equation is also conveniently written as

$$\sum_k^c n_k \, d\mu_k = -S \, dT + V \, dP \tag{7.7b}$$

Equation 7.7b corresponds to Equation 6.8, the Gibbs–Duhem equation, written in terms of the chemical potentials.

Furthermore, using the same procedure employed for the derivation of Equation 6.10, from Equation 7.7a, *for a mixture of fixed composition*, independent of the particular set of mole fractions considered, we obtain

$$d\mu_i = -\bar{s}_i \, dT + \bar{v}_i \, dP \tag{7.8}$$

This equation is one of the stepping-stones for the calculation of chemical potentials. At constant pressure and temperature, it may seem to indicate that the chemical potential of a compound i in a mixture is constant and independent of composition. On the other hand, we know that like any partial property, chemical potential is a function of the composition of the mixture. The fact is that Equation 7.8 was derived *for the case of fixed composition*, no matter what composition it would be. Thus, the composition dependence is hidden in the functionality of the partial molar properties $\bar{v}_i$ and $\bar{s}_i$. Therefore, the constant value of the chemical potential at constant pressure and temperature does depend on the composition of the system. This point is of outmost importance.

From the recognition that the chemical potential is the partial molar Gibbs energy, one can write Equation 6.15 for the chemical potential as

$$\mu_i = g - \sum_{k\neq i} x_k \left(\frac{\partial g}{\partial x_k}\right)_{T,P,x_r \atop r\neq i,\, r\neq k} \tag{7.9}$$

Equations can be transformed in so many ways that sometimes it is difficult to recognize them. One good example is the case of the *Euler's or integral form of the fundamental equation*, Equation 7.10. Combining Equation 4.2 with Equations 7.2b and 3.6b, we write

$$\sum_k^c n_k \, \mu_k = (U + PV) - TS$$

or, rearranging,

$$U = TS - PV + \sum_k^c n_k \, \mu_k \tag{7.10}$$

It is interesting to observe that from the three terms contributing to the total internal energy in Equation 7.10, the enthalpy and the Helmholtz functions retain only two of the three pieces of information contained in the internal energy,

$$H = U + PV = TS + \sum_k^c n_k \, \mu_k$$

$$A = U - TS = -PV + \sum_k^c n_k \, \mu_k$$

The Gibbs energy, on the other hand, retains only the compositional contribution, and thus it is the appropriate function to use when studying molecular interactions.

$$G = H - TS = U + PV - TS = \sum_k^c n_k \, \mu_k$$

It is also interesting to consider the differential form of Equation 7.10,

$$dU = T \, dS + S \, dT - P \, dV - V \, dP + \sum_k^c n_k \, d\mu_k + \sum_k^c \mu_k \, dn_k$$

or

$$dU = [T \, dS - P \, dV + \sum_k^c \mu_k \, dn_k] + [S \, dT - V \, dP + \sum_k^c n_k \, d\mu_k]$$

Comparison of this differential with Equation 4.7 shows that three terms from the second bracket of the right-hand side cancel out. These are the terms considered in the Gibbs–Duhem equation (7.7b).

MAXWELL RELATION IN TERMS OF THE CHEMICAL POTENTIAL

As the chemical potential is the partial molar Gibbs energy, from Equation 6.11 we write

$$\left(\frac{\partial \mu_j}{\partial n_i}\right)_{T,P,n_{k \neq i}} = \left(\frac{\partial \mu_i}{\partial n_j}\right)_{T,P,n_{k \neq j}} \tag{7.11}$$

This relation is of particular use for the study of electrolyte solutions, and it is seldom emphasized in the literature.

8 The Principles of Physical and Chemical Equilibrium

In this chapter, we discuss the conditions for equilibrium in multicomponent closed systems that can be formed by one or more phases (open subsystems) in direct contact. The numbers of moles of each compound present in each subsystem can change for one of the following reasons:

1. The subsystem is open and exchanges matter with its surroundings.
2. The subsystem is closed and undergoes chemical reactions.
3. The subsystem is open and undergoes chemical reactions.

The first case is a typical problem of phase equilibrium in which two nonreacting phases of different composition are put in direct contact in an overall closed system. The second case is typical of a study of chemical equilibrium. The third case is when two reacting phases are put in direct contact and allowed to reach phase and chemical equilibrium. These three cases will be discussed separately.

Before going into detail, it is necessary to agree on a change in the nomenclature. For the study of partial molar properties in mixtures of constant composition, it was convenient to use the subscript n to indicate that the numbers of moles did not change. For a closed system undergoing chemical reactions, the numbers of moles of the compounds will change, although the total mass of the system will be constant. Thus, to indicate a closed system from here on, we use the subscript m.

EXTREMUM PRINCIPLE FOR FUNCTIONS OTHER THAN ENTROPY

In Chapter 2, Equation 2.16 expressed in mathematical terms the extremum principle in terms of entropy. It is desirable to have this general principle expressed in terms of U, A, H, and G. For this, combining Equations 2.1d and 2.15 for an actual process in a closed system,

$$dU = TdS - T\delta S_{irr} + \delta W \tag{8.1}$$

For a process at constant entropy and at constant volume, the first and third terms of the right-hand side are null. As the second term is always negative for a real process, the internal energy decreases and it is a minimum at equilibrium. Thus, the mathematical expression of the extremum principle in terms of internal energy for a process at constant entropy and volume in a closed system takes the form

$$dU_{S,V,m} \leq 0 \tag{8.2}$$

By a similar argument, using Equations 4.1 and 3.6b, respectively, it is shown that in a closed system, A is at minimum at equilibrium for a process at constant temperature and volume, while H is at minimum for a process at constant entropy and volume.

$$dA_{T,V,m} \leq 0 \tag{8.3}$$

and

$$dH_{S,V,m} \leq 0 \tag{8.4}$$

Due to its importance for G, we repeat the argument here. From Equations 3.6b and 4.2 for a closed system, at constant temperature and pressure, we write

$$dG_{T,P,m} = [dU + PdV - TdS]_{T,P,m}$$

Combining this equation with Equation 8.1, and observing that at constant pressure the product PdV is equal to the actual irreversible work $\delta W = -P_E dV$, we write

$$dG_{T,P,m} = - TdS_{T,P,m} \leq 0 \tag{8.5}$$

Thus, for a process at constant temperature and pressure in a closed system, the Gibbs energy always decreases and it has its minimum value at equilibrium for the particular set of values of pressure and temperature. The key point to notice here is that once the system is at equilibrium, it is totally immaterial by which process it attained the equilibrium state at T, P, and m. In mathematical terms, for the condition of G being a minimum with respect to the variation of a generalized independent variable X *at equilibrium* we write

$$\left(\frac{\partial G}{\partial X} \right)_{T,P,m} = 0 \tag{8.6}$$

We are now in a position to present the analysis of the three possible processes by which a subsystem, forming a part of a closed system, can change its number of moles.

PHASE EQUILIBRIUM WITHOUT CHEMICAL REACTIONS

We consider here two nonreacting open subsystems α and β, in direct contact, forming part of an overall closed system of mass, m, which are left to reach equilibrium at a temperature, T, and a pressure, P. As the internal energy, the entropy, and the volume of the two subsystems in contact are additive, and the pressure and temperature of both subsystems at equilibrium are the same, the Gibbs energy of the overall closed system is the sum of the Gibbs energies of the subsystems. At equilibrium,

$$G = G^\alpha + G^\beta$$

Thus, for a differential change in the masses of the subsystems at constant temperature and pressure we have

$$dG_{T,P,m} = dG^\alpha_{T,P} + dG^\beta_{T,P}$$

For each subsystem, from Equation 7.2b,

$$dG_{T,P} = \sum_k^c n_k \, d\mu_k + \sum_k^c \mu_k \, dn_k$$

According to the Gibbs–Duhem equation, Equation 7.7b, at constant pressure and temperature the first term of the right-hand side is null. Thus, with all generality for each subsystem we write

$$dG_{T,P} = \sum_k^c \mu_k \, dn_k \tag{8.7}$$

For the overall closed system at constant pressure and temperature, adding the differential Gibbs energy changes for each subsystem and noting that necessarily $dn_i^\beta = -dn_i^\alpha$, we have

$$dG_{T,P,m} = \sum_k^c \left(\mu_k^\alpha - \mu_k^\beta\right) dn_k^\alpha$$

From this expression and Equation 8.6, for a differential change in the number of moles of any of the compounds i at equilibrium,

$$\left(\frac{\partial G}{\partial n_i^\alpha}\right)_{T,P,m} = \mu_i^\alpha - \mu_i^\beta = 0$$

or

$$\mu_i^\alpha = \mu_i^\beta \text{ for a compound } i \text{ in } all \text{ the phases in which it is present} \tag{8.8a}$$

If there are more than two phases present in the closed system, a similar analysis applied to each pair of phases in direct constant, for the case of π phases, leads to

$$\mu_i^\alpha(T,P,x^\alpha) = \mu_i^\beta(T,P,x^\beta) = \cdots = \mu_i^\pi(T,P,x^\pi) \tag{8.8b}$$

As indicated, the chemical potentials of compound i in all phases in which it is present are equal and are at the same temperature and pressure, but in each phase they are at the particular composition of the phase.

We observe that the equality of the chemical potential of a pure compound in two phases in equilibrium was deduced previously from direct application of the first and second laws. According to Equation 7.2a, for a pure compound the chemical potential is equal to the molar Gibbs energy, and as shown by Equation 5.14b for any two phases of a pure compound in equilibrium, the molar Gibbs energies are equal.

CHEMICAL EQUILIBRIUM IN A SINGLE-PHASE CLOSED SYSTEM

In a single-phase closed system undergoing chemical reactions at constant temperature and pressure, we write Equation 8.7 as

$$dG_{T,P,m} = \sum_{k}^{c} \mu_k \, dn_k \tag{8.7a}$$

where in this case, the subscript m has been added to the left-hand side to emphasize that this is valid for a closed system.

In a chemical reaction, the number of moles of *reactants*, R, consumed and the number of moles of *products*, P, produced are related by *stoichiometric coefficients*, λ:

$$\sum_{j}^{reactants} \lambda_j R_j = \sum_{k}^{products} \lambda_k P_k$$

Thus, the change in the number of moles of any two reactants is proportional to the ratio of their stoichiometric coefficients. The same happens for any two products. On the other hand, the ratio of the change in the number of moles of a reactant and a product is in the ratio of their stoichiometric coefficients affected with a minus sign, because the number of moles of reactant decreases as the number of mole of product increases. Thus, it is simpler to define the *stoichiometric number*, v_i, as

$$v_i = \lambda_i \text{ for products}$$

and

$$v_i = -\lambda_i \text{ for reactants}$$

and rewrite the stoichiometric equation in the form

$$\sum_{k}^{products} v_k P_k + \sum_{j}^{reactants} v_j R_j = 0$$

Hence, for any two compounds, reactants or products,

$$\frac{dn_1}{dn_2} = \frac{\nu_1}{\nu_2}$$

or, equivalently, for any two compounds participating in the reaction, reactants or products,

$$\frac{dn_1}{\nu_1} = \frac{dn_2}{\nu_2} = \cdots = d\xi$$

The parameter ξ is known as the *reaction coordinate*. Therefore, we write

$$dn_i = \nu_i d\xi \tag{8.9}$$

Equation 8.7a can be written as

$$dG_{T,P,m} = \left(\sum_{k}^{c} \nu_k \mu_k \right) d\xi$$

or, at equilibrium, from Equation 8.6,

$$\left(\frac{\partial G}{\partial \xi} \right)_{T,P,m} = \sum_{k}^{c} \nu_k \mu_k = 0 \tag{8.10}$$

This condition for chemical equilibrium applies to all reactions that can occur in a closed system (Figure 8.1).

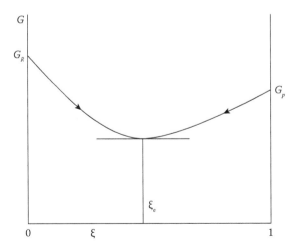

FIGURE 8.1   Gibbs function of a closed system in which a chemical reaction takes place.

Although the chemical potential for compound i at equilibrium has a single value for all possible reactions, the stoichiometric number of compound i in different reactions is not necessarily the same. Thus, the *thermodynamic condition for chemical equilibrium* is more precisely written in the form

$$\sum_{k}^{c} \nu_{k,r}\, \mu_k = 0 \qquad (8.11)$$

In this expression, the subscript r indicates that there is one such relation for each *independent reaction* that can occur in the system.

The *number of independent chemical reactions* possible (NIR) is obtained by writing the equation of formation of all compounds considered to be present, starting from their constituent elements. By algebraic combination of these imaginary reactions of formation, all elements, which are considered not to be present as elements, are eliminated. The set of equations resulting from this process gives the number of independent reactions possible. Although the set of final equations depends on the algebra, the number of such equations should be the same for all sets. In general, this number is larger or equal to the number of compounds and elements present as elements minus the number of constituent elements. It is highly recommended to eliminate the elements not present as elements in an organized way by selecting one equation containing one such element and using it to eliminate that particular element from all other equations.

PHASE AND CHEMICAL EQUILIBRIUM IN A CLOSED SYSTEM

In this case, Equations 8.8 and 8.11 will apply simultaneously. As the reactions will be at equilibrium in each of the phases present, with the same stoichiometric numbers and the same values for the chemical potentials of the compounds, the chemical equilibrium is conveniently solved in the phase where calculations are simpler. This is usually the gas phase, when present.

As shown in this chapter, the conditions for phase equilibrium in simple systems or chemical equilibrium in reacting systems are the direct consequence of the minimum in the Gibbs energy at equilibrium. This condition, in turn, is a direct consequence of the increase in entropy of the system due to the irreversibility of real processes. The generality of the thermodynamic treatment presented up to this point is further discussed in Chapter 10.

9 The Phase Rule and the Duhem Theorem

When solving a problem of phase or chemical equilibrium, it is necessary to know if there is enough information or it would be necessary to specify some variables before evaluating the remaining variables. As in any multivariable system of equations, if the number of variables is equal to the number of independent equations, the system can be solved. If there are more equations than variables, the system is overdetermined and, in the general case, no solution exists. On the other hand, if there are fewer relations than variables, the system is underdefined. The difference between the number of variables and the number of relations between them gives the available degrees of freedom (DF), that is, the number of variables that need to be specified in order to solve for all others.

For simplicity, we discuss separately the case of a closed nonreacting system with two or more phases in equilibrium and the case of a closed reacting system in which phases are in physical and chemical equilibrium.

PHASE RULE FOR NONREACTING SYSTEMS

The chemical potential of a compound in a homogeneous mixture is a function of temperature, pressure, and the composition of the mixture. Thus, in mathematical terms, each equality of the chemical potential of compound i included in Equation 8.8a or 8.8b represents one equation relating the variables of both phases in contact. These variables are the mole fractions x of the compounds in each phase, the temperature, T, and the pressure, P.

Thus, if π phases are present, Equation 8.8 indicates that there are $(\pi - 1)$ independent equations per compound relating the variables involved. For c compounds, the total number of equations available is $c\,(\pi - 1)$.

With respect to the variables involved, we recall that the mole fractions in each phase add to unity, so there are only $(c - 1)$ independent mole fractions for a phase with c components. Assuming that all c compounds are present in all the phases, as the pressure and the temperature are the same for all phases, the total number of independent variables is $[(c - 1)\,\pi + 2]$.

We recall here that in the case we are considering, there are no chemical reactions; thus,

$$DF = [(c - 1)\,\pi + 2] - c\,(\pi - 1)$$

or

$$DF = c - \pi + 2 \tag{9.1}$$

If one compound is not present in one phase, that is, if the value of one composition variable is known to be zero, there is one variable less and also one equation less, as the equality of chemical potentials of that compound does not enter into the count. Thus, the number of degrees of freedom remains unchanged.

PHASE RULE FOR REACTING SYSTEMS

In this case, we consider a single phase. If there is a number of independent reactions (*NIR*) that are possible in the system, the number of variables will not change, but the number of relations between the variables will increase by exactly the number *NIR* of independent equilibrium conditions (Equation 8.10). Hence, in this case,

$$DF = [(c - 1) \pi + 2] - [c (\pi - 1) + NIR]$$

$$DF = (c - NIR) - \pi + 2 \tag{9.2}$$

If one reaction goes to completion at equilibrium and some reactants that were initially in the system are totally consumed, both the reaction and the compounds totally consumed can be ignored.

COMMON CASES OF APPLICATION OF THE PHASE RULE

It is useful to keep in mind the number of variables that it is necessary to specify to define completely the intensive state of a system in equilibrium. For a pure compound or for a mixture of known composition in one phase, the values of two intensive variables are required. One single intensive variable completely defines the state of a pure compound in two phases. For a binary system in two phases, there are two degrees of freedom. Thus, if the composition of one of the phases is known, either the pressure or the temperature can be fixed arbitrarily, but not both. In chemical engineering calculations for vapor–liquid equilibrium in binary systems, if the liquid composition is known, the calculation of the vapor phase composition is called a "bubble" calculation. Similarly, the calculation of the liquid phase composition from information on the vapor phase composition is called a "dew" calculation. If the pressure is fixed and the temperature is unknown, the problem is further indicated as a *T* calculation, and vice versa for the pressure calculation. Thus, a *bubble P* calculation is the case of knowing the liquid composition and the temperature and having to obtain the vapor phase composition and pressure.

DUHEM THEOREM

One interesting case in chemical engineering is the "flashing" of a liquid mixture of known composition in a low-pressure drum to obtain a liquid phase and a vapor phase in equilibrium. This practical process is run in a continuous way, and the mass flow rate and composition of the feed are known. The *Duhem theorem* gives the number of intensive variables necessary to fix, so the mass flow rates and compositions of the vapor and liquid product streams are uniquely determined.

The surprising result is that even requiring an extra extensive variable to determine the extensive state of each phase, the number of intensive variables that is necessary to fix is only two, independent of the number of components of the mixture. Thus, if the composition and mass flow rate of the feed are known, fixing the temperature and pressure in the drum completely determines the system. Duhem's theorem can be demonstrated by a simple count of the number of variables and relations involved. In addition to the intensive variables considered by the phase rule, it is necessary to know one additional extensive variable per phase present to define the extensive state of each of them.

Thus, the total number of variables in this case for π phases is

$$(\text{Number of variables}) = [(c - 1)\, \pi + 2] + \pi = c\, \pi + 2$$

On the other hand, in the case considered, there are also additional relations between the variables. In the absence of chemical reactions, at steady state each compound that enters in the feed must leave the system in the outflow streams. The equations relating the masses of each compound in the three streams are known as material balances. There are as many material balances as compounds in the system. Hence,

$$(\text{Number of relations}) = (\pi - 1)\, c + c = c\, \pi$$

Thus, $DF = 2$

PHASE RULE WITH ADDITIONAL CONSTRAINTS

Additional constraints not considered in the cases discussed above will discount degrees of freedom. If a pure compound is present as a phase in a system of c compounds, this will discount $(c - 1)$ variables. In some complex cases, it is better to obtain the degrees of freedom by direct counting of variables and restrictions. For example, a rigid and semipermeable membrane separating two phases in equilibrium will cause the pressure to not be the same in both phases, and some compositions may have a zero value in some of the phases. Even if a compound is present in more than one phase but cannot permeate through the membrane, the equality of potentials would not apply for that compound and one relationship is lost.

10 Generality of the Thermodynamic Treatment for More Complex Systems

Simple systems discussed in the first seven chapters were formed by nonreacting compounds, did not include electrically charged species, and were considered to be large enough to make surface effects negligible. Chapter 8 introduced a first generalization of the thermodynamic treatment and included the discussion of reacting systems. Thus, we accept now that the internal energy of the system can change by the exchange of heat with the surroundings, by expansion or compression work due to the action of external forces, and also by the change in the number of moles of the components due to the fact that either the system is open or it undergoes chemical reactions, or both.

In this section, we discuss the possibility of having work other than expansion or compression work in a thermodynamic reversible process between two equilibrium states of the system. Equation 4.7 had the form

$$dU = TdS - PdV + \sum_{k}^{c} \mu_k \, dn_k \tag{4.7}$$

If there is a reversible work $\delta \hat{W}_{rev}$, different from expansion work, it needs to be incorporated. Thus, we write the generalized form of Equation 4.7 as

$$d\widehat{U} = TdS - PdV + \delta \hat{W}_{rev} + \sum_{k}^{c} \mu_k \, dn_k \tag{10.1}$$

In Equation 10.1, a special symbol is used for the internal energy as a reminder that the system under discussion is not a simple system. Otherwise, for all practical consequences, the algebra done for simple systems will remain unchanged. For example, the relation of the internal energy with the enthalpy of such a system follows Equation 3.6, written as

$$\hat{H} = \hat{U} + PV$$

Similarly, the Gibbs energy of the nonsimple system follows the definition given in Equation 4.2, namely,

$$\hat{G} = \hat{H} - TS = \hat{U} + PV - TS$$

The differential form of this expression,

$$d\hat{G} = d\hat{U} + PdV + VdP - TdS - SdT$$

combined with Equation 10.1, gives

$$d\hat{G} = -SdT + VdP + \delta\hat{W}_{rev} + \sum_{k}^{c} \mu_k \, dn_k \qquad (10.2)$$

In most practical applications, the pressure and temperature are kept constant and we write

$$d\hat{G}_{T,P} = \delta\hat{W}_{rev} + \sum_{k}^{c} \mu_k \, dn_k \qquad (10.3)$$

We consider two examples of application of Equation 10.3 here.

EQUILIBRIUM INVOLVING IONS

Ions are *charged particles* produced by dissolving salts in water or other appropriate solvent. The word *ion* means "traveler" in Greek. Dissolving sodium chloride in water, for example, and applying an electric potential on two electrodes immersed in the solution, negatively charged chloride species and positively charged sodium species will travel toward the appropriate electrodes. Sodium ions will travel toward the negatively charged electrode, or *cathode*, and chloride ions toward the positively charged electrode, or *anode*. Thus, these positively and negatively charged ions are called *cations* and *anions*, respectively. As an electric current will pass from one electrode to the other when a salt is in solution, the salts generating ions are called *electrolytes*.

Electrochemical systems are complex and the thermodynamic description of electrochemical processes is far from simple. The problem starts with the difficulty of defining unambiguously the electrostatic potential of an electrolyte solution. As with the internal energy or the enthalpy, in thermodynamics we normally do not know the "value" of a property, but we are satisfied with being able to determine the difference in its value for a particular process or condition.

Thus, to proceed, we accept that a phase α of an electrolyte solution has an electrostatic potential Φ^α, with respect to an arbitrary zero. This potential is determined by the concentrations of electrolytes present, the temperature, and to a minor extent for liquid solutions, the pressure of the system. The extra work required in this case is the electrical work necessary to move a charge $\delta\Omega_i$ from a state at zero potential to a state at potential Φ of the solution. The addition of an ion to the solution adds energy to its internal energy. As with the expansion or compression work, the thermodynamic work is the work done by the external force; hence,

$$\delta \hat{W}_{rev,i} = \Phi \delta \Omega_i$$

The units used for the potential are volts. As the charge is measured in coulombs, that is, joules per volt, the work is expressed in joules. Each ion i has a characteristic charge Z_i. Sodium ion has charge +1, chloride ion has charge -1, and so on. The Faraday constant F is equal to 9.6484 JV^{-1} mol^{-1}, which is the product of the charge of an electron, that is, 1.6022 10^{-19} J/V, times Avogadro's number, that is, 6.0220 10^{23} mol^{-1}. Hence, the charge of dn_i moles of ion i is

$$\delta \Omega_i = Z_i \, F dn_i$$

and the electric work per dn_i moles of ion i is

$$\delta \hat{W}_{rev,i} = \Phi Z_i \, F \, dn_i$$

Introducing this result in Equation 10.3, we write

$$d\hat{G}_{T,P} = \sum_k^c (\mu_k + \Phi Z_k F) \, dn_k$$

The *electrochemical potential* of ion i is then defined by

$$\hat{\mu}_i = \mu_i + \Phi Z_i F \tag{10.4}$$

Thus, the above equation reduces to the familiar form of Equation 8.7, namely,

$$d\hat{G}_{T,P} = \sum_k^c \hat{\mu}_k \, dn_k \tag{10.5}$$

and in total parallel with Equation 7.1, we write

$$\hat{\mu}_i = \left(\frac{\partial \hat{G}}{\partial n_i} \right)_{P,T,\Phi,n_{j \neq i}} \tag{10.6}$$

From here on, *mutatis mutandis*, the algebra is not any different than the one for noncharged species. For *phase equilibrium of ions in solution*, the equilibrium condition takes the form

$$\hat{\mu}_i^\alpha = \hat{\mu}_i^\beta = \cdots = \hat{\mu}_i^\pi \tag{10.7}$$

Writing this equation, in terms of chemical potentials, for an ion in two phases, we get

$$\mu_i^\alpha + \Phi^\alpha \, Z_i F = \mu_i^\beta + \Phi^\beta \, Z_i F$$

or

$$\mu_i^\alpha - \mu_i^\beta = \left[\Phi^\beta - \Phi^\alpha \right] Z_i \, F$$

This expression is better written as

$$\frac{1}{Z_i} \left[\mu_i^\alpha - \mu_i^\beta \right] = \Delta \Phi \, F \tag{10.8}$$

Equation 10.8, in which $\Delta \Phi$ is known as the *Donnan potential*, shows that the right-hand side is independent of the ion considered, and consequently, different ions in solutions at equilibrium meet the condition

$$\frac{1}{Z_i} \left[\mu_i^\alpha - \mu_i^\beta \right] = \frac{1}{Z_j} \left[\mu_j^\alpha - \mu_j^\beta \right] \tag{10.9}$$

This relation is known as the *Donnan equilibrium* condition for ions in solutions in contact through membranes.

Similarly, the case of "chemical equilibrium" for a single closed phase is found when a salt is dissolved in an appropriate solvent such as water, and it dissociates into its ions. A chemical equilibrium is established between the undissociated salt in solution and its ions. Strong electrolytes, like NaCl or KOH, dissociate completely, while weak electrolytes like alanine or acetic acid, will dissociate only partially. Another case of ionic equilibrium is metal in equilibrium with its ions in an acid solution. For all these cases, the condition of chemical equilibrium, obtained by the same algebra as the one used to obtain Equation 8.10 for the case of nonelectrolytes, gives

$$\sum_k \nu_k \hat{\mu}_k = 0 \tag{10.10}$$

Using Equation 10.4, we write

$$\sum_k \nu_k \mu_k + F \Phi \sum_k \nu_k Z_k = 0$$

In this equation, the second summation is null, as it is the difference of the charges of all the cations and the charges of all the anions. The sign of the charges is implicit in Z_i.

$$\sum_k \nu_k Z_k = 0 \tag{10.11}$$

Equation 10.11 is known as the *electroneutrality condition for all solutions of electrolytes*. Thus, finally for chemical equilibrium of electrolytes in solution, we write

$$\sum_k \nu_k \mu_k = 0 \qquad (10.12)$$

Equation 10.12 is identical to Equation 8.11 for nonelectrolytes. The important point to observe here is that in Equation 10.9 for phase equilibrium and also in Equation 10.12 for chemical equilibrium in electrolyte solutions, the loosely defined phase electrostatic potential Φ cancels out. In some sense, this potential has played the same role as the entropy in Chapter 5; it was used as a stepping-stone to derive equations of practical use. The use of these relations will be further explored in Chapter 18 dealing with electrolyte solutions.

ADSORPTION OF CHEMICAL SPECIES ON A SURFACE

As another example of the generality of the thermodynamic treatment, we discuss here the basic principle of equilibrium in the case of *adsorption of a solute from a solution over a solid surface*. The solution can be in the gas or liquid phase. As in the case of the previous example, we will not go into the details of the application here.

In the case of adsorption equilibrium, there are three phases present: a fluid phase that can be a solution or a pure compound, a solid phase, and an adsorbed phase. While the first two phases are volumetric in nature, the adsorbed phase is considered to be a two-dimensional phase covering the surface of the solid. At equilibrium, for any chemical species present in the fluid phase and in the adsorbed phase, the chemical potential of the compound will be the same in both phases:

$$\mu_i^{ads} = \mu_i$$

The symbol without superscript in this expression refers to the chemical potential in the fluid phase. The work of adsorption is the work to extend the adsorbing surface area A of the adsorbent to receive more adsorbate. The surface tension σ is defined as the force per unit of frontal surface length exerted by the surface; thus, the adsorption work to increase the adsorbed area by a differential dA is

$$\delta \hat{W}_{rev} = \sigma dA$$

Introducing this expression into Equation 10.3, for the adsorbed phase we obtain

$$d\hat{G}_{T,P}^{ads} = \sigma dA + \sum_k^c \mu_k \, dn_k^{ads} \qquad (10.13)$$

Here, using the equality of chemical potentials for compound i, the superscript *ads* is omitted. We now use the same method employed to derive Equation 6.5, namely, the fact that if the number of moles of *all* compounds adsorbed is multiplied by a factor λ, the area covered and the value of the Gibbs function will also be changed by the same factor. Thus,

$$d\left[\lambda \hat{G}_{T,P}^{ads}\right] = \sigma d[\lambda A] + \sum_{k}^{c} \mu_k d\left[\lambda n_k^{ads}\right]$$

Differentiation of this expression with respect to λ gives

$$\hat{G}_{T,P}^{ads} = \sigma A + \sum_{k}^{c} \mu_k n_k^{ads}$$

Thus,

$$d\hat{G}_{T,P}^{ads} = \sigma dA + A d\sigma + \sum_{k}^{c} \mu_k dn_k^{ads} + \sum_{k}^{c} n_k^{ads} d\mu_k$$

Comparison of this expression with Equation 10.13 shows that, necessarily,

$$A d\sigma + \sum_{k}^{c} n_k^{ads} d\mu_k = 0 \qquad (10.14)$$

Equation 10.14 is the Gibbs–Duhem equation for the adsorbed phase.

In the work of adsorption, it is common to define the *Gibbs adsorption*, Γ_i, as the number of moles of compound i adsorbed per unit surface area of adsorbent.

$$\Gamma_i = \frac{n_i^{ads}}{A} \qquad (10.15)$$

Combining Equations 10.14 and 10.15, we obtain

$$d\sigma = -\sum_{k}^{c} \Gamma_k d\mu_k \qquad (10.16)$$

Variations of Equation 10.16 are commonly used in adsorption studies. For example, for the case of adsorption on a solid surface of a solute 2 from a solution in a solvent 1, in which the solvent is *not* adsorbed by the solid surface, that is, $\Gamma_1 = 0$, from Equation 10.16 we get

$$\Gamma_2 = -\frac{d\sigma}{d\mu_2} \qquad (10.17)$$

This expression is normally used as the starting point in physical chemistry texts.

As a final example of the large number of applications to be found by extension of the basic thermodynamic treatment, we present here one expression useful for the study of small systems where the surface effects become important.

Consider a droplet of a solution in a chamber containing vapors of the solvent and air. The number of moles of solute in the droplet is constant, while the number of moles of solvent can be changed by evaporation or condensation of solvent in the droplet. In this case, as there is no change in the number of moles of solute, Equation 10.13 takes the form

$$d\hat{G}_{T,P}^{ads} = \sigma_s \, dA + \mu_s \, dn_s$$

The subscript s has been used to indicate that the surface tension in this case is that of the solvent in the droplet. Assuming a spherical droplet of radius r, a change in the surface area of sphere is related to a change in its volume by

$$dA = 2\frac{dV}{r}$$

In addition, for a dilute solution the partial molar volume of the solvent can be approximated by its pure compound molar volume v_s and we can write

$$dV = v_s \, dn_s$$

Thus,

$$d\hat{G}_{T,P}^{ads} = \left[2\sigma_s \frac{v_s}{r} + \mu_s\right] dn_s$$

and the chemical potential of the solvent in the droplet, following Equation 10.6, is

$$\hat{\mu}_s = \left(\frac{\partial \hat{G}}{\partial n_s}\right)_{P,T,\sigma_s} = \left[2\sigma_s \frac{v_s}{r} + \mu_s\right] \tag{10.18}$$

Clearly, the change in the chemical potential of the solvent is important in small systems, and as the radius of the droplet increases, the difference of chemical potentials becomes negligible.

The purpose of this chapter has been to show some selected examples in which the same standard ideas of the first nine chapters can be applied to discuss more complex systems. We have followed the developments to the point where the equations obtained are normally used as a starting point for applications.

11 Ideal Gas and Ideal Gas Mixtures

CHEMICAL POTENTIAL OF A COMPOUND IN AN IDEAL GAS MIXTURE

As discussed in Chapter 8, the study of phase and chemical equilibrium requires knowledge of the chemical potentials of the components of a mixture at the temperature, T, of the system. A convenient starting point to calculate chemical potentials is Equation 7.8, which indicates that at constant pressure and temperature the chemical potentials of the compounds in the mixture have fixed values. This result is in agreement with the conclusion obtained in Chapter 9, from the application of the phase rule to a single phase multicomponent mixture of fixed composition. This system has two degrees of freedom, and fixing the temperature and pressure, the system is completely specified. We also recall that, in principle, in Equation 7.8 the composition dependence of the chemical potential is hidden in the functionality of the partial molar properties $\bar{v}_i$ and $\bar{s}_i$. Thus, at constant temperature Equation 7.8 reads

$$d\mu_{i,T} = \bar{v}_i \, dP \qquad (11.1)$$

This equation would be directly applicable if the partial molar volume of compound i in the mixture as a function of pressure could be obtained from Equation 6.2 or Equation 6.15. This would be the case if an equation of state (EOS) giving the volume of the mixture as a function of composition and pressure at the temperature of interest were available. In practice, the use of available volume-explicit EOSs is restricted to gases of nonpolar pure compounds at low pressure and relatively high temperatures. Thus, a better way to proceed is to start the analysis considering the case of a compound in a multicomponent ideal gas mixture for which the volumetric behavior can be described without approximations. From Equation 1.1 or Equation 5.27, for n moles of mixture we write

$$V^* = \frac{\left[\sum\limits_{k}^{c} n_k \right] RT}{P}$$

The partial molar volume of compound i in an ideal gas mixture, from Equation 6.2, is

$$\bar{v}_i^* \equiv \left(\frac{\partial V^*}{\partial n_i} \right)_{T,P,n_{j \neq i}} = \frac{RT}{P} \qquad (11.2)$$

Comparison of Equation 11.2 with Equation 5.27 shows that the partial molar volume of a component of an ideal gas mixture at temperature, T, and pressure, P, is equal to the molar volume of the pure compound as ideal gas under the same conditions of pressure and temperature.

$$\bar{v}_i^* = v_i^{0,*} = \frac{RT}{P}$$

This is true for an ideal gas mixture only. We observe that n_i moles of pure compound i as ideal gas, at the temperature and pressure of the mixture, occupy a total volume that is different from the total volume of the gas mixture.

$$V_i^{0,*}(P,T) = \frac{n_i\,RT}{P} \neq V^*$$

In an ideal gas, molecules are mere points in space and do not interact in any way with one another. Each compound i in the ideal gas mixture occupies the whole volume V^*. If this is the case, as all other components of the ideal mixture also exert pressure on the walls, the pressure exerted on the wall of the container by a compound i cannot be the total pressure P. Both conditions cannot hold simultaneously. The n_i moles of compound i in the ideal gas mixture exert a *partial pressure* p_i^* on the walls of the container when the total volume of the mixture is V^*. Thus,

$$p_i^* = \frac{n_i\,RT}{V^*} = \frac{n_i}{\sum n_k} P = y_i\,P$$

In this equation, we have used the symbol y_i for the mole fraction of compound i in a gas phase mixture in exact parallel with Equation 6.7. Addition over all components of the mixture gives the total pressure P exerted by the ideal gas mixture. For compound i in a real gas mixture, that is, a gas mixture whose volumetric behavior does not follow Equation 5.27, we define its partial pressure by

$$\bar{p}_i \equiv y_i\,P \qquad\qquad\qquad (11.3)$$

Two points need clarification here. First, the partial pressure of a compound in a *real* gas mixture is not the pressure that compound i would exert on the wall if it occupies the same volume at the temperature T. This is only true for a component of an ideal gas mixture. Second, the partial pressure is not a partial molar property. The partial pressure of a component in a gas mixture is simply defined by Equation 11.3. Returning to the ideal gas case, combining Equations 11.1 and 11.2, we write

$$d\mu_{i,T}^* = \frac{RT}{P}dP = RT\,d\ln P$$

Integrating between two pressures P_1 and P_2 for a compound as ideal gas at temperature T, we write

$$\mu_i^*(T,P_2) = \mu_i^*(T,P_1) + RT \ln \frac{P_2}{P_1}$$

It is conventional to consider state 1 as the pure compound ideal gas at the temperature T of the system and at *one-unit pressure*, whatever the units of pressure being used. Hence, for the pure compound as ideal gas at pressure P and temperature T of the mixture, we write

$$\mu_i^{0,*}(T,P) = \mu_i^{0,*}(T,1) + RT \ln P$$

and for the compound i, in an ideal gas mixture at total pressure P and temperature T,

$$\mu_i^*(T,P,y) = \mu_i^{0,*}(T,1) + RT \ln(y_i P) \tag{11.4}$$

PROPERTY CHANGES IN THE MIXING OF IDEAL GASES

For the process of mixing pure compounds as ideal gases at a pressure P and a temperature T, to form an ideal gas mixture at the same pressure and temperature in a closed system, by analogy to Equations 6.18 and 6.19, for the molar Gibbs energy change, we write

$$\Delta g_m^* = \sum_{k=1}^{c} y_k \left(\bar{g}_k^* - g_k^{0,*} \right)$$

Because according to Equation 7.1 the chemical potential is the partial molar Gibbs energy, from Equation 11.4,

$$\Delta g_m^* = \sum_{k=1}^{c} y_k (\mu_k^* - \mu_k^{0,*}) = RT \sum_{k=1}^{c} y_k [\ln(y_k P) - \ln P]$$

or

$$\Delta g_m^* = RT \sum_{k=1}^{c} y_k \ln y_k \tag{11.5}$$

The Gibbs–Helmholtz equation, Equation 7.6b, written in terms of molar property changes, has the form

$$\left(\frac{\partial \left(\Delta g_m / RT \right)}{\partial T} \right)_{P,n} \equiv - \frac{\Delta h_m}{RT^2}$$

Thus, as the right-hand side of Equation 11.5 becomes independent of temperature after division by RT,

$$\Delta h_m^* = \sum_{k=1}^{c} y_k \left(\bar{h}_k^* - h_k^{0,*} \right) = 0 \tag{11.6}$$

In addition, as Equations 11.2 and 5.27 show that the partial molar volume of each component in the mixture is equal to its pure compound molar volume,

$$\Delta v_m^* = \sum_{k=1}^{c} y_k \left(\bar{v}_k^* - v_k^{0,*} \right) = 0 \tag{11.7}$$

Two more conclusions can be reached, for the constant pressure mixing process,

$$\Delta u_m^* = \sum_{k=1}^{c} y_k \left(\bar{u}_k^* - u_k^{0,*} \right) = \Delta h_m^* - P \Delta v_m^* = 0 \tag{11.8}$$

and, finally, as the process is isothermal,

$$\Delta g_m^* = \Delta h_m^* - T \Delta s_m^*$$

Thus,

$$\Delta s_m^* = - R \sum_{k=1}^{c} y_k \ln y_k \tag{11.9}$$

Note that when forming an ideal gas mixture at pressure P and temperature T in a closed system, starting with the pure compounds as ideal gases at the same pressure and temperature, the process occurs without any volume change or heat effect. However, the entropy does change. As discussed in Chapter 6, the fact that the process is isochoric and isenthalpic indicates that the ideal gas mixture is ideal. On the other hand, ideal mixtures are not necessarily in the ideal gas state. They can be a real gas mixture, a liquid, or even a solid mixture. An additional important conclusion is that for the ideal gas case, the composition dependence of the chemical potential shows in the argument of the logarithmic term of Equation 11.4. This observation is the basis for the treatment of real mixtures discussed in Chapter 12.

12 Equilibrium in Terms of Fugacity and Activity

DEFINITION OF FUGACITY AND FUGACITY COEFFICIENT

For the treatment of mixtures in gaseous or liquid phases for which a pressure-explicit equation of state would be available with the adjustable parameters dependent on composition, one could think of using Equation 11.1 to calculate chemical potentials. In fact, the partial molar volume, as defined by Equation 6.2, can be then evaluated using the following transformation:

$$\bar{v}_i \equiv \left(\frac{\partial V}{\partial n_i}\right)_{T,P,n_{j\neq i}} = -\left(\frac{\partial P}{\partial n_i}\right)_{T,V,n_{j\neq i}} \left(\frac{\partial V}{\partial P}\right)_{T,n} = \left(\frac{\partial P}{\partial n_i}\right)_{T,V,n_{j\neq i}} V \kappa_T$$

In the first equality to the right, after the identity, we have applied Equation 4.28, and in the second equality, we have applied the definition of the coefficient of isothermal compressibility, Equation 5.2. Although this treatment is feasible, it is cumbersome and discouraging. One alternative, which will be developed in Chapter 13, is to evaluate the chemical potential from the Helmholtz function using Equation 4.16. We include these considerations here to emphasize the ingenious step taken by G. N. Lewis by introducing the definition of fugacity. This step is comparable to the introduction of the concepts of entropy or of chemical potential. G. N. Lewis proposed to start from what was known for ideal gas mixtures for which the isothermal differential form of Equation 11.4 is

$$d\mu_{i,T}^*(P,y) = RT\, d\ln[y_i P] \tag{12.1}$$

In this expression, appearing naturally from the treatment of the ideal gas mixture, the composition dependence of the chemical potential is in the argument of the logarithmic term. In a major leap forward, Lewis defined a function of composition, temperature, and pressure f_i, called *fugacity*, which for any component i in a mixture, real or ideal, gives the isothermal change of the chemical potential of compound i by the relation

$$d\mu_{i,T}(P,x) \equiv RT\, d\ln f_i(x,T,P) \tag{12.2}$$

In this identity, which is valid for any phase gas or liquid, for generality the mole fraction has been designated by x. From Equation 11.4 for compound i in an ideal gas mixture and in the pure compound state, we have, respectively,

$$f_i^* = x_i P \tag{12.3a}$$

and

$$f_i^{0,*} = P \tag{12.3b}$$

As the pressure decreases tending to zero, any system *at equilibrium* approaches the behavior of the ideal gas system, and the limiting value of the fugacity tends to the value of the "partial pressure" of the component in the ideal gas mixture; hence,

$$\lim_{P \to 0} f_i = x_i P \tag{12.4}$$

This limiting behavior suggests the convenience of defining the *fugacity coefficient*, ϕ_i, of compound i by

$$\phi_i(T,P,x) \equiv \frac{f_i(T,P,x)}{x_i P} \tag{12.5}$$

and we rewrite Equation 12.4 as

$$\lim_{P \to 0} \phi_i = 1 \tag{12.6}$$

The relations for a pure compound equivalent to Equations 12.5 and 12.6 are

$$\phi_i^0(T,P) \equiv \frac{f_i^0(T,P)}{P} \tag{12.7}$$

and

$$\lim_{P \to 0} \phi_i^0 = 1 \tag{12.8}$$

Integrating Equation 12.2 between the state of compound i as pure compound ideal gas at one unit pressure and its state in a mixture at pressure P, after rearrangement, gives

$$\mu_i(T,P,x) = \mu_i^{0,*}(T,1) + RT \ln f_i(T,P,x) \tag{12.9}$$

PHASE EQUILIBRIUM IN TERMS OF FUGACITIES

Thus, for phase equilibrium of compound i in phases α and β, the equality of chemical potentials, $\mu_i^\alpha = \mu_i^\beta$, takes the form

$$\mu_i^{0,*}(T,1) + RT \ln f_i^\alpha \left(T,P,x^\alpha\right) = \mu_i^{0,*}(T,1) + RT \ln f_i^\beta \left(T,P,x^\beta\right)$$

or, simply,

$$f_i^\alpha \left(T, P, x^\alpha \right) = f_i^\beta \left(T, P, x^\beta \right) \tag{12.10}$$

Thus, *for phase equilibrium calculations, with all generality, the equality of fugacities replaces the equality of chemical potentials.* At this point, one may question the advantage of having defined the fugacity and fugacity coefficient of a compound in a mixture, in comparison with the direct use of chemical potentials. One advantage is encountered in the treatment of dilute solutions. As shown by Equation 12.1 for an ideal gas mixture, as the concentration of compound i goes to zero, its chemical potential goes to minus infinity, while according to Equation 12.3, its fugacity goes to zero. On the other extreme, as the compound gets more concentrated in the ideal gas mixture, its fugacity goes to the value of the pressure of the system. This physical interpretation of fugacity helps in making estimates for iterative calculations. Other advantages will become apparent. Isothermal integration of Equation 12.2 for compound i between some arbitrarily chosen state (a) and its state in a mixture at pressure P gives

$$\mu_i(T,P,x) = \mu_i^{(a)}\left(T,P^{(a)},x^{(a)}\right) + RT \ln \frac{f_i(T,P,x)}{f_i^{(a)}\left(T,P^{(a)},x^{(a)}\right)} \tag{12.11}$$

One possibility is to consider the arbitrary state as the pure compound at the temperature of the system and at a well-defined real or imaginary state characterized by its pressure $P^{(\theta)}$. This particular reference state is called the *standard state* of the compound. In this case, we write

$$\mu_i(T,P,x) = \mu_i^{(\theta)}\left(T,P^{(\theta)}\right) + RT \ln \frac{f_i(T,P,x)}{f_i^{(\theta)}\left(T,P^{(\theta)}\right)} \tag{12.12}$$

DEFINITION OF ACTIVITY

The ratio of fugacities of the right-hand side of Equation 12.12 is called the *activity of compound* i *with respect to the standard state* θ. This function is normally used in the study of chemical equilibrium.

$$a_{i,\theta}(T,P,x) \equiv \frac{f_i(T,P,x)}{f_i^{(\theta)}\left(T,P^{(\theta)}\right)} \tag{12.13}$$

Depending on the problem studied, it may be more convenient to use other units for measuring concentration and other standard states. A value of the activity of a compound without a clear specification of the standard state used for its evaluation is meaningless.

In simplified nomenclature, we rewrite Equation 12.12 as

$$\mu_i = \mu_i^{(\theta)} + RT \ln a_{i,\theta} \tag{12.12a}$$

CHEMICAL EQUILIBRIUM IN TERMS OF ACTIVITY

For the case of chemical equilibrium, then, we write the condition of equilibrium, Equation 8.11, as

$$\sum_{k}^{c} \nu_k \mu_k^{(\theta)} + RT \sum_{k}^{c} \nu_k \ln a_{k,\theta} = 0$$

or

$$\sum_{k}^{c} \nu_k \ln a_{k,\theta} = -\frac{\displaystyle\sum_{k}^{c} \nu_k \mu_k^{(\theta)}}{RT} \tag{12.14a}$$

For all practical purposes, the above expression is the final result needed. Observe that due to the implicit sign of the stoichiometric numbers, the sum of the right-hand side is the sum of the standard values of the molar Gibbs energies of the products minus the sum of the standard values of the molar Gibbs energies of the reactants; that is, this sum represents the *standard Gibbs energy change of reaction*. From Equation 7.2b, we write

$$\Delta G_T^\theta \equiv \sum_{k}^{c} \nu_k \mu_k^{(\theta)} \tag{12.14b}$$

where the subscript T indicates that the value of the Gibbs energy change of reaction is at the temperature of the system and the superscript θ indicates that this value is based on well-defined standard states for the chemical potentials. In addition, as the coefficients multiplying the logarithms can be written as exponents of the argument of the logarithms and the sum of logarithms is equal to the logarithm of the product of the terms, we get

$$\sum_{k}^{c} \nu_k \ln a_{k,\theta} = \sum_{k}^{c} \ln a_{k,\theta}^{\nu_k} = \ln \left[\prod_{k} a_{k,\theta}^{\nu_k} \right]$$

For simplicity, the symbol K_T is used for the product in the above expression, and noting that the stoichiometric numbers are positive for products and negative for reactants, we revert to the use of the stoichiometric coefficients that are always positive and write

$$K_T \equiv \prod_{k} a_{k,\theta}^{\nu_k} = \frac{\displaystyle\prod_{k}^{\text{products}} a_{Pk,\theta}^{\lambda_{Pk}}}{\displaystyle\prod_{j}^{\text{reactants}} a_{Rj,\theta}^{\lambda_{Rj}}} \tag{12.15}$$

This expression is called the *thermodynamic equilibrium constant*. Thus, Equation 12.14a takes the simple form

$$\ln K_T = -\frac{\Delta G_T^\theta}{RT} \tag{12.16}$$

ENTHALPY CHANGE OF REACTION

The consistent use of Equation 12.16 requires that the same standard state for a compound is used in the calculation of the activities for Equation 12.15 and for the calculation of the *standard Gibbs energy change of reaction* in Equation 12.14b. Because in practice these two terms are calculated independently, as a reminder, we have kept θ as a subscript for the activities and as a superscript for the Gibbs energy change.

Finally, the temperature dependence of the equilibrium constant is obtained combining Equation 12.16 with the Gibbs–Helmholtz equation, Equation 7.6b,

$$\left(\frac{\partial \ln K_T}{\partial T}\right)_{P,n} \equiv \frac{\Delta H_T^\theta}{RT^2} \tag{12.17}$$

where

$$\Delta H_T^\theta \equiv \sum_k^c \nu_k \, h_k^{(\theta)} \tag{12.18}$$

is the *standard enthalpy change of reaction* at the temperature of the system.

13 Calculation of Fugacities from Equations of State

The determination of the value of the fugacity of a compound as a function of temperature, pressure, and mixture composition is crucial for the study of physical equilibrium, as indicated by Equation 12.10. Similarly, the values of activity required by Equation 12.14a for the study of chemical equilibrium can be obtained directly from Equation 12.13 once the values of fugacities are known. The key equation for the calculation of the values of fugacities is Equation 12.5, written as

$$f_i = y_i \, \phi_i P \qquad (13.1)$$

For a pure compound, Equation 12.7 takes the form

$$f_i^0 = \phi_i^0 P \qquad (13.2)$$

For generality, in these equations we have not indicated the independent variables for each function. As the above equations are perfectly general and apply to any kind of mixture, for the mole fraction of compound i we use either the symbol y_i or x_i, as required. Since for a pure compound the fugacity of saturated liquid is equal to the fugacity of the saturated vapor, for this particular case, the method for calculating fugacity coefficients discussed here is directly applicable for a saturated liquid phase. For liquid compounds at pressures different than the saturation pressure, a correction must be applied. In this chapter, we consider the evaluation of fugacities from equations of state (EOSs) that use temperature, pressure, and composition as independent variables, $v = v \, (T, P, x)$, and from EOSs using temperature, molar volume, and composition as independent variables, $P = P \, (T, v, x)$.

USE OF VOLUME-EXPLICIT EOSs: $v = v \, (T, P, x)$

Combining Equations 11.1 and 12.2, we obtain

$$d \ln f_i = \frac{\overline{v}_i}{RT} dP \qquad (13.3)$$

For compound i, in an ideal gas mixture at the same temperature T, pressure P, and composition x of the real mixture,

$$d \ln f_i^* = \frac{\overline{v}_i^*}{RT} dP$$

or, using Equations 11.2 and 12.3a,

$$d \ln[x_i P] = \frac{1}{P} dP \tag{13.4}$$

Subtracting Equation 13.4 from Equation 13.3 and using Equation 12.5, we obtain

$$d \ln \phi_i = \frac{1}{RT} \left[\bar{v}_i - \frac{RT}{P} \right] dP$$

Integrating this expression from zero pressure to P and remembering that as $P \rightarrow 0$, $\phi_i \rightarrow 1$,

$$\ln \phi_i (T, P, x) = \frac{1}{RT} \int_0^P \left[\bar{v}_i - \frac{RT}{P} \right] dP \tag{13.5}$$

For compound i in an *ideal gas* mixture, from Equation 11.2,

$$\phi_i^* = 1 \tag{13.6}$$

In addition, writing Equation 13.5 for a pure compound i at temperature T and pressure P,

$$\ln \phi_i^0 (T, P) = \frac{1}{RT} \int_0^P \left[v_i^0 - \frac{RT}{P} \right] dP \tag{13.7}$$

Although Equation 13.7 is of limited practical use due to the lack of EOSs with pressure as an independent variable, it is of considerable theoretical interest. The fact that we do not have expressions in terms of these independent variables does not imply that this relation is not valid in general terms. One interesting variation of this equation is obtained introducing the definition of the compressibility factor, Equation 5.16,

$$\ln \phi_i^0 (T, P) = \int_0^P \frac{\left[z_i^0 - 1 \right]}{P} dP \tag{13.7a}$$

In addition to its applicability for a pure compound, this equation is equally applicable to a mixture of constant composition. As discussed in Chapter 9, a single-phase mixture of constant composition behaves as a pure compound and has two degrees of freedom. Thus, for a mixture of constant composition we can write

$$\ln \phi_m (T, P) = \int_0^P \frac{[z_m - 1]}{P} dP \tag{13.7b}$$

Equation 13.7b is a *de facto* definition of the fugacity coefficient of a gas mixture of constant composition. Taking the analogy one step further, and interpreting the mixture of constant composition in a single phase as a new pure compound, we can define the fugacity of a gas mixture by

$$\phi_m \equiv \frac{f_m}{P} \tag{13.8}$$

Returning to the case of a compound in a mixture and subtracting Equation 13.7 from Equation 13.5, we write

$$\ln \frac{\phi_i(T, P, x)}{\phi_i^0(T, P)} = \frac{1}{RT} \int_0^P \left[\overline{v}_i - v_i^0\right] dP \tag{13.9}$$

An immediate conclusion obtained from this equation is that, using Equation 6.22, for compound i in an ideal mixture, be it gas, liquid, or solid, the right-hand side is null and *the fugacity coefficient of the compound in the ideal mixture at any composition is equal to the fugacity coefficient of the pure compound at the same pressure and temperature:*

$$\phi_i^{id}(T, P, x) = \phi_i^0(T, P) \tag{13.10}$$

Thus, for a compound in an ideal mixture, from Equations 13.1 and 13.10,

$$f_i^{id}(T, P, x) = x_i \phi_i^0(T, P) P \tag{13.11}$$

It is important to note that as per Equation 13.9, Equations 13.10 and 13.11 are valid only if the ideal mixture behavior holds in all the range from zero pressure to the pressure P of the mixture. Equation 13.11 is especially useful for first estimates for the fugacity in iterative calculations.

USE OF PRESSURE-EXPLICIT EOSs: $P = P(T, v, x)$

EOSs having as independent variables the volume, the temperature, and the composition of the fluid are the most common ones in practice. It is unfortunate that the derivation of equations for this case is the more elaborate one. Although algebra can be done by computer applications, it is important to understand the principles involved in the derivation of equations, and thus we will deal with the algebra following a conceptual approach.

First, we observe that in this case, the integration required needs to be done over volume. For the real system of fixed composition x, at a volume V and temperature T, the pressure P is given by Equation 5.16, written as

$$P = z \frac{\left(\sum n_k\right) RT}{V} \tag{5.16a}$$

For a system as ideal gas at the same total volume, temperature, and composition, according to Equation 5.27 the pressure will have a value P^* given by

$$P^* = \frac{\left(\sum n_k\right)RT}{V} \tag{5.27a}$$

Thus,

$$P^* = \frac{P}{z}$$

Independent of the set of variables considered, Equation 12.9 is written as

$$\mu_i(T, P, x) = \mu_i(T, V, x) = \mu_i^{0,*}(T, 1) + RT \ln f_i$$

while for the ideal gas system, it has the form

$$\mu_i^*(T, P^*, x) = \mu_i^*(T, V, x) = \mu_i^{0,*}(T, 1) + RT \ln(x_i P^*) = \mu_i^{0,*}(T, 1) + RT \ln(x_i P) - RT \ln z$$

Subtracting the ideal gas equation from the equation for the real system, using Equation 12.5 and rearranging,

$$\ln \phi_i = \frac{\mu_i - \mu_i^*}{RT} - \ln z \tag{13.12}$$

In this equation, we have not indicated the variables for each function, as the two possibilities are clearly shown in the two preceding equations. The next step is to obtain the chemical potential using Equation 4.16. For this, from Equation 4.15,

$$d A_T = - P \, dV$$

and, at the same value of the total volume,

$$d A_T^* = -P^* dV = -\frac{\left(\sum_k n_k\right)RT}{V} dV$$

$$d\left[A_T - A_T^*\right] = -\left[P - \frac{\left(\sum_k n_k\right)RT}{V}\right] dV$$

Thus, integrating between zero pressure, or infinite volume, and the pressure and total volume of the system, and observing that at the lower integration limit both integrants are equal to zero, we write

$$A_T - A_T^* = -\int_{\infty}^{V}\left[P - \frac{\left(\sum_k n_k\right)RT}{V}\right]dV = +\int_{V}^{\infty}\left[P - \frac{\left(\sum_k n_k\right)RT}{V}\right]dV$$

This equation combined with Equation 4.16 gives

$$\mu_i - \mu_i^* = \int_{V}^{\infty}\left[\left(\frac{\partial P}{\partial n_i}\right)_{T,V,n_{j\neq i}} - \frac{RT}{V}\right]dV$$

Hence, Equation 13.12 can be written as

$$\ln\phi_i = \frac{1}{RT}\int_{V}^{\infty}\left[\left(\frac{\partial P}{\partial n_i}\right)_{T,V,n_{j\neq i}} - \frac{RT}{V}\right]dV - \ln z \qquad (13.13)$$

We emphasize that in this equation, the compressibility factor z refers to the mixture and that the partial derivative of the analytical equation for pressure with respect to the number of moles of compound i is unrelated to any partial molar quantity. As EOSs are normally written in terms of the molar volume, it is convenient to rewrite this equation as

$$\ln\phi_i = \frac{1}{RT}\int_{v}^{\infty}\left[n\left(\frac{\partial P}{\partial n_i}\right)_{T,V,n_{j\neq i}} - \frac{RT}{v}\right]dv - \ln z \qquad (13.13a)$$

However, *the differentiation of the analytical expression for the pressure with respect to the number of moles of compound i must be done at constant total volume V and not constant molar volume v = V/n.*

For the derivation of an expression for a pure compound fugacity coefficient from an EOS using as independent variables pressure, temperature, and composition, it is simpler to start with the differential form of Equation 13.7a,

$$d\ln\phi = [z - 1]\frac{dP}{P}$$

From Equation 5.15, we obtain

$$dP = \frac{RT}{v}dz - z\frac{RT}{v^2}dv$$

Thus, dividing by P, using again Equation 5.16 and rearranging,

$$\frac{dP}{P} = \frac{dz}{z} - \frac{dv}{v}$$

Hence, we can now write

$$d \ln \phi = [z-1]\left[\frac{dz}{z} - \frac{dv}{v}\right] = dz - d \ln z - [z-1]\frac{dv}{v}$$

Integrating this expression from zero pressure, at which both the fugacity coefficient and the compressibility factor are unity and the molar volume goes to infinity, to the pressure and volume of the system,

$$\ln \phi = [z-1] - \ln z - \int_{\infty}^{v} [z-1]\frac{dv}{v} \tag{13.14}$$

This expression is useful to calculate the fugacity coefficient of a pure compound or of a mixture of constant composition.

14 Fugacity of a Mixture and of Its Components

FUGACITY AND FUGACITY COEFFICIENT OF A MIXTURE

The definitions of the fugacity coefficient and fugacity of a mixture by Equations 13.7b and 13.8, respectively, bring up an interesting extension to the idea of treating a mixture of constant composition as a pseudo–pure compound. For a pure compound, the chemical potential is equal to the molar Gibbs energy of the compound. Thus, we can rewrite Equation 12.9 as

$$g = RT \ln f + g^*(T, 1)$$

In this equation, we have omitted all distinctions between pure compound and mixture of constant composition. Translating this equation to the latter case, we would write

$$g_m = RT \ln f_m + g_m^*(T, 1) \tag{14.1}$$

On the other hand, for a compound i in the mixture, from Equation 12.9,

$$\mu_i = RT \ln f_i + \mu_i^*(T, 1)$$

and according to Equation 7.2a,

$$g_m = \sum_k^c x_k \, \bar{g}_k = \sum_k^c x_k \, \mu_k$$

or

$$g_m = RT \sum_k^c x_k \ln f_k + RT \sum_k^c x_k \mu_k^*(T, 1)$$

Hence, from Equations 14.1 and 7.2a,

$$g_m = RT \ln f_m + g_m^*(T, 1) = RT \sum_k^c x_k \ln f_k + RT \sum_k^c x_k \, \mu_k^*(T, 1)$$

or

$$\ln f_m = \sum_k^c x_k \ln f_k - \frac{1}{RT}\left[g_m^*(T,1) - \sum_k^c x_k \,\mu_k^*(T,1) \right]$$

But, from Equation 11.5 we know that

$$\left[g_m^*(T,1) - \sum_k^c x_k \,\mu_k^*(T,1) \right] = RT\left[\sum_k^c x_k \ln x_k \right]$$

So, we finally write

$$\ln f_m = \sum_k^c x_k \ln \frac{f_k}{x_k} \tag{14.2}$$

and, as an additional step, subtracting $\ln P$ from both sides, we obtain

$$\ln \phi_m = \sum_k^c x_k \ln \phi_k \tag{14.3}$$

FUGACITY COEFFICIENTS AS PARTIAL PROPERTIES

Comparison of Equations 14.2 and 14.3 with Equation 6.6 indicates that $\ln\left(f_i/x_i \right)$ behaves as a partial molar property for $\ln f_m$, and $\ln \phi_k$ behaves as a partial molar property for $\ln \phi_m$. Thus, if $\ln \phi_m$ is obtained from an equation of state for the mixture using Equation 13.7b, then, following Equation 6.2, the fugacity coefficient for component i of the mixture can be obtained as

$$\ln \phi_i \equiv \left(\frac{\partial n \ln \phi_m}{\partial n_i} \right)_{T,P,nj\neq i} \tag{14.4}$$

or, following Equation 6.15,

$$\ln \phi_i = \ln \phi_m - \sum_{k\neq i} x_k \left(\frac{\partial \ln \phi_m}{\partial x_k} \right)_{\substack{T,P,x_r \\ r\neq i,\, r\neq k}} \tag{14.5}$$

GIBBS–DUHEM EQUATION IN TERMS OF FUGACITY COEFFICIENTS

The recognition that the logarithm of the fugacity of a compound in a mixture behaves as a partial molar property of the logarithm of the fugacity of the mixture has other useful implications. First, the fugacity coefficients of the components in a mixture are not all independent, but they are related by the *Gibbs–Duhem equation*, Equation 6.9, written as

$$\sum_{k}^{c} x_k \, d\ln\phi_k = \left(\frac{\partial \ln\phi_m}{\partial T}\right)_{P,n} dT + \left(\frac{\partial \ln\phi_m}{\partial P}\right)_{T,n} dP \qquad (14.6)$$

and also, from Equation 6.11,

$$\left(\frac{\partial \ln\phi_j}{\partial n_i}\right)_{T,P,n_{k\neq i}} = \left(\frac{\partial \ln\phi_i}{\partial n_j}\right)_{T,P,n_{k\neq j}} \qquad (14.7)$$

The derivatives of the right-hand side of Equation 14.6, giving the variation of the logarithm of the fugacity coefficient with temperature and pressure, can be obtained from Equation 14.1 written as

$$\ln\phi_m = \frac{g_m - g_m^*(T,1)}{RT} - \ln P^{\dagger}$$

For the derivative with respect to pressure, using Equation 4.18 we obtain

$$\left(\frac{\partial \ln\phi_m}{\partial P}\right)_{T,n} = \frac{v_m}{RT} - \frac{1}{P} = \frac{z_m - 1}{P} \qquad (14.8)$$

and for the derivative with respect to temperature, using Equation 7.6b,

$$\left(\frac{\partial \ln\phi_m}{\partial T}\right)_{P,n} = -\frac{[h_m - h_m^*(T,1)]}{RT^2} \qquad (14.9)$$

In this latter equation, which is seldom used, as most cases of study are under isothermal conditions, $h_m^*(T,1)$ can be written simply as $h_m^*(T)$ since the enthalpy of the ideal gas is independent of pressure. Equations 14.8 and 14.9 are also valid for the variation of the fugacity coefficient of a pure compound with pressure and temperature. For a compound i in a mixture, by the same argument used to derive Equation 6.10, from Equation 14.6 we obtain

$$d\ln\phi_k = \left(\frac{\partial \ln\phi_k}{\partial T}\right)_{P,n} dT + \left(\frac{\partial \ln\phi_k}{\partial P}\right)_{T,n} dP \qquad (14.10)$$

---

† This expression can be written as

$RT\ln\phi_m = g_m(T,P,x) - \left[g_m^*(T,1,x) + RT\ln P\right] = g_m(T,P,x) - g_m^*(T,P,x) = g_m^{res}(T,P,x)$, where the residual value of the molar Gibbs energy $g_m^{res}(T,P,x)$ is the contribution of the intermolecular forces to the property of the fluid.

and applying the same argument to Equations 14.8 and 14.9,

$$\left(\frac{\partial \ln \phi_i}{\partial P}\right)_{T,n} = \frac{\bar{v}_i}{RT} - \frac{1}{P} \tag{14.11}$$

and for the derivative with respect to temperature, using Equation 7.6,

$$\left(\frac{\partial \ln \phi_i}{\partial T}\right)_{P,n} = -\frac{\left[\bar{h}_i - h_i^*(T,1)\right]}{RT^2} \tag{14.12}$$

GIBBS–DUHEM EQUATION IN TERMS OF FUGACITIES

Again, in this case, Equation 14.12 finds limited use. Sometimes it is desirable to have the *Gibbs–Duhem equation in terms of fugacities* instead of fugacity coefficients. This form is better obtained from Equation 14.6. Replacing Equation 12.5 in the left-hand side and Equations 14.8 and 14.9 in the right-hand side of Equation 14.6, we obtain

$$\sum_k^c x_k\, d\ln f_k - \sum_k^c x_k\, d\ln x_k - \left(\sum_k^c x_k\right) d\ln P = -\frac{\left[h_m - h_m^*(T,1)\right]}{RT^2} dT + \left(\frac{v_m}{RT} - \frac{1}{P}\right) dP$$

In this expression, it is interesting to notice that

$$\sum_k^c x_k\, d\ln x_k = \sum_k^c x_k \frac{dx_k}{x_k} = \sum_k^c dx_k = d\left(\sum_k^c x_k\right) = 0$$

and, in addition,

$$\left(\sum_k^c x_k\right) d\ln P = \frac{1}{P} dP$$

Thus,

$$\sum_k^c x_k\, d\ln f_k = -\frac{\sum_k^c x_k\left[\bar{h}_k - h_k^{0,*}(T,1)\right]}{RT^2} dT + \frac{\sum_k^c x_k \bar{v}_k}{RT} dP \tag{14.13}$$

Equation 14.13 is the general form of the Gibbs–Duhem equation in terms of fugacities. Grouping all terms of Equation 14.13 under a single summation on the left-hand side,

$$\sum_k^c x_k \left[d\ln f_k + \frac{\sum_k^c x_k\left[\bar{h}_k - h_k^{0,*}(T,1)\right]}{RT^2} dT - \frac{\bar{v}_k}{RT} dP \right] = 0$$

For this equation to be valid in all of the composition range, necessarily,

$$d \ln f_i = -\frac{\left[\bar{h}_i - h_i^{0,*}(T,1) \right]}{RT^2} dT + \frac{\bar{v}_i}{RT} dP \qquad (14.14)$$

From Equation 14.14, then,

$$\left(\frac{\partial \ln f_i}{\partial P} \right)_{T,n} = \frac{\bar{v}_i}{RT} \qquad (14.15)$$

and

$$\left(\frac{\partial \ln f_i}{\partial T} \right)_{P,n} = -\frac{\left[\bar{h}_i - h_i^{0,*}(T,1) \right]}{RT^2} \qquad (14.16)$$

Equations 14.15 and 14.16 can also be obtained directly from Equations 14.11 and 14.12, respectively. For a pure compound, Equations 14.15 and 14.16 take the form

$$\left(\frac{\partial \ln f_i^0}{\partial P} \right)_T = \frac{v_i^0}{RT} \qquad (14.17)$$

and

$$\left(\frac{\partial \ln f_i^0}{\partial T} \right)_P = -\frac{\left[h_i^0 - h_i^{0,*}(T,1) \right]}{RT^2} \qquad (14.18)$$

It is important to emphasize that the expressions for fugacity and fugacity coefficients for all compounds forming a mixture must be derived from the same equation of state. This ensures that these expressions are thermodynamically consistent and satisfy both the Gibbs–Duhem equation and Maxwell's relations.

To close, we consider a particular case of Equation 14.13 that is of practical importance. Most studies of liquid mixtures are performed under isothermal conditions, and the first term of the right-hand side of this equation cancels out. In addition, for liquid mixtures the second term of the right-hand side is negligible because the volume of the liquid phase is orders of magnitude smaller than the product RT.

Hence, *for isothermal studies of liquid mixtures the Gibbs–Duhem equation in terms of fugacities* is usually written as

$$\sum_k^c x_k d \ln f_k = 0 \qquad (14.19)$$

15 Fugacities, Activities, and Activity Coefficients in Liquid Mixtures of Nonelectrolytes

LEWIS'S CONVENTION FOR ACTIVITY COEFFICIENTS

The methods to calculate fugacities described in Chapter 13 assumed the availability of equations of state valid from zero pressure to the pressure of the system at the temperature and composition of the system. While for systems in the gas phase this is a reasonable assumption, for condensed mixtures, this is not always the case. The volumetric behavior of an aqueous solution of sugar, for example, is not easily described from the ideal gas to the liquid phase using a single equation of state. However, the fact that we do not have an analytical expression to handle the algebra does not mean that the thermodynamic relations between the thermodynamic properties are not valid. Although in this chapter we develop relations of general use, we concentrate mostly on cases when the pure compounds forming a mixture exist as liquids at the temperature T and pressure P of the system. Hence, using Equation 13.9,

$$\ln \frac{\phi_i(T,P,x)}{\phi_i^0(T,P)} = \frac{1}{RT} \int_0^P \left[\bar{v}_i - v_i^0\right] dP \tag{13.9}$$

For convenience, we define

$$\ln \gamma_{i,0}(T,P,x) \equiv \frac{1}{RT} \int_0^P [\bar{v}_i - v_i^0] dP \tag{15.1}$$

and using Equations 12.5 and 12.7, we write

$$\gamma_{i,0}(T,P,x) = \frac{\phi_i(T,P,x)}{\phi_i^0(T,P)} = \frac{f_i}{x_i f_i^0} \tag{15.2a}$$

or

$$f_i(T,P,x) = x_i \gamma_{i,0}(T,P,x) f_i^0(T,P) \tag{15.2b}$$

Or dividing both sides by $x_i P$ and using Equations 13.1 and 13.2,

$$\phi_i = \gamma_{i,0}\phi_i^0 \qquad (15.2c)$$

Equation 15.2c is an important link between the thermodynamic treatments based on equation of state and those based on excess functions, and it has been used to open Part II, "Mixtures," of this book.

In addition, *the activity of compound* i *with respect to the standard state of pure compound at the same pressure and temperature of the system*, indicated by the subscript 0, according to Equation 12.13, is

$$a_{i,0}(T,P,x) \equiv \frac{f_i(T,P,x)}{f_i^0(T,P)} = x_i\gamma_{i,0} \qquad (15.3)$$

The fact that the mole fraction of compound i multiplied by the factor $\gamma_{i,0}$, defined by Equation 15.1, gives its activity suggests calling $\gamma_{i,0}$ the activity coefficient of compound i. This convention, in which both the activity coefficient and the activity are unity for the pure compound, is known as the *Lewis convention*. Other useful conventions are discussed in Chapter 16. In preparation for the more general treatment of ideal mixtures presented in Chapter 16, in Chapter 12 we introduced the general symbol θ as a way to identify the particular standard state chosen for the analytical treatment of equations. Comparison of Equation 15.3 with Equation 12.13 shows that Lewis's convention refers to the particular case when the standard state used for the definition of the activity of compound i is chosen as the pure compound at the temperature of the system.

Although we have indicated temperature, pressure, and composition as the variables for the activity coefficient, its definition by Equation 15.1 indicates that the main variables are composition and temperature only. The integration of the right-hand side of Equation 15.1 can be divided in the integration from pressure zero to the saturation pressure of the compound and the integration from the saturation pressure to the pressure of the system. The first of these integrals is independent of the pressure of the system, while for the second integral, the volumes refer to liquid volumes that, for all practical purposes, are independent of pressure.

As discussed in the derivation of Equation 13.11, for compound i in an ideal mixture from zero pressure to the pressure P of the system, the right-hand side of Equation 15.1 is null and the activity coefficient of the compound in the ideal mixture at any composition is equal to unity,

$$\gamma_{i,0}^{id}(T,P,x) = 1 \qquad (15.4)$$

Thus, in Lewis's convention, for a compound in an ideal mixture, Equation 15.3 gives

$$f_i^{id,0}(T,P,x) = x_i\, f_i^0(T,P) \qquad (15.5)$$

For the activity coefficient of a compound in an actual solution, in Lewis's convention, we write more formally,

$$\lim_{x_i \to 1} \gamma_{i,0} = 1 \qquad (15.6)$$

Deviations from Lewis's ideal behavior are said to be positive if the activity coefficient is larger than unity and negative if the activity coefficient is smaller than unity. The approximation of ideal mixture is excellent for dilute solutions and also for mixtures of close members of the same homologous series at any concentration. In these cases, we will have negligible heat and volume effects by mixing the compounds. As the difference in the hydrocarbon chain length increases between members of the same homologous series, weak negative deviations appear. Negative deviations are also found in mixtures of components from different homologous series when either hydrogen bonding or electron donor–acceptor complexes can be formed between the compounds. However, *most mixtures of compounds from different homologous series present positive deviations from ideality. Positive deviations increase with the difference in polarity between the compounds.*

The fact that dilute mixtures are formed with negligible volume change or heat effects suggests a generalization of the concept of ideal mixture, as discussed below.

HENRY'S CONVENTION FOR THE ACTIVITY COEFFICIENT OF A SOLUTE

Lewis's convention for the activity coefficient is a convenient one when the compounds in the mixture all exist as pure liquids at the temperature of the system. When one of the compounds is a solid at the temperature of the system, Lewis's convention can be used but needs some additional considerations. Let us consider a case in which *the solvent, component 1, is almost pure and the solute, component 2, is at high dilution.* For the solvent, in Lewis's convention,

$$f_1^{dil} \cong x_1 f_1^0 \qquad (15.7)$$

From the isothermal form of Equation 14.13, as given by Equation 14.19, neglecting pressure effects for liquid mixtures,

$$x_1 d \ln f_1^{dil} + x_2 d \ln f_2^{dil} = 0$$

Thus, as f_1^0 is independent of composition,

$$d \ln f_2^{dil} = -\frac{x_1}{x_2} d \ln x_1 = -\frac{dx_1}{x_2} = \frac{dx_2}{x_2} = d \ln x_2$$

or

$$d \ln \left(\frac{f_2^{dil}}{x_2} \right) = 0$$

This result indicates that in the dilute region, the ratio of the fugacity of the solute to its mole fraction is a constant independent of composition. Designating this constant value by $H_{2,1}$, known as *Henry's constant for solute 2 dissolved in solvent 1*, we write

$$f_2^{dil} = x_2 H_{2,1} \tag{15.8}$$

This expression, in fact, defines an ideal behavior for the solute, and we can define an *activity coefficient in Henry's convention*, $\gamma_{i,H}$, which is unity at infinite dilution of the solute. Thus, for a concentrated solution of the solute i in solvent r, we write

$$f_i = x_i \gamma_{i,H} H_{i,r} \tag{15.9}$$

with

$$\lim_{x_i \to 0} \gamma_{i,H} = 1 \tag{15.10}$$

We observe that for the definition of the activity coefficient in Henry's convention, it is particularly important to identify the nature of the solvent r in which the solute is at infinite dilution. The value of Henry's constant depends on the solvent in which the solute is at infinite dilution. For a particular solvent r, the value of Henry's constant is then the limiting value of the ratio of the fugacity of the compound to its mole fraction as the mole fraction tends to zero. Because the ratio tends to a limit of zero over zero at infinite dilution, applying the L'Hôpital theorem, we obtain

$$H_{i,r} = \lim_{x_i \to 0} \frac{f_i}{x_i} = \lim_{x_i \to 0} \frac{df_i}{dx_i} \tag{15.11}$$

Thus, Henry's constant is the value of the slope of the fugacity as a function of mole fraction for the solute at high dilution in solvent r. In a graphical representation of the fugacity of the solute as a function of its mole fraction, the value of Henry's constant would be found where the tangent to the fugacity curve at its origin, $x_i = 0$, cuts the vertical axis at $x_i = 1$ (Figure 15.1).

GENERALIZED DEFINITION OF ACTIVITY

As both conventions have much in common, we can write a single generalized expression for both of them as

$$f_i = x_i \gamma_{i,\theta} f_i^\theta \tag{15.12}$$

Thus, the activity of compound i with respect to the generalized standard state θ of the compound is given by Equation 12.13, written as

$$a_{i,\theta} \equiv \frac{f_i}{f_i^\theta} = x_i \gamma_{i,\theta} \tag{15.13}$$

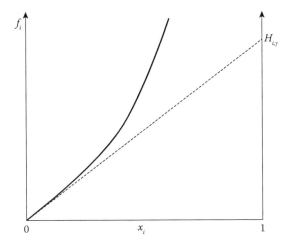

FIGURE 15.1   Determination of Henry's constant.

For a component i in an ideal mixture, in these two conventions,

$$f_i^{id,\theta} = x_i f_i^{\theta} \tag{15.14}$$

and the activity of the compound in the ideal mixture is equal to the mole fraction, and becomes unity for the pure compound in its standard state.

The main difference between both conventions is that in the Lewis convention, the reference state used to normalize the activity coefficient to unity is the same as the standard state that gives unit activity for the compound in a real or ideal mixture, while in Henry's convention, this is not the case. In Henry's convention, the reference state used to normalize the activity coefficient to unity is the state of the solute at infinite dilution in the particular solvent, while the standard state that gives unit activity is the imaginary state of the solute in an ideal mixture (in Henry's sense), extrapolated to a mole fraction of the solute equal to unity. *Thus, in general we distinguish between a reference state used to normalize the activity coefficient to unity and a standard state in which the activity of the compound in an ideal mixture would be unity.*

RELATIONS BETWEEN THE ACTIVITY COEFFICIENTS IN LEWIS'S AND HENRY'S CONVENTIONS

For the case of a compound that can be a solute or a solvent, depending on the situation, it may be convenient to use Lewis's or Henry's conventions. As the fugacity of the compound is independent of the convention used to define the activity coefficient, we can write

$$f_i = x_i \gamma_{i,0} f_i^0 = x_i \gamma_{i,H} H_{i,r}$$

or

$$\ln \gamma_{i,0} = \ln \gamma_{i,H} + \ln \frac{H_{i,r}}{f_i^0} \tag{15.15}$$

Taking the limit of this expression when the mole fraction of component i tends to zero in the particular solvent (state of infinite dilution), using Equations 15.10 and 15.14 and rearranging,

$$\ln H_{i,r} = \ln f_i^0 + \lim_{x_i \to 0} \left(\ln \gamma_{i,0} \right) \tag{15.16}$$

This equation is important because it allows us to evaluate Henry's constant from a knowledge of the fugacity of the pure compound, Equation 13.2, and the value of the activity coefficient in Lewis's convention at infinite dilution in the particular solvent. Replacing Equation 15.16 back into Equation 15.15,

$$\ln \gamma_{i,H} = \ln \gamma_{i,0} - \lim_{x_i \to 0} \left(\ln \gamma_{i,0} \right) \tag{15.17}$$

Similarly, taking the limit of Equation 15.15 when x_i tends to unity, using Equation 15.6 and rearranging,

$$\ln \gamma_{i,0} = \ln \gamma_{i,H} - \lim_{x_i \to 1} \left(\ln \gamma_{i,H} \right) \tag{15.18}$$

This equation is not particularly useful, and it is included here for completeness only. The important point is that having analytical expressions for the activity coefficient of the compounds as a function of composition in one convention, it is possible to obtain the analytical expressions in the other convention.

GENERALIZED EQUATIONS FOR AN IDEAL MIXTURE

The Gibbs energy change obtained by forming an ideal mixture starting from all its components in their respective standard states can be written as

$$\Delta g_m^{id,\theta} = \sum_1^c x_k \left(\bar{g}_k^{id,\theta} - g_k^\theta \right) = RT \sum_1^c x_k \ln \frac{f_k^{id,\theta}}{f_k^\theta} = RT \sum_1^c x_k \ln x_k \tag{15.19}$$

For the enthalpy change of mixing in forming the ideal mixture, then, from Equation 7.6,

$$\Delta h_m^{id,\theta} = \sum_1^c x_k \left(\bar{h}_k^{id,\theta} - h_k^\theta \right) = -RT^2 \left(\frac{\partial \left(\Delta g_m^{id,\theta} \big/ RT \right)}{\partial T} \right)_{P,n} = 0$$

Thus,

$$\bar{h}_k^{id,\theta} = h_k^\theta \tag{15.20}$$

Similarly, from Equation 4.18, as Equation 15.17 indicates that $\Delta g_m^{id,\theta}$ is independent of pressure, we write

$$\Delta v_m^{id,\theta} = \sum_1^c x_k \left(\bar{v}_k^{id,\theta} - v_k^\theta \right) = \left(\frac{\partial \Delta g_m^{id,\theta}}{\partial P} \right)_{T,n} = 0$$

Thus,

$$\bar{v}_k^{id,\theta} = v_k^\theta \tag{15.21}$$

In addition, we can also write

$$\Delta u_m^{id,\theta} = \sum_1^c x_k \left(\bar{u}_k^{id,\theta} - u_k^\theta \right) = \Delta h_m^{id,\theta} - P\Delta v_m^{id,\theta} = 0$$

from where

$$\bar{u}_k^{id,\theta} = u_k^\theta$$

Recapitulating, a generalized ideal mixture, real or hypothetical, is formed from its pure compounds at their respective standard state without any heat effect or volume change:

$$\Delta h_m^{id,\theta} = 0 \tag{15.22}$$

and

$$\Delta v_m^{id,\theta} = 0 \tag{15.23}$$

while from Equation 4.2, the entropy change of forming the ideal mixture from the individual compounds in their standard states is

$$\Delta s^{id,\Theta} = \frac{\Delta h^{id,\Theta} - \Delta g^{id,\Theta}}{T}$$

$$\Delta s_m^{id,\theta} = -R \sum_1^c x_k \ln x_k \tag{15.24}$$

PROPERTIES OF IDEAL MIXTURES IN LEWIS'S CONVENTION

In Lewis's convention, ideal mixtures will form starting from the pure compounds at the same pressure and temperature of the mixture without changes in volume or heat effects. Thus, all the above equations are valid. As studies of liquid mixtures in the Lewis convention are the most common, we repeat these equations here:

$$\Delta v_m^{id,0} = \sum_k x_k \left(\overline{v}_k^{id,0} - v_k^0 \right) = 0$$

and this relation implies that

$$\overline{v}_i^{id,0} = v_i^0$$

Similarly, from

$$\Delta h_m^{id,0} = \sum_k x_k \left(\overline{h}_k^{id,0} - h_k^0 \right) = 0$$

we obtain $\overline{h}_i^{id,0} = h_i^0$

For the Gibbs energy change,

$$\Delta g_m^{id,0} = RT \sum_k x_k \ln \frac{f_k^{id,0}}{f_k^0} = RT \sum_k x_k \ln \frac{x_k f_k^0}{f_k^0} = RT \sum_k x_k \ln x_k$$

$$\Delta s^{id,0} = \frac{\Delta h^{id,0} - \Delta g^{id,0}}{T} = -R \sum_k x_k \ln x_k$$

As these equations are just particular cases of Equations 15.22 through 15.24, there is no need to assign them a particular equation number here.

USE OF MOLALITY AS A MEASURE OF CONCENTRATION

Molality, that is, the moles of solute per 1000 g of solvent, is a convenient way to express the concentration of dilute solutions. In this case, the fugacity of solute i at a molality m_i is writen in Henry's convention as

$$f_i = \tilde{m} \gamma_{i,H}^{(m)} H_{i,r}^{(m)} \tag{15.25}$$

In this expression, $\tilde{m}$ is the dimensionless molality of the solute, that is, the value of its molality divided by 1 [mol of i /1000 g of solvent]. This later value is the concentration of the standard state for the ideal solution in Henry's sense. In addition, we have introduced the superscript (m) in order to distinguish the activity coefficient and the standard fugacity used here from those used in Henry's convention with mole fraction as measure of concentration. Again here, taking the limit of zero molality and applying L'Hôpital's theorem, we obtain

$$H_{i,r}^{(m)} = \lim_{\tilde{m} \to 0} \frac{f_i}{\tilde{m}} = \lim_{\tilde{m} \to 0} \frac{df_i}{d\tilde{m}} \tag{15.26}$$

Conversion between the two sets of activity coefficients in Henry's convention is direct. As the fugacity is independent of the system chosen to express concentration, we have

$$f_i = x_i \gamma_{i,H} H_{i,r} = \tilde{m}_i \gamma_{i,H}^{(m)} H_{i,r}^{(m)}$$

Thus,

$$\ln \gamma_{i,H} = \ln\left(\frac{\tilde{m}_i}{x_i}\right) + \ln \gamma_{i,H}^{(m)} + \ln\left(\frac{H_{i,r}^{(m)}}{H_{i,r}}\right) \tag{15.27}$$

Simple algebra shows that

$$x_i = \frac{\tilde{m}_i}{\tilde{m}_i + \sum_{k \neq 1} \tilde{m}_k + 1000/M_s}$$

or

$$\frac{\tilde{m}_i}{x_i} = \frac{1000}{M_s} + \tilde{m}_i + \sum_{k \neq i} \tilde{m}_k$$

where M_s is the molar mass of the solvent. Thus, as the normalization of the activity coefficient in Henry's convention is taken at the limit when the number of moles of i, n_i, tends to zero in pure solvent, the dimensionless molalities of all other possible solutes are identically zero and the expression above tends to the constant value of the first term on the right-hand side, while $\ln\gamma_{i,H}$ and $\ln\gamma_{i,H}$ tend to zero as n_i goes to zero. Hence,

$$\frac{H_{i,r}^{(m)}}{H_{i,r}} = \frac{M_s}{1000} \tag{15.28}$$

Equation 15.28 gives a simple relation between both values of Henry's constants. Replacing these terms back in Equation 15.27 and rearranging, we finally write

$$\ln \gamma_{i,H} = \ln \gamma_{i,H}^{(m)} + \ln\left(1 + 0.001 M_s \sum_k m_k\right) \tag{15.29}$$

where this time the sum is over all solutes, including solute i.

16 Activity Coefficients and Excess Properties

In order to evaluate the activity coefficients of components of a liquid mixture of nonelectrolytes from experimental data, it is necessary to know the value of the fugacity of the pure compounds in the liquid state at the pressure and temperature of the system.

EVALUATION OF THE FUGACITY OF A PURE COMPOUND LIQUID AT P AND T OF THE SYSTEM

For a pure compound that exists as a saturated liquid at the temperature T of the system, from Equation 14.17, we write

$$\ln\left(\frac{f_i^{0,L}}{f_i^{0,s}}\right) = \frac{1}{RT}\int_{P_i^s}^{P} v_i^{0,L} dP \approx \frac{v_i^{0,L}(P - P_i^s)}{RT}$$

or

$$f_i^{0,L} = f_i^{0,s} \exp\left[\frac{v_i^{0,L}(P - P_i^s)}{RT}\right]$$

But from Equation 13.2,

$$f_i^{0,s} = \phi_i^{0,s} P_i^s$$

Thus,

$$f_i^{0,L} = \phi_i^{0,s} \exp\left[\frac{v_i^{0,L}(P - P_i^s)}{RT}\right] P_i^s \tag{16.1}$$

The fugacity coefficient of a pure compound, saturated at the temperature, T, of the system, can be calculated from an equation of state for the saturated vapor phase using either Equation 13.7 or Equation 13.13. The molar liquid volume can be obtained from data for liquid densities, and the saturation pressure can be obtained from an equation of the type of Equation 5.17. The exponential term in Equation 16.1 is usually referred to as the *Poynting correction*. Considering a typical molar liquid volume of 100 ml and a temperature of 300 K, the Poynting correction for 10 atm of

117

difference between the pressure, P, of the system and the saturation pressure is about 1.05. For 100 atm of pressure difference, the Poynting correction is about 1.5. In the rare cases of negative deviations from ideality, in which the pressure of the system is lower than the saturation pressure, P_i^s, of the compound, the Poynting correction is a value slightly smaller than unity.

On the other hand, experimental evidence shows that the compressibility factor of pure compounds is less than unity at reduced temperatures below unity. Thus, according to Equation 13.7a, the fugacity coefficients at saturation for nonpolar compounds are less than unity. Approximate estimates based on generalized correlations suggest that at a reduced temperature of 0.7, the fugacity coefficient is slightly less than 1; it is about 0.8 at a reduced of temperature 0.8, and it decreases to about 0.75 at a reduced temperature of 0.9 and to 0.6 near the critical point. At moderate conditions, and also at high-pressure and reduced temperatures close to unity, both the fugacity coefficient and the Poynting correction tend to compensate. At reduced temperatures below 0.8 and moderate pressure, their product is less than unity, and at high pressure and low temperatures, their product is higher than unity. As a first approximation, then, the fugacity of the pure liquid is closer to the saturation pressure of the pure compound than to the total pressure of the system.

$$f_i^{0,L} \approx P_i^s$$

Similar equations can be written for a compound that exists as a saturated solid at the temperature, T, of the system.

EVALUATION OF ACTIVITY COEFFICIENTS FROM PHASE EQUILIBRIUM STUDIES

The key equations for the study of phase equilibria are obtained from the equality of fugacities at equilibrium, Equation 12.10. In the case of vapor–liquid equilibrium, for the fugacity of the compound in the vapor phase we use Equation 13.1, and for the fugacity of the compound in the liquid phase, we use Equations 15.2b and 16.1,

$$y_i \phi_i^V P = x_i \gamma_{i,0} \phi_i^{0,s} \exp\left[\frac{v_i^{0,L}\left(P - P_i^s\right)}{RT}\right] P_i^s \tag{16.2}$$

In Equation 16.2, we have used the conventional symbols y_i and x_i for the mole fraction in the vapor and liquid phases, respectively. In this equation, all quantities other than the activity coefficient can be determined independently. The fugacity coefficients are obtained from equations of state, while the compositions, the pressure, and the temperature are measured directly. The density and saturation pressure of the pure compound are obtained from the literature. For practical work, the following shorthand notation is useful:

$$y_i P = x_i \gamma_{i,0}\left(P_i^s\right)' \tag{16.2a}$$

In Equation 16.2a, we have used the following definition for a "corrected" value of the vapor pressure of the pure compound:

$$\left(P_i^s\right)' \equiv \left(\frac{\phi_i^{0,s}}{\phi_i^V}\right)\exp\left[\frac{v_i^{0,L}\left(P - P_i^s\right)}{RT}\right]P_i^s \tag{16.3}$$

As discussed in the paragraph following Equation 16.1, at reduced temperatures below 0.8 and pressures of a few atmospheres, at which ϕ_i^V is nearly equal to 1, the whole factor correcting the value of the vapor pressure is close to unity. Thus, as the normal boiling point of organic liquids is normally below a reduced temperature of 0.7, the correction factor is usually ignored and Equation 16.2 is simply used as

$$y_i P = x_i \gamma_{i,0} P_i^s \tag{16.4}$$

This expression is known as the modified Raoult's equation. For historical reasons, Equation 16.4 with $\gamma_{i,0} = 1$ is called Raoult's law.

Although there are analytical methods based on group contributions to estimate the values of the activity coefficients in a liquid mixture, their evaluation comes from experimental measurements carried on binary systems. Complete sets of data have been collected and are available in the literature. For the case of two compounds that *exist as liquids at the temperature of the system*, each liquid pure compound is degassed until its vapor pressure coincides with the known value at the temperature of interest. Then, one compound is added to a thermostated cell and the other compound is added in steps, registering the total pressure once the temperature is stabilized after each addition. Both the vapor and the liquid phases are then analyzed, and the activity coefficient of each compound is determined from Equation 16.2 or, at moderate conditions, Equation 16.4. We observe that the activity coefficient of each compound in the mixture can be evaluated independently. The problem is, however, that the activity coefficients are not all independent. For thermodynamic consistency, they should satisfy the Gibbs–Duhem equation, as discussed below. For calculation of phase equilibrium, Equation 16.2 can be used without further assumptions if the compound i exists as liquid at the temperature of the system. Otherwise, as discussed in Chapter 13, the use of an equation of state is preferred.

For calculation of phase equilibrium, Equation 16.2 can be used without further assumptions if the compound i exists as liquid at the temperature of the system. Otherwise, as discussed in Chapter 13, the use of an equation of state is preferred.

For the case of liquid–liquid equilibrium, in which a compound i is present in two liquid phases, α and β, at mole fractions x_i^α and x_i^β. From Equation 12.10, using Equation 15.2b, we see that the pure compound fugacity cancels out and the equilibrium equation takes the simple form

$$x_i^\alpha \gamma_{i,0}^\alpha = x_i^\beta \gamma_{i,0}^\beta \tag{16.5}$$

In this case, vapor–liquid equilibrium in an isothermal cell containing the two liquid phases in contact with a gas phase allows the determination of both $\gamma_{i,0}^\alpha$ and $\gamma_{i,0}^\beta$.

GIBBS–DUHEM EQUATION IN TERMS OF ACTIVITY COEFFICIENTS IN LEWIS'S CONVENTION

From Equation 15.2, we write

$$\ln f_i = \ln x_i + \ln \gamma_{i,0} + \ln f_i^0$$

Combining this expression with the Gibbs–Duhem equation in terms of fugacities, Equation 14.13, we obtain

$$\sum_k^c x_k d \ln x_k + \sum_k^c x_k d \ln \gamma_{k,0} + \sum_k^c x_k d \ln f_k^0$$

$$= -\frac{\displaystyle\sum_k^c x_k [\bar{h}_k - h_k^{0,*}(T)]}{RT^2} dT + \frac{\displaystyle\sum_k^c x_k \bar{v}_k}{RT} dP$$

As shown in the derivation of Equation 14.13, the first term on the left-hand side of this expression is null. The third term on the left-hand side, according to Equation 14.13, has the form

$$\sum_k^c x_k d \ln f_k^0 = -\frac{\displaystyle\sum_k^c x_k [h_k^0 - h_k^{0,*}(T)]}{RT^2} dT + \frac{\displaystyle\sum_k^c x_k v_k^0}{RT} dP$$

Thus,

$$\sum_k^c x_k d \ln \gamma_{k,0} = -\frac{\displaystyle\sum_k^c x_k [\bar{h}_k - h_k^0]}{RT^2} dT + \frac{\displaystyle\sum_k^c x_k [\bar{v}_k - v_k^0]}{RT} dP \qquad (16.6)$$

Equation 16.6 is the Gibbs–Duhem equation in terms of activity coefficients in Lewis's convention. As in previous cases, we can now group all the terms under a single summation. For this summation to be null at all concentrations, the factor multiplying the mole fractions must be null:

$$d \ln \gamma_{i,0} = -\frac{[\bar{h}_i - h_i^0]}{RT^2} dT + \frac{[\bar{v}_i - v_i^0]}{RT} dP \qquad (16.7)$$

The change of the activity coefficient with temperature is not of much interest, as the activity coefficients are usually measured at the temperature of interest. For the pressure variation, we obtain

$$\left(\frac{\partial \ln \gamma_{i,0}}{\partial P}\right)_T = \frac{\left[\bar{v}_i - v_i^0\right]}{RT} \tag{16.8}$$

Although, due to the incompressibility of the liquid phase, for most practical situations the activity coefficients are quite insensitive to changes in pressure, for completeness, we note here that the pressure chosen for its definition is rather arbitrary. In the treatment above, we have chosen the pressure P of the system.

In some cases, the standard state is chosen as the saturated liquid pure compound. In this case, the fugacity of the pure compound, f_i^{0s}, depends only on temperature. From Equation 15.2, it is clear that as the fugacity of the compound is independent of the choice of standard state, a change of the latter will require a change of the activity coefficient to a value $\gamma_{i,0s}$ satisfying the relation

$$\gamma_{i,0s}\, f_i^{0s}(T,P_i^s) = \gamma_{i,0}\, f_i^0(T,P)$$

Thus,

$$d \ln \gamma_{i,0s} = d \ln \gamma_{i,0} + d \ln f_i^0(T,P) - d \ln f_i^{0s}(T,P_i^s)$$

From this equality, using Equations 16.7 and 14.17 and observing that the last term on the right-hand side is independent of pressure, we obtain

$$\left(\frac{\partial \ln \gamma_{i,0s}}{\partial P}\right)_T = \frac{\bar{v}_i}{RT} \tag{16.9}$$

The fact that for liquid mixtures the typical magnitude of the terms on the right-hand side of both Equations 16.8 and 16.9 is of the order of 10^{-3} indicates that the activity coefficients are quite insensitive to pressure changes.

The Gibbs–Duhem equation, Equation 16.6 in Lewis's convention, is used for the testing the thermodynamic consistency of values of activity coefficients obtained in isothermal experiments. For these conditions in binary systems, Equation 16.6 takes the form

$$x_1 d \ln \gamma_1 + x_2 d \ln \gamma_2 = \frac{\Delta v_m}{RT} dP \tag{16.10}$$

Thus, for a change in composition,

$$x_1 \frac{d \ln \gamma_{1,0}}{dx_1} + x_2 \frac{d \ln \gamma_{2,0}}{dx_1} = \frac{\Delta v_m}{RT} \frac{dP}{dx_1}$$

where we have used Equation 6.18 for the volume change of mixing. This stringent test of slopes has not found much application in practice and alternative procedures are of common use.

In the paragraph under Equation 14.18, we remarked that when all fugacity coefficients are obtained from a single equation of state, they satisfy the Gibbs–Duhem equation. Following the same pattern, it is convenient to combine all the independently measured activity coefficients in a single function, as discussed in the next section. When a single analytical expression for the excess Gibbs energy is used to reduce the experimental values of the activity coefficients of all components of the mixture, the smoothed values produced necessarily satisfy the Gibbs–Duhem equation.

EXCESS GIBBS ENERGY AND OTHER EXCESS FUNCTIONS IN ANY CONVENTION

For generality, we use here the symbol θ to indicate the standard state chosen to define the ideal mixture behavior. For the particular case of the Lewis convention, the symbol 0 replaces θ. The excess Gibbs energy function, G_θ^E, is defined as the difference between the Gibbs energy of the actual mixture and the Gibbs energy of the ideal mixture, with respect to well-defined standard states of the compounds, both mixtures taken at the same composition, pressure, and temperature.

$$G_\theta^E \equiv G(T,P,x) - G^{id,\theta}(T,P,x) \tag{16.11a}$$

or, in molar terms,

$$g_\theta^E \equiv g(T,P,x) - g^{id,\theta}(T,P,x) \tag{16.11b}$$

Following the same procedure used in the derivation of Equation 6.10, we write for the partial molar excess properties,

$$\bar{g}_{i,\theta}^E \equiv \bar{g}_i - \bar{g}_i^{id,\theta} \tag{16.11c}$$

Analogous relations hold for the excess volume, excess enthalpy, excess entropy, and so forth.

From Equation 4.2, we write

$$g_\theta^E \equiv h_\theta^E - Ts_\theta^E \tag{16.12a}$$

In addition, following again the same method used to derive Equation 6.10, relations between partial molar excess properties are obtained. For example,

$$\bar{g}_{i,\theta}^E \equiv \bar{h}_{i,\theta}^E - T\bar{s}_{i,\theta}^E \tag{16.12b}$$

and, from Equation 7.6, the Gibbs–Helmholtz form,

$$h_\theta^E = -RT^2 \left(\frac{\partial\left(g_\theta^E/RT\right)}{\partial T} \right)_{P,n} \tag{16.13a}$$

$$\overline{h}_{i,\theta}^{E} = -RT^2 \left(\frac{\partial\left(\overline{g}_{i,\theta}^{E}/RT\right)}{\partial T} \right)_{P,n} \tag{16.13b}$$

As discussed in Chapter 26, these expressions are useful in the development of models for the representation of mixtures.

ACTIVITY COEFFICIENTS AND THEIR RELATION TO EXCESS FUNCTIONS IN ANY CONVENTION

The activity coefficients are directly related to the excess Gibbs energy. From Equation 16.9c, we write

$$\overline{g}_{i,\theta}^{E} \equiv \overline{g}_i - \overline{g}^{id,\theta} = \mu_i - \mu_i^{id,\theta} = RT \ln \frac{f_i}{f_i^{id,\theta}}$$

With the purpose of having general equations, not limited to Lewis's convention, we rewrite Equations 15.2 and 15.9 as

$$f_i = x_i \gamma_{i,\theta} f_i^{\theta} \tag{16.14}$$

Also, from Equation 15.14,

$$f_i^{id,\theta} = x_i f_i^{\theta} \tag{16.15}$$

Thus,

$$\overline{g}_{i,\theta}^{E} = RT \ln \gamma_{i,\theta} \tag{16.16}$$

From this expression and Equation 16.13b,

$$\overline{h}_{i,\theta}^{E} = -RT^2 \left(\frac{\partial \ln \gamma_{i,\theta}}{\partial T} \right)_{P,n} \tag{16.17}$$

and from Equations 6.12b, 16.16, and 16.17,

$$\overline{s}_{i,\theta}^{E} = -R \ln \gamma_{i,\theta} - RT \left(\frac{\partial \ln \gamma_{i,\theta}}{\partial T} \right)_{P,n} \tag{16.18}$$

From these expressions, the most important is Equation 16.16. It shows that *the logarithm of the activity coefficient is the partial molar property of the excess Gibbs energy over the product* RT. Thus, according to Equations 6.5 and 6.2, respectively, we write

$$\frac{g_{\theta}^{E}}{RT} = \sum_k x_k \ln \gamma_{k,\theta} \tag{16.19}$$

and

$$\ln \gamma_{i,\theta} = \left(\frac{\partial \left(n g_\theta^E / RT \right)}{\partial n_i} \right)_{T,P,nj \neq i} \tag{16.20}$$

or, alternatively,

$$\ln \gamma_{i,\theta} = \frac{g_\theta^E}{RT} - \sum_{k \neq i} x_k \left(\frac{\partial \left(g_\theta^E / RT \right)}{\partial x_k} \right)_{T,P,xj; j \neq i, j \neq k} \tag{16.21}$$

Equations 16.19 and 16.20, or Equation 16.21 if preferred, are the key equations for smoothing experimentally measured data of activity coefficients and producing sets of values of these activity coefficients that necessarily satisfy the Gibbs–Duhem equation. Individually measured activity coefficients are smoothed using a parametric form for the left-hand side of Equation 16.19. This function can be suggested by a model of solutions or be a simple polynomial form. Once adjustable parameters for the excess Gibbs energy function have been fitted to minimize the deviation of the fit from the experimental data, values of activity coefficients are generated using Equation 16.20 or Equation 16.21. These smoothed values necessarily satisfy the Gibbs–Duhem equation.

THERMODYNAMIC CONSISTENCY OF ACTIVITY COEFFICIENTS IN LEWIS'S CONVENTION

As discussed above, results obtained from the measurement of the individual activity coefficients in binary mixtures can be made to be thermodynamically consistent by obtaining the values of the excess Gibbs function using Equation 16.19, fitting these resulting values with an analytical function containing a few adjustable parameters, and using Equation 16.20 to generate "smoothed" consistent values of the activity coefficients. More rigorous studies prefer to test directly how consistent are the values of the individual activity coefficients obtained from the experiments at each composition. Two tests based on the combination of the Gibbs–Duhem equation with the definition of excess Gibbs energy have been developed for this purpose. Equation 16.19 in the Lewis convention for a binary system takes the form

$$\frac{g_0^E}{RT} = x_1 \ln \gamma_{1,0} + x_2 \ln \gamma_{2,0}$$

Then,

$$d \left(\frac{g_0^E}{RT} \right) = x_1 d \ln \gamma_{1,0} + x_2 d \ln \gamma_{2,0} + (\ln \gamma_{1,0}) dx_1 + (\ln \gamma_{2,0}) dx_2$$

The first two terms on the right-hand side of this expression are to the terms appearing in the Gibbs–Duhem equation, Equation 16.10; thus, for isothermal data, considering that $dx_2 = -dx_1$, we write

$$d\left(\frac{g_0^E}{RT}\right) = \frac{\Delta v_m}{RT}dP + \left[\ln\left(\frac{\gamma_{1,0}}{\gamma_{2,0}}\right)\right]dx_1$$

or

$$\ln\left(\frac{\gamma_{1,0}}{\gamma_{2,0}}\right) = \frac{d\left(g_0^E/RT\right)}{dx_1} - \frac{\Delta v_m}{RT}\frac{dP}{dx_1} \tag{16.22}$$

This differential test is less stringent than the one based on Equation 16.10, but still only high-precision data can satisfy it. Data are normally tested by the integral form of this equation. Integrating between $x_1 = 0$ and $x_1 = 1$, neglecting the small contribution of the second term on the right-hand side, and noting that in both limits the value of g_0^E is zero,

$$\int_0^1 \ln\left(\frac{\gamma_{1,0}}{\gamma_{2,0}}\right)d\,x_1 = 0 \tag{16.23}$$

Activity coefficients that do not satisfy this integral test, sometimes referred to as area test, are normally rejected. Equation 16.23 indicates that the surface area under the curve $\ln(\gamma_{1,0}/\gamma_{2,0})$ in a plot versus x_1 should add to zero. Thus, the surface area under this curve will have a section above the horizontal axis and an equivalent section below the horizontal axis (Figure 16.1).

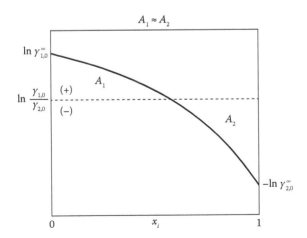

FIGURE 16.1   Consistency test for activity coefficients.

With respect to Equation 16.23, one interesting feature to notice is that in the plot of $\ln(\gamma_{1,0}/\gamma_{2,0})$ versus x_1, at the limit $x_1 \to 0$, $\gamma_{2,0} \to 1$ and $\gamma_{1,0}$ tends to a large value at infinite dilution that we denote by $\gamma_{1,0}^{\infty}$. Similarly, at the limit $x_1 \to 1$, $\gamma_{1,0} \to 1$ and $\gamma_{2,0}$ tends to a large value at infinite dilution denoted by $\gamma_{2,0}^{\infty}$. Thus, the curve of $\ln(\gamma_{1,0}/\gamma_{2,0})$ versus x_1 has a value $\ln \gamma_{1,0}^{\infty}$ at $x_1 = 0$, crosses the x-axis at some intermediate composition, and reaches the value $[-\ln \gamma_{2,0}^{\infty}]$ at $x_1 = 1$. This feature of the function $\ln(\gamma_{1,0}/\gamma_{2,0})$ will be used in the study of azeotropy. In Chapter 17, we first discuss when a mixture will split in two or more phases and when it will be stable in a single phase.

EXCESS GIBBS ENERGY IN LEWIS'S CONVENTION IN TERMS OF FUGACITY COEFFICIENTS

Due to its interest in the development of mixing rules for equations of state, we digress briefly here and observe that combining Equation 16.19 for the Lewis convention with the activity coefficients given by Equation 15.2a, we obtain

$$\frac{g_0^E}{RT} = \sum_k x_k \ln \gamma_{k,0} = \sum_k x_k \ln \frac{\phi_k}{\phi_k^0} = \sum_k x_k \ln \phi_k - \sum_k x_k \ln \phi_k^0$$

and using Equation 14.3, we write

$$\frac{g_0^E}{RT} = \ln \phi_m - \sum_k x_k \ln \phi_k^0 \tag{16.24}$$

Equation 16.24 is an equivalent form of Equation 15.2c.

17 Mixture Behavior, Stability, and Azeotropy

BASIC RELATIONS

As compounds mix to some degree in the liquid phase, the Gibbs energy of the system decreases toward a minimum at equilibrium. The Gibbs energy change obtained by forming a mixture starting from all the pure compounds can be written as

$$\Delta g_m \equiv g - g^0 = g - \sum_k x_k \mu_k^0 \tag{17.1}$$

The terms of Equation 17.1 can be written more explicitly as

$$\Delta g_m = \sum_1^c x_k \left(\bar{g}_k - g_k^0 \right) = \sum_1^c x_k \left(\mu_k - \mu_k^0 \right) = RT \sum_1^c x_k \ln \frac{f_k}{f_k^0} \tag{17.2}$$

For a binary system,

$$\frac{\Delta g_m}{RT} = x_1 \ln \frac{f_1}{f_1^0} + x_2 \ln \frac{f_2}{f_2^0} \tag{17.3}$$

Thus, expressing the fugacities in terms of the pure compound standard state, that is, defining the ideal behavior in Lewis's sense, from Equation 15.2 we write

$$\frac{\Delta g_m}{RT} = \left(x_1 \ln x_1 + x_2 \ln x_2 \right) + \left(x_1 \ln \gamma_{1,0} + x_2 \ln \gamma_{2,0} \right) \tag{17.4a}$$

The first bracket of the right-hand side gives the contribution of the ideal mixture, Equation 15.19, and as both mole fractions are less than unity, its value is always negative. This suffices to prove that an ideal mixture is always a stable mixture. The second bracket gives the contribution of nonideality of the mixture, and it can be identified with the excess Gibbs energy in Lewis's sense, Equation 16.19. Thus,

$$\frac{\Delta g_m}{RT} = \frac{\Delta g_m^{id,0}}{Rt} + \frac{g_0^E}{RT} \tag{17.4b}$$

POSITIVE AND NEGATIVE DEVIATIONS FROM IDEAL BEHAVIOR IN LEWIS'S SENSE

For negative deviations from ideality in Lewis's sense, $\gamma_{i,0} < 1$, the second bracket of the right-hand side of Equation 17.4a is also negative and makes the mixture still more stable than an ideal solution. For positive deviations from ideality in Lewis's sense, $\gamma_{i,0} > 1$, the second bracket of the right-hand side of Equation 17.4a is positive. If this contribution is large enough to overcome the contribution of the ideal mixture term, it produces instability, that is, liquid–liquid phase separation. A "good solvent" for a given compound should give negative deviations, ideal mixing, or at most weak positive deviations.

Although molecular effects are complex and resist generalizations, it is possible to establish some rules of thumb and be prepared to find exceptions. For the case of organic compounds, there are two main characteristics of their molecules that help in anticipating the behavior of their mixtures:

1. Length and shape of the hydrocarbon skeleton
2. Functional groups present

The effect of the functional group is so dominant that we classify the organic molecules in homologous families. In this respect, water can be considered the shortest alcohol, that is, an alcohol without a hydrocarbon chain. When mixing compounds, the group–group interactions govern the physical and chemical behavior of the mixture. Some group–group interactions are so strong as to cause a chemical reaction and the formation of new compounds. Good examples are the cases of groups COOH and OH, COO and H_2O, and CO and NH_2. We restrict our discussion here to nonreactive cases.

Large differences in the lengths of the hydrocarbon chains tend to give negative deviations. These are important in hydrocarbon–hydrocarbon systems, but they tend to be overshadowed by the larger effect of functional groups when these are present. The effect of the shape of the hydrocarbon skeleton is minor, and in most practical situations, it can be ignored. However, the effect of unsaturation in hydrocarbon structures may be important since unsaturated aliphatic compounds and also aromatic compounds can participate in electron donor–acceptor complexes and thus favor mixing.

BERG CLASSIFICATION OF LIQUIDS AND MIXTURE BEHAVIOR

Berg [1] proposed a classification of liquids in five groups that we extend here to six by differentiating between saturated and unsaturated hydrocarbons. The classification is based mostly on the capacity of the molecules to form hydrogen bonds and electron donor–acceptor complexes, but it also roughly orders the compounds on a scale of decreasing polarity. The modified Berg's classification is as follows:

- *Group 1:* Molecules forming tridimensional networks of H bonds. These are highly associated liquids, such as water, polyalcohols, polyphenols, hydroxyacids, amino alcohols, and amides.

- *Group 2:* Molecules having both active hydrogen atoms and electron donor atoms, such as O, N, and F. These are associated liquids, such as nitro-compounds and nitriles with hydrogen atoms in the α position (i.e., in the adjacent C atom) with respect to the N atom, nitromethane and acetonitrile, phenols, acids (including HCN, HF, HCl, and HNO_3), alcohols, and primary and secondary amines.
- *Group 3:* Molecules having electron donor atoms (O, N, and F) but not having an active hydrogen atom. Nitriles and nitrocompounds without α-hydrogen, esters, ketones, aldehydes, esters, tertiary amines, and so forth, are in this group.
- *Group 4:* Molecules having an active hydrogen atom but not an electron donor atom, such as O, N, or F. These are polychlorinated compounds with Cl and H in the same or in adjacent carbon atoms.
- *Group 5:* Unsaturated hydrocarbons, aliphatic or aromatic.
- *Group 6:* Saturated hydrocarbons.

From group 3 down, mixtures of compounds from the same group are close to ideal with slight positive deviations. Within a same group, positive deviations increase as the difference between the molecules increases. For example, a mixture of two ketones will be more ideal than a mixture of one ketone and one nitrile. Long hydrocarbon chains in both molecules of the same group tend to reduce the nonideality. If the molecules have similar vapor pressures, their mixtures can exhibit maximum-pressure azeotropes.

When mixing compounds of different groups from group 3 down, there are two cases deserving special mention. Members of group 3 mixed with members of group 4 will always give negative deviations from ideality. If their vapor pressures are similar, they can exhibit minimum-pressure azeotropes. Second, polar solvents from group 3 (and also from group 2), when mixed with members of group 5, give electron donor–acceptor complexes and thus favor solubility.

Groups 1 and 2 contain the most difficult "personalities" among the chemical species. When members of groups 1 or 2 are mixed with members of groups 4, 5, or 6 they always exhibit positive deviations that can be strong enough to cause liquid phase separation. When members of groups 1 and 2 are mixed with members of group 3, there can be either weak positive or weak negative deviations. And this is also the situation when mixing members of group 1 among themselves, members of group 2 among themselves, and members of group 1 with members of group 2. Negative deviations that lead to azeotropic behavior are found in mixtures of water with strong acids or other associating liquids of group 2 and also in mixtures of organic acids or phenols with amines or alcohols.

While negative deviations from ideality in Lewis's sense produce stable mixtures, strong positive deviations may cause instability, leading to the separation of the mixture into two liquid phases. Typical examples of a binary system formed by two partially miscible compounds are the cases of butanol and water and the more extreme case of petroleum and water. However, even in the case of strong positive deviations, two liquid compounds are miscible to a minor extent. Starting from one of the liquid compounds pure (solvent), it is always possible to dissolve in it a small concentration of the second liquid compound (solute). However, if the concentration of the solute is increased above a certain limit, the liquid mixture may split into two separate liquid phases. Each liquid phase will be rich in one of the components and lean in the other. In ternary or higher systems, the additional components will distribute between the two liquid phases formed.

PARTIAL MISCIBILITY

In the mixing process of two completely miscible liquid compounds, or in the miscibility regions of two partially miscible liquid compounds, that is, in the range where the mixture is stable, the Gibbs energy decreases toward equilibrium; thus, for the formation of a stable binary mixture the Gibbs energy change is necessarily negative,

$$\Delta g_m < 0$$

Figure 17.1 presents a plot $(\Delta g_m/RT)$ versus x showing that for a system in which the two compounds are totally miscible in all of the composition range, the slope of the curve $\Delta g_m/RT$ versus x increases continuously with an increase in x. Thus,

$$\frac{d}{dx}\left[\frac{d\Delta g_m/RT}{dx}\right]_T \geq 0$$

So, the criterion for the stability of a binary mixture can be written as

$$\left[\frac{\partial^2(\Delta g_m/RT)}{\partial x^2}\right]_T \geq 0 \tag{17.5}$$

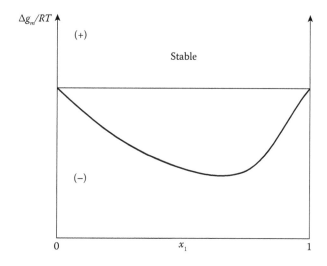

FIGURE 17.1   Stable system.

where x refers to x_1 or to x_2. When Equation 17.5 is satisfied over all of the concentration range, the compounds are miscible in all proportions. Binary systems having strong positive deviations from ideality in Lewis's sense may exhibit immiscibility if Equation 17.5 fails to be satisfied. If this happens, the mixture will be unstable and split into two phases. This behavior is depicted in Figure 17.2, where between points 1 and 2, the slope of the curve of $(\Delta g_m/RT)$ versus x decreases with composition. In this case, the Gibbs energy of the system is less when there are two phases present instead of one.

For a binary liquid mixture, Equation 17.1 takes the form

$$\Delta g_m \equiv g - x_1\mu_1^0 - x_2\mu_2^0 \qquad (17.1a)$$

Taking the second derivative with respect to either x_1 or x_2, for the condition of stability we obtain

$$\left[\frac{\partial^2 \Delta g_m}{\partial x^2}\right]_T = \left[\frac{\partial^2 g}{\partial x^2}\right]_T \geq 0 \qquad (17.6)$$

This result shows a very important conclusion; that is, both the plots of Δg_m versus x and g versus x convey the same information regarding a possible instability of the system. If at any concentration range the second derivative of the function becomes negative, that is, the curve presents an inflection point, the mixture becomes unstable and separates into two phases.

In Chapter 6, we discussed the relations between molar properties and partial molar properties, taking the molar volume as an example, but observing that the conclusions were also valid for all other molar and partial molar quantities. Hence, by analogy to what was shown in Figure 6.1, Figure 17.3 shows that a tangent line to

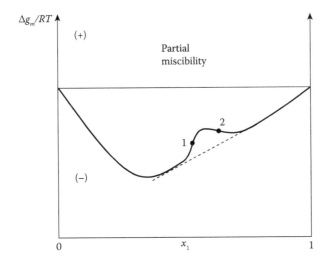

FIGURE 17.2 System showing instability gap k.

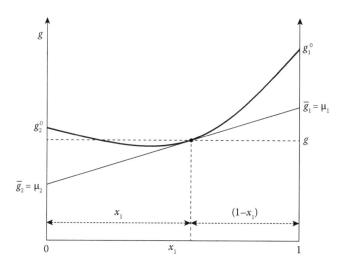

FIGURE 17.3   Stable binary system.

the curve of g versus x intercepts the vertical axis located at $x_1 = 0$ at a value equal to the chemical potential of compound 2, and the vertical axis located at $x_1 = 1$ at a value equal to the chemical potential of compound 1 at the particular composition where it touches the curve. If the curve of g versus x does not have any inflection point, the system presents complete miscibility in all the composition range. In this case, the slope of the curve with respect to x_1 will continuously increase with respect to an increase in x_1 or with respect to an increase of x_2. However, if Equation 17.6 is not satisfied in some composition range, as shown in Figure 17.4, the tangent will touch the curve in two points representing two liquid phases of different compositions having identical values for the chemical potential of compound 1 and also identical values for the chemical potential of compound 2. Thus, if the condition for stability is not satisfied, the mixture will split into two liquid phases of different compositions.

The points of contact of the tangent line with the curve for the molar Gibbs energy, describing the compositions of the liquid phases at equilibrium, are called binodal points, and the inflection points are called spinodal points. For modeling purposes, it is of interest to relate stability to the excess Gibbs energy in the Lewis convention. From Equations 17.4b and 17.5, we conclude that to show *instability* in some range of composition, the following condition must be met:

$$\frac{\partial^2 \Delta g_m}{\partial x^2} = \frac{\partial^2 \Delta g_m^{id,0}}{\partial x^2} + \frac{\partial^2 g^E}{\partial x^2} < 0$$

or

$$\frac{\partial^2 g^E}{\partial x^2} < -\frac{\partial^2 \Delta g_m^{id,0}}{\partial x^2}$$

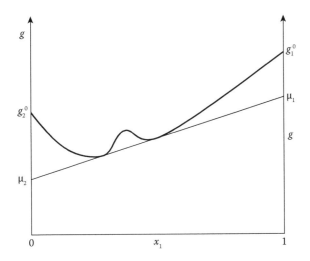

FIGURE 17.4 System showing instability gap.

Taking the second derivative of the contribution of the ideal mixture, Equation 15.19, with respect to either x_1 or x_2, we obtain

$$\frac{\partial^2 \Delta g_m^{id,0}}{\partial x^2} = \frac{RT}{x_1 x_2}$$

As the right-hand side of this expression is always positive, we confirm that the ideal mixture is stable in all of the composition range. The most interesting conclusion, however, is obtained when we combine this later result with that obtained for the second derivative of the excess Gibbs energy and write that for *instability*,

$$\frac{\partial^2 g^E}{\partial x^2} < -\frac{RT}{x_1 x_2} \tag{17.7}$$

Equation 17.7 shows that in order to represent liquid–liquid separation in a binary system, that is, in order to represent *instability*, an excess Gibbs energy function must have a negative value of its second derivative *in at least part of the composition range*. This is equivalent to say that in some part of the composition range, excess Gibbs energy function must present two inflection points in which its derivative changes sign. Notably, the two-parameter expression of Wilson [2], which is one of the most successful for the representation of vapor–liquid equilibrium, fails to meet this condition, and thus it is unable to represent liquid–liquid equilibrium. In fact, for a binary system, the Wilson equation for the excess Gibbs energy can be written as

$$g^E{}_{Wilson} = -RT \left[x_1 \ln (x_1 + x_2 \Lambda_{12}) + x_2 \ln (x_1 \Lambda_{21} + x_2) \right]$$

where Λ_{12} and Λ_{21} are temperature dependent but composition independent adjustable binary parameters. Taking the second derivative of this expression, one obtains

$$\frac{\partial^2 \left[g^E \right]_{Wilson}}{\partial x^2} = RT \left[x_1 \frac{(1 - \Lambda_{12})^2}{(x_1 + x_2 \Lambda_{12})^2} + x_2 \frac{(1 - \Lambda_{21})^2}{(x_1 \Lambda_{21} + x_2)^2} \right]$$

The right-hand side of this expression is always positive, no matter the values of the binary parameters Λ_{12} and Λ_{21}. Hence, the Wilson equation in its two-parameter form cannot satisfy Equation 17.7.

CONDITION FOR STABILITY OF LIQUID MIXTURES

Particularly useful forms of the stability condition are those written in terms of the fugacities of the compounds. For clarity, before giving the final forms, we present three alternative formulations of the Gibbs–Duhem equation for a binary liquid mixture under isothermal conditions. From Equation 14.19, for this case we write

$$x_1 d \ln f_1 + x_2 d \ln f_2 = 0$$

and for an infinitesimal change in composition of the mixture,

$$x_1 \frac{d \ln f_1}{dx_1} + x_2 \frac{d \ln f_2}{dx_1} = 0 \qquad (14.19a)$$

Replacing x_2 by $(1 - x_1)$ and rearranging, we write

$$x_1 \left[\frac{d \ln f_1}{dx_1} - \frac{d \ln f_2}{dx_1} \right] = \frac{d \ln f_2}{dx_2} \qquad (14.19b)$$

while replacing x_1 by $(1 - x_2)$ in Equation 14.19a, we obtain

$$x_2 \left[\frac{d \ln f_1}{dx_1} - \frac{d \ln f_2}{dx_1} \right] = \frac{d \ln f_1}{dx_1} \qquad (14.19c)$$

Taking the derivative of Equation 17.3 with respect to x_1,

$$\frac{d(\Delta g_m / RT)}{dx_1} = \left(\ln f_1 - \ln f_2 \right) + \left(x_1 \frac{d \ln f_1}{dx_1} + x_2 \frac{d \ln f_2}{dx_1} \right) - \left(\ln f_1^0 - \ln f_2^0 \right)$$

We recognize here that the second bracket of the right-hand side is null, according to Equation 14.19a above, and the third bracket is a constant, independent of composition. Thus, for the second derivative, the only important term is the first bracket. Hence,

$$\frac{d^2 \left(\Delta g_m / RT \right)}{dx_1^2} = \frac{d \ln f_1}{dx_1} - \frac{d \ln f_2}{dx_1}$$

Comparison of this result with inequality (Equation 17.5) shows that the *condition for stability* can be written as

$$\left(\frac{d \ln f_1}{dx_1} - \frac{d \ln f_2}{dx_1} \right) > 0 \tag{17.8a}$$

or, equivalently, as $dx_1 = -dx_2$,

$$\left(\frac{d \ln f_1}{dx_1} + \frac{d \ln f_2}{dx_2} \right) > 0 \tag{17.8b}$$

Combining the inequality in Equation 17.8a with Equations 14.19b and 14.19c of the Gibbs–Duhem equation written above, for stability it is necessary that

$$\frac{d \ln f_1}{dx_1} > 0 \tag{17.9}$$

and

$$\frac{d \ln f_2}{dx_2} > 0 \tag{17.10}$$

According to the inequalities in Equations 17.9 and 17.10, in a stable liquid mixture the fugacity of a compound always increases with an increase in its concentration.

AZEOTROPY

For the purposes of this discussion, we start with Equation 16.4, which is valid at moderate conditions of temperature and pressure. Adding the equations for both compounds in a binary mixture, we obtain

$$P = x_1 \gamma_{1,0} P_i^s + x_2 \gamma_{2,0} P_2^s \tag{17.11}$$

For an ideal mixture, then,

$$P^{id} = x_1 P_i^s + x_2 P_2^s = P_2^s + \left(P_1^s - P_2^s \right) x_1$$

Thus, at constant temperature, in a plot of P versus x_1, the equilibrium pressure of an ideal mixture will be a straight line. On the other hand, if the activity coefficients are larger than unity, that is, for positive deviations from ideality, the equilibrium pressure will be a curve above the straight line. If the activity coefficients are smaller than unity, that is, negative deviations from ideality, the equilibrium pressure will be a curve below the straight line. In an ideal solution, the different kinds of molecules do not distinguish each other and they mix without heat effects or volume change. In the case of positive deviations, the two compounds attract each other with forces that are weaker than those existing in an ideal solution. Similarly, in the case of negative deviations, the two compounds attract each other with forces that are stronger than those existing in an ideal solution. There are extreme cases in which at constant temperature the equilibrium pressure of the mixture passes through a maximum for positive deviations or a minimum for negative deviations. The composition at which this extremum occurs is called the *azeotropic point* of the mixture. An azeotropic mixture behaves as a pure compound and at constant pressure boils at constant temperature. From Equation 14.13, for the vapor phase of a binary mixture in an isothermal system,

$$y_1 d \ln f_1^V + y_2 d \ln f_2^V = \frac{\overset{V}{v_m}}{RT} dP$$

or for an incremental change of composition in the liquid phase,

$$y_1 \frac{d \ln f_1^V}{dx_1} + y_2 \frac{d \ln f_2^V}{dx_1} = \frac{v_m^V}{RT} \frac{dP}{dx_1}$$

Similarly, for the liquid phase,

$$x_1 \frac{d \ln f_1^L}{dx_1} + x_2 \frac{d \ln f_2^L}{dx_1} = \frac{v_m^L}{RT} \frac{dP}{dx_1}$$

Although for the liquid phase the right-hand side can be considered null and neglected, we keep the term here, as this simplification is not applicable for the gas phase, and it makes no difference for the treatment that follows. Because at equilibrium the fugacities of a compound in the vapor and liquid phases for each compound are equal to each other, we omit the superscripts V and L. Thus, taking the difference of these two equations, we write

$$(y_1 - x_1) \frac{d \ln f_1}{dx_1} + (y_2 - x_2) \frac{d \ln f_2}{dx_1} = \left(\frac{\Delta v_m}{RT} \right) \frac{dP}{dx_1}$$

As the mole fractions in both the liquid phase and the vapor phase add to unity, we express the mole fractions of compound 2 in terms of compound 1 and write

$$(y_1 - x_1) \left[\frac{d \ln f_1}{dx_1} - \frac{d \ln f_2}{dx_1} \right] = \left(\frac{\Delta v_m}{RT} \right) \frac{dP}{dx_1}$$

At a maximum or a minimum of the equilibrium pressure, the term on the right-hand side is equal to zero, and according to Equation 17.8a, the square bracket on the left-hand side is always positive for a stable liquid mixture; thus, necessarily,

$$y_1^{az} = x_1^{az}$$

As the composition of the vapor phase becomes equal to the composition of the liquid phase, at the azeotropic point the mixture behaves as a pure compound and boils at a constant temperature. In fact, the word *azeotrope* comes from the Greek term (*zeo* = "to boil," *trope* = "something changed," a = "no," which means "no change on boiling"), indicating precisely this behavior. A liquid mixture that has an azeotropic point but its initial composition is different from the azeotropic composition will evolve toward the azeotropic composition as it boils, and once this composition is reached, it will continue boiling at constant temperature as if it were a pure compound. Boiling at constant pressure, a maximum-pressure azeotropic mixture has a boiling point that is lower than the boiling points of the pure compounds, that is, it presents a minimum in the representation of the boiling temperature versus composition. The reverse is obviously true for a mixture presenting a minimum-pressure azeotropic behavior.

ACTIVITY COEFFICIENTS AND AZEOTROPY

At the azeotropic point, as the composition of the vapor phase is the same as the composition of the liquid phase, we write Equation 16.4 as

$$P^{az} = \gamma_{i,0}^{az} P_i^s$$

or

$$\gamma_{i,0}^{az} = \frac{P^{az}}{P_i^s}$$

Thus, for a binary system that presents an azeotropic point,

$$\ln \frac{\gamma_{1,0}^{az}}{\gamma_{2,0}^{az}} = \ln \frac{P_2^s}{P_1^s}$$

As discussed after the presentation of Equation 16.23, the term $\ln(\gamma_{1,0}/\gamma_{2,0})$ has the value $\ln \gamma_{1,0}^{\infty}$ at $x_1 = 0$ and $\left[-\ln \gamma_{2,0}^{\infty} \right]$ at $x_1 = 1$. Thus, in order to have an azeotropic point, the logarithm of the ratio of the vapor pressures of the compounds should be within these limits. If it happens that

$$-\ln \gamma_{2,0}^{\infty} \le \ln \frac{P_2^s}{P_1^s} \le \ln \gamma_{1,0}^{\infty} \tag{17.12}$$

the binary system will present an azeotropic point.

From a thermodynamic point of view, azeotropic behavior is due to a combination of nonideality of the liquid mixture and nearly equal pure compound vapor pressure values (or boiling point temperatures) of the components of the mixture.

HETEROAZEOTROPES

When two compounds are partially miscible or almost immiscible in the liquid phase, vapor liquid equilibrium is established between the two liquid phases, say phase α rich in compound 1 and phase β rich in compound 2, and a single vapor phase. The heterogeneous liquid phase will then be in equilibrium with a vapor phase of fixed composition and will boil at a constant temperature that is lower than the boiling points of either of the compounds. This fact is in agreement with the behavior of homoazeotropes originated by positive deviations from ideality in Lewis's sense. At equilibrium, the fugacity of each of the compounds in the vapor phase will be equal to its fugacity in the liquid phase in which it is diluted, and also to its fugacity in the liquid phase in which it is concentrated. For convenience, we choose the latter. In phase α, the mole fraction and the activity coefficient of compound 1 are both close to unity, and the same is true for compound 2 in phase β. Thus, using Equation 16.2a, we write

$$y_1^{az} P^{az} = x_1^\alpha \gamma_1^\alpha \left(P_1^s\right)' = \left(1 - x_2^\alpha\right)\gamma_1^\alpha \left(P_1^s\right)'$$

and

$$y_2^{az} P^{az} = x_2^\beta \gamma_2^\beta \left(P_2^s\right)' = \left(1 - x_1^\beta\right)\gamma_2^\beta \left(P_2^s\right)'$$

On the right-hand side of both equations above, we have opted for expressing the composition of each phase in terms of the solubility of compound 1 in phase β, x_1^β, and the solubility of compound 2 in phase α, x_2^α. Adding over both compounds,

$$P^{az} = \left(1 - x_2^\alpha\right)\gamma_1^\alpha \left(P_1^s\right)' + \left(1 - x_1^\beta\right)\gamma_2^\beta \left(P_2^s\right)' \tag{17.13}$$

Hence,

$$y_1^{az} = \frac{\left(1 - x_2^\alpha\right)\gamma_1^\alpha \left(P_1^s\right)'}{\left(1 - x_2^\alpha\right)\gamma_1^\alpha \left(P_1^s\right)' + \left(1 - x_1^\beta\right)\gamma_2^\beta \left(P_2^s\right)'}$$

or

$$y_1^{az} = \frac{\left(P_1^s\right)'}{\left(P_1^s\right)' + \left[\left(1 - x_1^\beta\right)\gamma_2^\beta \big/ \left(1 - x_2^\alpha\right)\gamma_1^\alpha\right]\left(P_2^s\right)'} \tag{17.14}$$

with a similar expression for y_2^{az}.

The assumption of total immiscibility, that is, $x_1^\alpha = 0$ and $x_1^\beta = 0$, provides first approximations that usually are quite close to the final results.

$$P^{az} \approx P_1^s + P_2^s$$

and

$$y_i^{az} \approx \frac{P_i^s}{P_1^s + P_2^s}$$

Considering these two approximate expressions, it is easy to see that as the vapor pressures are a function of temperature, the composition of the azeotropic vapor phase is a function of the boiling temperature of the heteroazeotrope. In addition, it is also seen that the boiling temperature of the heteroazeotrope can be obtained using expressions for the temperature dependence of the vapor pressures of the pure compounds and iterating values of temperature until the azeotropic equilibrium pressure equals the prevailing pressure of the system. The normal boiling point of the azeotropic mixture is obtained when the prevailing pressure is set equal to 1 atm. Obviously, with the appropriate correction factors, these conclusions are also applicable to Equations 17.13 and 17.14.

PHASE EQUILIBRIUM DIAGRAMS

Although with the advent of computers graphical methods have lost their importance as tools to solve problems, plots retain their educational value. Good figures help us to think, retain concepts, and even program a computer. For binary vapor liquid equilibrium, typical diagrams depict in a single figure the curves of the equilibrium pressure P versus x_1 and versus y_1, at a fixed temperature T, or the curves of the equilibrium temperature T versus x_1 and versus y_1 at a fixed pressure P. A typical vapor–liquid equilibrium diagram for the isothermal case is presented in Figure 17.5, and the isobaric case is shown in Figure 17.6.

For a mixture of overall composition z_1, points b and d are the bubble and dew points, respectively. In both figures, the solid line is the bubble line, that is, the line joining the bubble points, and the segmented line is the dew line, that is, the line joining the dew points. For pressures above the bubble line in Figure 17.5, the mixture is in liquid phase, and for pressures below the dew line, the mixture is in vapor phase.

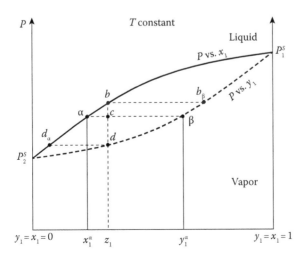

FIGURE 17.5 Isothermal vapor–liquid equilibrium diagram.

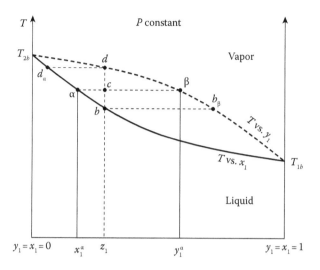

FIGURE 17.6   Isobaric vapor–liquid equilibrium diagram.

Decompressing at constant temperature a liquid mixture of overall composition z_1, the bubble point pressure is the value of the pressure at which the first bubble of vapor appears. A further decrease in pressure will cause an increase in the vapor phase with the corresponding decrease of the liquid phase. At the dew point, the last drop of liquid phase disappears. Starting with the vapor phase and increasing the pressure at constant temperature, at the dew point the first drop of liquid appears, and at the bubble point, the last bubble of vapor disappears.

Similarly, for temperatures below the bubble line in Figure 17.6, the mixture is in liquid phase, and for temperatures above the dew line, the mixture is in vapor phase. Changes in temperature at constant pressure will result in the appearance or disappearance of one of the phases at the dew and bubble points. The horizontal line joining the equilibrium compositions of coexisting liquid and vapor phases at fixed temperature and pressure is called a tie line. Thus, in Figures 17.5 and 17.6, the line b–b_β is the bubble point tie line and the line d_α–d is the dew point tie line for a mixture of overall composition z_1. At an intermediate point between the bubble point and the dew point of a mixture, such as point c in the figures, a saturated liquid phase coexists at equilibrium with a saturated vapor phase. Taking as a basis of calculation 1 mole of mixture of composition z_1, at point c the system will be formed by L moles of liquid phase of composition x_1^α and V moles of vapor phase of composition y_1^β with $(L + V) = 1$. Since in 1 mole of mixture of composition z_1 there are z_1 moles of compound 1, by material balance of compound 1,

$$z_1 = Vy_1^\beta + Lx_1^\alpha = (1 - L)y_1^\beta + Lx_1^\alpha$$

or

$$L = \frac{y_1^\beta - z_1}{y_1^\beta - x_1^\alpha}$$

Hence,

$$V = 1 - L = \frac{z_1 - x_1^\alpha}{y_1^\beta - x_1^\alpha}$$

$$\left(\frac{L}{V}\right) = \frac{y_1^\beta - z_1}{z_1 - x_1^\alpha} \qquad (17.15)$$

As can be seen in Figures 17.5 and 17.6, the relative amounts of phases L and V are inversely proportional to the segments of the tie line joining the point at the mixture composition z_1 with the corresponding equilibrium compositions x_1^α and y_1^β. This is the so-called *lever rule*. If the temperature of the mixture is increased at a constant pressure P, as shown in Figure 17.6, the liquid will start boiling at point b and the boiling point will continually increase until point d is reached. Both the relative amounts of the two phases and their composition will continuously change until all the mixture is in the vapor phase. When the temperature of the system is higher than the critical temperature of one of the compounds, that particular compound is in "supercritical condition." In this case, as depicted in Figure 17.7, the bubble curve will not reach the pure compound 1 axis. One interesting effect observed in this case is that decompression of a saturated vapor does not always produce a superheated gas. Figure 17.7 depicts the case of a mixture with a composition slightly to the right of critical point c, which is initially at a high pressure and being decompressed until a saturated vapor phase is reached. If the mixture is further decompressed, condensation occurs and a saturated liquid phase will coexist with the saturated vapor phase. Further decompression will cause the relative amount of liquid phase to increase at first, and decrease afterwards until a saturated vapor phase is reached again. Then the fluid passes to the superheated region. This unusual behavior is known as retrograde condensation.

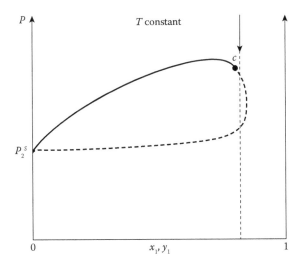

FIGURE 17.7   Isothermal vapor–liquid equilibrium at $T > T_c$ of component 1.

In Figures 17.5 and 17.6, we observe that the compositions of the liquid and vapor phases in equilibrium are never equal to each other. Not all binary mixtures behave this way. As discussed before, binary mixtures having strong positive deviations from ideality in Lewis's sense may present a maximum in the curve of P versus x_1 at a fixed temperature T. At this maximum pressure, $y_1 = x_1$, the "azeotropic point," the mixture boils at constant pressure and temperature, as if it were a pure compound. Figures 17.8 and 17.9 illustrate maximum boiling pressure (minimum boiling temperature) behavior. Similarly, binary mixtures having strong negative deviations

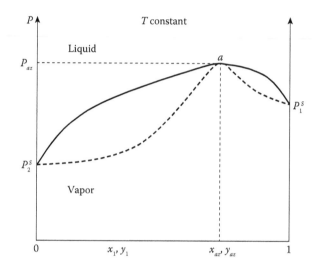

FIGURE 17.8   Isothermal azeotropic system.

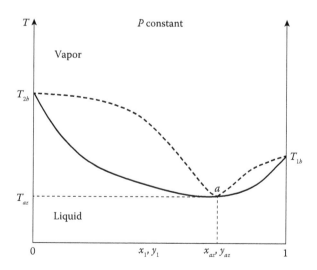

FIGURE 17.9   Isobaric azeotropic system.

from ideality in Lewis's sense, that is, mixtures of compounds forming either a hydrogen bond or an electron donor–acceptor complex, may present a minimum in the equilibrium pressure at constant temperature, that is, a maximum boiling temperature at constant pressure.

Liquid mixtures exhibiting strong positive deviations from ideality in Lewis's sense eventually may show (liquid–liquid) phase separation. As liquid–liquid equilibrium concentrations are a weak function of pressure, temperature versus mole fraction, T–x, diagrams are preferred to P–x diagrams. In this case, the equilibrium pressure is seldom of concern.

Figure 17.10 shows the binodal curve, that is, the mutual solubility curve joining the binodal points described in Figure 17.4, as a function of temperature for the most common case of a system presenting an upper consolute or upper critical temperature (UCT). For this kind of system, the compounds are totally miscible at high temperature, and at low temperatures two liquid phases coexist at equilibrium. The less common case of a system showing a lower consolute temperature (LCT) is shown in Figure 17.11. Some rare systems, like the nicotine–water system, present a closed solubility loop exhibiting a lower and an upper critical solution temperature. At the critical point, the second and third derivatives of the molar Gibbs energy with respect to the mole fraction are equal to zero. Similarly to vapor–liquid equilibrium, for liquid–liquid equilibrium of a mixture of an overall composition z_1, the amounts L_α and L_β of the phases present are in the proportion given by the lever rule. The solubilities x_1^β and x_2^α are fixed once the temperature and the pressure are fixed. If a vapor phase coexists with the two liquid phases, fixing the temperature or the pressure, according to the phase rule, all other variables are fixed.

Figure 17.12 depicts the case of two liquid phases in equilibrium with a vapor phase at a temperature T and equilibrium pressure P_{eq}, which is of interest in steam distillation. At equilibrium, the compositions of the three phases are fixed

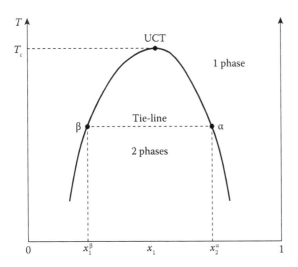

FIGURE 17.10   Two-phase isobaric system with UCT.

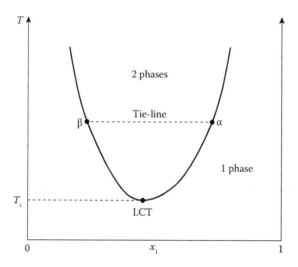

FIGURE 17.11   Two-phase isobaric system with LCT.

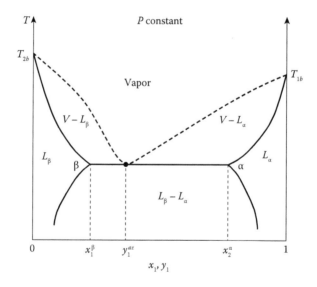

FIGURE 17.12   Isobaric heteroazeotrope.

at x_1^β, x_2^α, and y_1^{az}. If this three-phase system exchanges heat with its surroundings, neither the temperature, the pressure, nor the compositions of the phases will change while there are three phases present. Only the relative amount of the phases will change until one of the phases disappears. This state of the system is called a heteroazeotropic point. If at constant pressure the temperature increases, one liquid phase will disappear, and if the temperature decreases, the vapor phase will disappear. The vapor phase in equilibrium with two liquid phases is always an azeotropic phase, in the sense that its composition will not change until one of the phases disappear. For all practical purposes, a diagram at constant pressure of two solid solutions S_α and S_β in equilibrium with a liquid phase L looks identical to Figure 17.12. For this case, the point of coexistence of the three phases is called a eutectic point instead of an azeotropic point.

Figure 17.13 shows the case of solid–liquid equilibrium when an intermediate compound (1–2) is formed from pure compounds 1 and 2.

The case presented in Figure 17.13 considers that there are no solid solutions of either compound 1 or compound 2 with the intermediate compound (1–2). In cases of having more than one intermediate compound, the diagram becomes more complex.

In all the cases discussed above, we have considered positive deviations from ideality in Lewis's sense. These are by far the most common cases found in practice. Figures 17.14 and 17.15 present typical P versus x_1 diagrams for negative deviations with no minimum-pressure azeotrope and with a minimum-pressure azeotrope, respectively. Binary systems formed by two volatile compounds showing negative deviations from ideality in Lewis's sense never split into two liquid phases. If the solute is nonvolatile, at high concentration it may saturate the solution and separate into a solid phase.

For ternary systems, triangular diagrams are used. These diagrams are particularly useful in cases of liquid–liquid equilibria of two highly immiscible solvents containing a common solute. Figure 17.16 shows how to read the compositions (mole fractions or mass fractions) in a ternary diagram. All points lying on a line parallel to the side opposite to a pure compound vertex have the composition in that compound equal to the value of the intercept of the line with the axis of that compound.

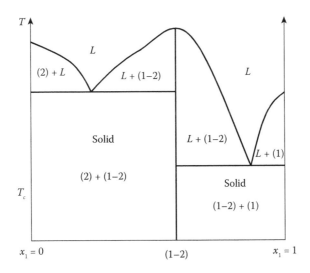

FIGURE 17.13   Isobaric solid–liquid equilibrium with the formation of an intermediate.

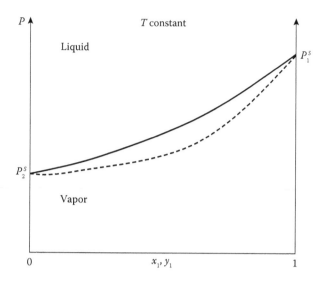

FIGURE 17.14   Isothermal vapor–liquid equilibrium system with negative deviations from ideality.

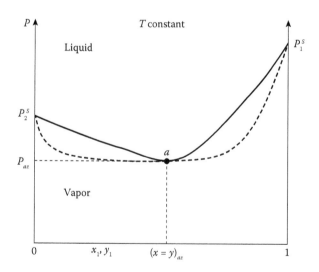

FIGURE 17.15   Isothermal azeotropic vapor–liquid equilibrium system with negative deviations from ideality.

An important property of ternary diagrams is that all points lying on a straight line that joins a pure compound vertex with the opposite side have the same proportion of the other two compounds, as shown in the side intercepted. This property is useful when adding a third compound to a binary mixture. The dotted line in Figure 17.16 shows a ternary mixture in which compounds B and C are in the proportion 1:2. At any point of the diagram, a line passing parallel to a side opposite to a particular vertex will intercept the side corresponding to the compound of that vertex at the compound composition in the mixture.

Figure 17.17 shows the case of solvents A and C, which have an immiscibility gap, and a solute B that is miscible with both solvents in all proportions. The binodal curve and one tie line are shown. Figure 17.18 shows the rare case of two partially miscible pairs.

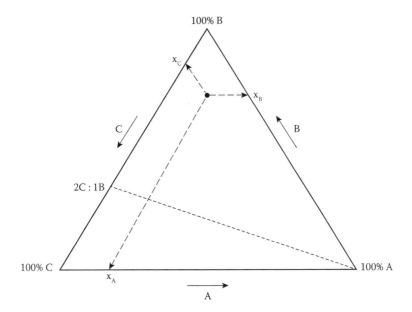

FIGURE 17.16   Reading a triangular diagram.

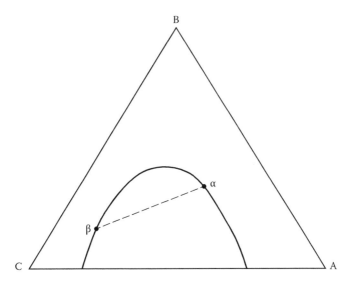

FIGURE 17.17   One immiscibility gap.

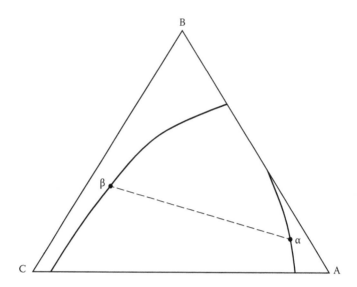

FIGURE 17.18   Two miscibility gaps.

REFERENCES

1. Berg, L. 1969. Selecting the agent for distillation processes. *Chem. Eng. Prog.* 65(9): 52–57.
2. Wilson, G. M. 1964. Vapor-liquid equilibrium. XI. A new expression for the excess free energy of mixing. *J. Am. Chem. Soc.* 86(2): 127–130.

18 The Thermodynamics of Aqueous Electrolyte Solutions

As discussed in Chapter 10, when a salt is dissolved in water or in other appropriate solvent, the molecules dissociate into ions. In aqueous solutions, strong electrolytes, that is, those formed from a strong acid neutralized with a strong base, will dissociate almost completely into ions, while weak electrolytes will dissociate only partially. In a media of lower dielectric constant than water, such as furfural, acetonitrile, alcohols, chloroacetic acid, dioxane, acetone, acetic acid, or in their mixtures with water, conductivity measurements show that all electrolytes are increasingly weak; that is, they are partially associated, as the solvent moves down in the scale of dielectric constants. Thus, the classification of strong electrolytes as strong acids, bases, and their salts (chlorides, fluorides, sulfates of sodium, potassium, magnesium, copper, zinc, etc.) is only valid in aqueous media. On the other hand, weak electrolytes such as acetic acid or chloroacetic acid in concentrated aqueous solutions can associate to such a high degree as to change the properties of water as solvent. The dielectric constant of air is so low that there are no ions present in the vapor phase over a solution of a volatile electrolyte. All molecules are fully associated. In mercury or sodium lamps, ions exist in the vapor phase under a voltage difference and in the absence of air.

BASIC RELATIONS

With these considerations in mind, without specifying the extent of the actual dissociation in aqueous solution, for 1 mole of an electrolyte E that in total dissociation would give ν_+ cations C^{Z+} and ν_- anions A^{Z-}, we write

$$C_{\nu_+} A_{\nu_-} = \nu_+ C^{Z_+} + \nu_- A^{Z_-}$$

For a single electrolyte, the electroneutrality condition reduces to

$$\nu_+ Z_+ + \nu_- Z_- = 0 \tag{18.1}$$

In this equation, the sign of the charge is implicit in Z_i. For clarity, it is better to have the signs of the charges explicit and write

$$\nu_+ Z_+ = \nu_- |Z_-| \tag{18.1a}$$

Variations of this relation that are often used in the literature without further explanation may be confusing at first, so we write some of them in detail.

$$\frac{1}{Z_+} = \frac{v_+}{v_-|Z_-|} = \frac{v_+}{v_+ Z_+} \tag{18.1b}$$

$$\frac{1}{|Z_-|} = \frac{v_-}{v_+ Z_+} \tag{18.1c}$$

One important form is obtained by multiplying Equation 18.1a first by $|Z_-|$ to obtain

$$v_+|Z_+ Z_-| = v_-(Z_-)^2$$

and then multiplying Equation 18.1a by Z_+ to obtain

$$v_+(Z_+)^2 = v_-|Z_+ Z_-|$$

Taking the difference of these two expressions and rearranging, we get

$$|Z_+ Z_-| = \frac{v_+(Z_+)^2 + v_-(Z_-)^2}{v} \tag{18.1d}$$

with

$$v \equiv v_+ + v_- \tag{18.2}$$

According to Equation 12.15, the equilibrium constant for the ionic dissociation in terms of activities takes the form

$$K_T = \frac{a_+^{v_+} a_-^{v_-}}{a_E} \tag{18.3}$$

For the dissociation of the electrolyte E, according to Equation 12.16, the value of the equilibrium constant is obtained from the standard Gibbs energy change:

$$K_T = \exp\left[\frac{\mu_E^\theta - v_+\mu_+^\theta - v_-\mu_-^\theta}{RT}\right] \tag{18.4}$$

Some treatments of electrolyte solutions have proposed to use mole fractions as a measure of composition. For all practical purposes, the use of molality is simpler and gives a better range of values. As an example, the solubility of common salt (NaCl) in water at 298 K is 360 g kg^{-1} of water or 6.16 moles per 55.51 moles of water. Thus, at saturation, that is, the maximum concentration of salt possible at this temperature, the mole fraction of each ion is 0.100, while the molality is 6.16. Having decided to use molality as the measure of concentration, the next step is to choose the standard states for the activity coefficients to be used in the evaluation of the activities by

$$a_i = \tilde{m}_i \gamma_i = \nu_i \tilde{m} \gamma_i \qquad (18.5)$$

Thus, for the cation,

$$a_+ = \tilde{m}_+ \gamma_+ = \nu_+ \tilde{m} \gamma_+ \qquad (18.5a)$$

and for the anion,

$$a_- = \tilde{m}_- \gamma_- = \nu_- \tilde{m} \gamma_- \qquad (18.5b)$$

In this expression, as the activities are dimensionless, $\tilde{m}_i$ is the dimensionless molality of the ion i and $\tilde{m}$ is the dimensionless molality of the electrolyte solute; that is, the value of the molality divided by 1[mole of i/1000g of solvent]. Similarly to the case of the use of molality for nonelectrolytes discussed in Chapter 15, the reference state for the activity coefficient of the ions is their state at infinite dilution, and their standard state is the ideal solution in Henry's sense at 1[mole of i/1000g of solvent]. At the reference state, the activity coefficient of an ion is normalized to unity.

$$\lim_{m_i \to 0} \gamma_i = 1 \qquad (18.6)$$

MEAN IONIC ACTIVITY COEFFICIENT

At the standard state, the activity of an ion is equal to unity (dimensionless). This is so because in its standard state the ion is in an ideal solution at unit molality. The normalization of the activity coefficients of the ions to unity at their state in an infinitely dilute solution is of great importance. At this state, the presence of any other ion is immaterial, be it a co-ion or a counterion. Thus, the same condition is valid independently of the nature of the electrolyte generating the ion. With this normalization, although its value is not known, the standard state potential of an ion in solution is fixed and well defined, and it is independent where the ion came from. The standard state for the electrolyte is chosen so that the constant K_T in Equation 18.4 is equal to unity.

$$\mu_E^\theta = \nu_+ \mu_+^\theta + \nu_- \mu_-^\theta$$

Again here, the value of μ_E^θ for the electrolyte is not known, but we know that for each electrolyte it has a fixed and well-defined value depending only on the temperature and the pair of ions forming the electrolyte. Hence, from Equation 18.3 we write

$$a_E = a_+^{\nu_+} a_-^{\nu_-} = \left(\tilde{m}_+ \gamma_+ \right)^{\nu_+} \left(\tilde{m}_- \gamma_- \right)^{\nu_-} = \tilde{m}^\nu \gamma_\pm^\nu \left(\nu_+^{\nu_+} \nu_-^{\nu_-} \right) \qquad (18.7)$$

where the mean ionic activity coefficient of the electrolyte $\gamma_\pm$ is defined as

$$\gamma_\pm^v \equiv \gamma_+^{v_+} \gamma_-^{v_-} \tag{18.8}$$

with $v \equiv v_+ + v_-$, as defined by Equation 18.2.

At infinite dilution, by normalization of the activity coefficients of the ions, we have

$$\lim_{m \to 0} \gamma_\pm = 1 \tag{18.9}$$

OSMOTIC COEFFICIENT

For a single electrolyte aqueous solution at constant temperature, neglecting pressure effects, the Gibbs–Duhem equation, which relates the changes in a_E with the changes in the activity of water a_W, takes the form

$$n_E d \ln a_E + n_W d \ln a_W = 0$$

where $n_E = \tilde{m}$ and $n_W = 1000/M_W$ are the number of moles of salt and water, respectively.

$$\tilde{m} d \ln a_E + \frac{1000}{M_W} d \ln a_W = 0 \tag{18.10}$$

The activity of water is sometimes given in terms of the osmotic coefficient of the solution, defined as

$$\varphi \equiv -\frac{1000}{M_W \sum_j (v\tilde{m})_j} \ln a_W \tag{18.11}$$

For an aqueous solution of a nonvolatile electrolyte, the activity of water is obtained directly by measuring the vapor pressure, P, of the solution at the temperature of interest. From Equation 16.4, we write

$$a_W = x_W \gamma_W = \frac{P}{P_W^S} \tag{18.12}$$

Here, P_W^S is the vapor pressure of pure water at the temperature of the system. For work at high pressure, the correction factors included in Equation 16.2a should be included in Equation 18.12. It is of interest to obtain the relations between the osmotic coefficient of a single electrolyte solution, φ, and the mean ionic activity coefficient of the electrolyte. For a single electrolyte in solution, Equation 18.11 takes the form

$$\ln a_W \equiv -\frac{M_W v\tilde{m}}{1000} \varphi \tag{18.11a}$$

Then,

$$d \ln a_W \equiv -\frac{M_W \nu \tilde{m}}{1000} d\varphi - \frac{M_W \nu}{1000} \varphi d\tilde{m}$$

Combining this expression with the Gibbs–Duhem equation, Equation 18.10, we get

$$d \ln a_E = \nu d\varphi + \nu \varphi \frac{d\tilde{m}}{\tilde{m}}$$

From Equation 18.7,

$$d \ln a_E = \nu \frac{d\tilde{m}}{\tilde{m}} + \nu d \ln \gamma_{\pm}$$

Thus, equating these two expressions and rearranging, the relation between the osmotic coefficient of a single electrolyte solution and the mean ionic activity coefficient takes the form

$$d \ln \gamma_{\pm} = d\varphi + \frac{(\varphi - 1)}{\tilde{m}} d\tilde{m} \qquad (18.13)$$

This differential relation can be used to obtain the mean ionic activity coefficient in terms of the osmotic coefficient and vice versa. Integrating between the limit at infinite dilution, where the mean ionic activity coefficient and the osmotic coefficient tend to unity, and a molality m,

$$\ln \gamma_{\pm} = (\varphi - 1) + \int_{0}^{\tilde{m}} \frac{(\varphi - 1)}{\tilde{m}} d\tilde{m} \qquad (18.14)$$

Because in Equation 18.14 the molality appears as a ratio, for simplicity, the tilde differentiating it from its dimensionless value is sometimes dropped. Rearranging Equation 18.13, we write

$$\tilde{m} d \ln \gamma_{\pm} = \tilde{m} d\varphi + \varphi d\tilde{m} - d\tilde{m} = d(\tilde{m}\varphi) - d\tilde{m}$$

or

$$d(\tilde{m}\varphi) = \tilde{m} d \ln \gamma_{\pm} + d\tilde{m}$$

Integrating between the same limits as before and rearranging,

$$\varphi = \frac{1}{\tilde{m}} \int_{0}^{\tilde{m}} \tilde{m} d \ln \gamma_{\pm} + 1 \qquad (18.15)$$

Again, as the molality appears as a ratio in this expression, sometimes the difference between the molality and its dimensionless form is ignored.

NEED FOR CONSIDERING THE EXISTENCE OF INDIVIDUAL IONS

One natural question arising at this point is why, for the thermodynamic treatment of strong electrolytes, one should go through all the trouble of considering independent ions in solution instead of defining an activity coefficient for the electrolyte as such by the relation

$$\hat{a}_E = \tilde{m}\gamma_E$$

The idea here is not to ignore the real existence of ions, which is evident from conductivity measurements, but to explore whether it is absolutely necessary to consider their existence for the thermodynamic treatment when it would be much simpler to consider the electrolyte as a single entity. The answer to this question is found by observing the behavior of the solution of an electrolyte in the very dilute region.

If we ignore the presence of ions and consider only the presence of molecules, we get from the above relation, in the very dilute region, when $\gamma_E \to 1$ and $\hat{a}_E^{dil} \approx \tilde{m}$, the following relation:

$$\frac{d \ln \hat{a}_E^{dil}}{d\tilde{m}} = \frac{1}{\tilde{m}}$$

In addition, using the Gibbs–Duhem equation, Equation 18.10, we can calculate the slope of $\ln \hat{a}_W^{dil}$ plotted versus the molality of salt $\tilde{m}$:

$$\frac{d \ln \hat{a}_W^{dil}}{d\tilde{m}} = -\frac{M_W}{1000}$$

On the other hand, if the presence of the ions is considered for thermodynamic treatment, in the dilute region, that is, when $\gamma_\pm \to 1$, Equation 18.7 gives

$$a_E^{dil} = \tilde{m}^\nu (\nu_+^{\nu_+} \nu_-^{\nu_-})$$

and from Equation 18.11 we obtain

$$\frac{d \ln a_W^{dil}}{d\tilde{m}} = -\frac{\nu M_W}{1000}$$

Figure 18.1 shows a plot of $(-\ln a_W)$ versus $\tilde{m}$ in the dilute region for NaCl ($\nu = 2$) in aqueous solution at 298.2 K, and the slope of the line is 0.036, as predicted by the consideration of the ions, and not 0.018, as would be the case ignoring the ions in the thermodynamic approach.

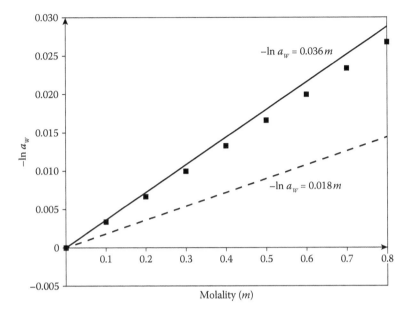

FIGURE 18.1 Comparison of the experimental dependence of $-\ln a_W$ as a function of molality for aqueous NaCl solution at 25°C with calculated lines: ——, calculated line with assumed presence of ions; ------, calculated line with the assumption of no ions present; ■, experimental data [1]. (Reprinted and partially modified from Wilczek-Vera, G. and Vera J. H. Peculiarities of the Thermodynamics of Electrolyte Solutions: A Critical Discussion. *Can.J.Chem.Eng.* 81:70–79, 2003. With permission from Wiley & Sons.)

BEHAVIOR OF THE ACTIVITY OF INDIVIDUAL IONS

At high dilution, the ions of one kind are not affected by any other ion present in the solution. One reasonable question is whether, as the solution gets concentrated, the activity coefficient of an ion in the presence of a particular counterion will be the same or different from the activity coefficient of the same ion with a different counterion, both solutions at the same concentration and temperature. This question can be stated more clearly considering Equation 18.8 for solutions at the same concentration of two salts with the same anion, but with different cation. Consider, for example, a solution of a sodium salt with an anion Y^- and a potassium salt with the same anion, both solutions at the same concentration and temperature. If the activity coefficients of the anion were independent of the cation present, we would have

$$\left[\frac{(\gamma_\pm^2)_{NaY}}{(\gamma_\pm^2)_{KY}}\right] \equiv \left[\frac{\gamma_{Na^+}\gamma_{Y^-}}{\gamma_{K^+}\gamma_{Y^-}}\right] = \left[\frac{\gamma_{Na^+}}{\gamma_{K^+}}\right]?$$

If this were true, the ratio of the mean ionic activity coefficients of two electrolytes with a common ion would have the same value independent of the nature of the common ion.

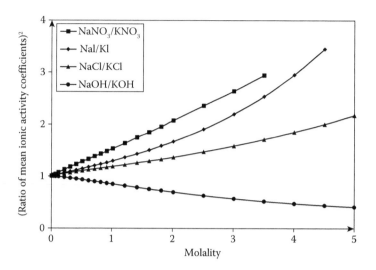

FIGURE 18.2 Purported ratios of $\gamma_{Na^+}/\gamma_{K^+}$ calculated for different salts. (Experimental data [2]; Plot: reprinted and partially modified with permission from Wilczek-Vera, G. and Vera J.H.: Peculiarities of the thermodynamics of electrolyte solutions: a critical discussion. *Can. J. Chem. Eng.* 2003. 81. 70–79. Copyright Wiley-VCH Verlag GmbH & Co. KGaA.)

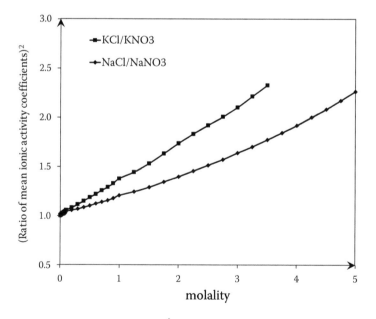

FIGURE 18.3 Purported ratios of $\gamma_{Cl^-}/\gamma_{NO_3^-}$ calculated for different salts. (Experimental data [2]; Plot: reprinted and partially modified with permission from Wilczek-Vera, G. and Vera J.H.: Peculiarities of the thermodynamics of electrolyte solutions: a critical discussion. *Can. J. Chem. Eng.* 2003. 81. 70–79. Copyright Wiley-VCH Verlag GmbH & Co. KGaA.)

Figure 18.2 shows the values of the left-hand side of this equation for aqueous solutions of electrolytes of the same cation with different anions at 298.2 K. Figure 18.3 shows the values of the left-hand side of this equation for aqueous solutions of electrolytes of the same anion with different cations at 298.2 K. These results show that at the same concentration, the activity coefficient of an ion is strongly dependent on the nature of the counterion present. At high dilution, the activity coefficient of an ion tends to be independent of the counterion, but as the solution becomes more concentrated, the ion behaves differently depending on its counterion.

DEBYE–HÜCKEL THEORY FOR DILUTE AQUEOUS ELECTROLYTE SOLUTIONS

Up to this point, we have made a deliberate effort to avoid introducing models or theories that are not part of the mathematical structure of thermodynamics. Models and theories change with time; the mathematics supporting the thermodynamic treatment of fluid systems does not. However, as a tool to discuss equations useful in the thermodynamic treatment of aqueous electrolyte solutions, we introduce here the expressions obtained by Debye and Hückel [4] for the treatment of dilute electrolyte solutions. This theory has stood the test of time. Debye and Hückel developed a model by considering the mathematics of inserting charged particles (ions) into a solvent. In their model, based on what is called the MacMillan–Mayer framework, the solvent is not a component of the mixture, but it is just a background media supporting the ions. This restriction presents a problem for the thermodynamic treatment of electrolyte solutions in mixed solvents, but for all practical purposes, it can be ignored in the work of aqueous solutions or other single-solvent systems. For the activity coefficient of a single ion in a single solvent, Debye obtained the expression

$$\ln \gamma_i = -\frac{A_D Z_i^2 I^{1/2}}{1 + B_{Di} I^{1/2}} + b_i I \tag{18.16}$$

where A_D is a theoretical parameter that depends on the solvent and the temperature of the system. Although in principle B_{Di} is a parameter related to the size of the ion, in practice it is usually set equal to unity. The parameter b_i is an adjustable parameter. The ionic strength I, written in terms of (dimensionless) molality, is defined as

$$I = \frac{1}{2} \sum_{j=1}^{all\,ions} \tilde{m}_j Z_j^2 \tag{18.17}$$

For a single electrolyte in solution, the ionic strength is given by

$$I = \frac{\tilde{m}}{2} \sum_{j=1}^{all\,ions} \nu_j Z_j^2 \tag{18.17a}$$

For 1:1 electrolytes such as HCl or NaCl, the ionic strength is numerically equal to the (dimensionless) molality. The Debye–Hückel expression for the mean ionic activity coefficient of the electrolyte obtained from Equation 18.8, assuming the same value of B_D for the cation and the anion, can be written as

$$\ln \gamma_\pm = -\frac{A_D|Z_+Z_-|I^{1/2}}{1+B_DI^{1/2}} + bI \tag{18.18}$$

with

$$b = \frac{(\nu_+b_+ + \nu_-b_-)}{\nu}$$

The absolute value of the product of the charges of the ions in Equation 18.18 arises from the relation of electrical neutrality for the electrolyte, Equation 18.1d. From Equation 18.10, the activity coefficient for water in a single electrolyte aqueous solution obtained from the Debye–Hückel theory is

$$\ln \gamma_W = \frac{2A_DM_W}{(10B_D)^3}\left[(1+B_DI^{1/2}) - \frac{1}{(1+B_DI^{1/2})} - 2\ln\left(1+B_DI^{1/2}\right)\right]$$
$$+ \ln\left[1+\frac{\nu M_W\tilde{m}}{1000}\right] - \frac{\nu M_W\tilde{m}}{1000}\left[1+\frac{bI}{2}\right] \tag{18.19}$$

LIMITING VALUES OF THE MEAN IONIC ACTIVITY COEFFICIENT AND THE OSMOTIC COEFFICIENT AT HIGH DILUTION

One immediate application of the Debye–Hückel equation is to obtain explicitly the limiting value of the osmotic coefficients as the ionic strength goes to zero. As it can be seen from the definition of the osmotic coefficient by Equation 18.11, and more clearly from the case of a single electrolyte in solution, as the molality of the solute goes to zero, the activity of water goes to unity and its logarithm goes to zero, so no clear limiting behavior for the osmotic coefficient is obtained. At high dilution, the limiting behavior obtained from the Debye–Hückel equation takes the form

$$\ln \gamma_\pm^\infty = -A_D|Z_+Z_-|I^{1/2} \tag{18.20}$$

Thus, according to Equation 18.17a, for a single electrolyte in solution,

$$\ln \gamma_\pm^\infty = -A_D|Z_+Z_-|\left[\frac{1}{2}\sum_{j=1}^{all\,ions}\nu_jZ_j^2\right]^{\frac{1}{2}}\tilde{m}^{1/2} = -A_\gamma\tilde{m}^{1/2} \tag{18.20a}$$

We have indicated by a subscript γ in the Debye–Hückel constant A that for any electrolyte other than type 1:1, the value of this constant is modified in this step. This fact is normally ignored in texts. Inserting Equation 18.20a into Equation 18.15, we observe that in the dilute region the osmotic coefficient is given by

$$\varphi^\infty = -\frac{A_\gamma}{\tilde{m}} \int_0^{\tilde{m}} \tilde{m}\, d(\tilde{m}^{1/2}) + 1 = -\frac{A_\gamma}{3}\tilde{m}^{1/2} + 1$$

Hence,

$$\lim_{\tilde{m}\to 0} \varphi = 1 \qquad (18.21)$$

As discussed in Chapter 20, this limiting behavior of the osmotic coefficient is well known for the case of the osmotic pressure in the study of nonelectrolyte solutions, but it is not immediately obvious for aqueous electrolyte solutions. In addition, for electrolyte aqueous solutions, this limiting behavior seems to create a problem with the value of the integrand at the lower limit of the integral in Equation 18.14.

$$\ln \gamma_\pm = (\varphi - 1) + \int_0^{\tilde{m}} \frac{(\varphi - 1)}{\tilde{m}}\, d\tilde{m} \qquad (18.14)$$

Again in this case, at the limit of infinite dilution the integrand seems to be undefined. However, the behavior of the osmotic coefficient in the dilute region just obtained shows that

$$\lim_{\tilde{m}\to 0}\left[\frac{\varphi - 1}{\tilde{m}^{1/2}}\right] = -\frac{A_\gamma}{3} \qquad (18.22)$$

This result suggests that a change in the denominator of the integrand of Equation 18.14 from $\tilde{m}$ to $\tilde{m}^{1/2}$ would solve the problem. This is achieved noting that

$$\frac{d\left(\tilde{m}^{1/2}\right)}{d\tilde{m}} = \frac{1}{2}\tilde{m}^{-1/2} = \frac{1}{2}\frac{\tilde{m}^{1/2}}{\tilde{m}}$$

or

$$2\frac{d\left(\tilde{m}^{1/2}\right)}{\tilde{m}^{1/2}} = \frac{d\tilde{m}}{\tilde{m}}$$

Hence, Equation 18.14 is written as

$$\ln \gamma_\pm = (\varphi - 1) + 2\int_0^{\tilde{m}} \frac{(\varphi - 1)}{\tilde{m}^{1/2}}\, d\tilde{m}^{1/2} \qquad (18.14a)$$

In this form, at the lower limit of integration the integrand has a well-defined value given by Equation 18.22. For this reason, experimentally measured values of the osmotic coefficient or of the mean ionic activity coefficient are fitted with advantage as a function of $\tilde{m}^{1/2}$ instead of $\tilde{m}$.

INDIRECT MEASUREMENT OF THE MEAN IONIC ACTIVITY COEFFICIENT

Equation 18.14a is the basic relation for calculating the mean ionic coefficient from the osmotic coefficient data. The vapor pressure of water (or solvent) is measured under isothermal conditions for different molalities of a single electrolyte in solution. The activity of water is then obtained from Equation 18.12, and the osmotic coefficient calculated from Equation 18.11a. The values of the osmotic coefficient are then correlated as a function of molality, usually in terms of $\tilde{m}^{1/2}$, and the smooth function is used with Equation 18.14a to obtain values of the mean ionic activity coefficient $\gamma_\pm$ as a function of molality.

DIRECT ELECTROCHEMICAL MEASUREMENT OF THE MEAN IONIC ACTIVITY COEFFICIENT

The mean ionic activity coefficient of an electrolyte in aqueous solution can be obtained measuring the potential difference between the responses of an ion-selective electrode (ISE) sensitive to the cation and an ISE sensitive to the anion. These ISEs, of which the pH meter is a precursor, are commercially available for most ions of practical interest. The voltage response $E_{i,k}$ of an ISE sensitive to ion i at molality $\tilde{m}_{i,k}$ in aqueous solution is given by the Nernst equation:

$$E_{i,k} = E_{i,0} + \frac{RT}{Z_i F}\ln(a_{i,k}) \tag{18.23}$$

In this equation, other than the usual symbols R, T, Z_i, and $a_{i,k}$ for the gas constant, the absolute temperature, the charge of the ion, and the activity of ion i at molality m_k, F is the Faraday constant and $E_{i,0}$ is the standard electrode potential. As the ISEs are transient instruments, the standard electrode potential has a constant value only over relatively short lapses of time. If an ISE is immersed in the same solution on different days, the standard electrode potential will have different values. Hence, either measurements are carried out in continuous runs in short periods of time, or if carried out on different days, the electrodes should be recalibrated as explained below. For clarity, we write the Nernst equation for the cation and for the anion as

$$E_{+,k} = E_{+,0} + \frac{RT}{Z_+ F}\ln(a_{+,k})$$

and

$$E_{-,k} = E_{-,0} - \frac{RT}{|Z_-| F}\ln(a_{-,k})$$

If both electrodes are immersed in the same solution at the same time and their responses are measured directly one against the other in a voltmeter, the voltage difference is

$$\Delta E_k = \Delta E_0 + \frac{RT}{F} \ln \left[a_{+,k}^{\frac{1}{Z_+}} a_{-,k}^{\frac{1}{|Z_-|}} \right] \tag{18.24a}$$

Now, using Equations 18.1b and 18.1c, we write

$$\left[a_{+,k}^{\frac{1}{Z_+}} a_{-,k}^{\frac{1}{|Z_-|}} \right] = \left[a_{+,k}^{v_+} a_{-,k}^{v_-} \right]^{\frac{1}{v_+ Z_+}}$$

But, according to Equation 18.7,

$$a_+^{v_+} a_-^{v_-} = \tilde{m}^v \gamma_\pm^v (v_+^{v_+} v_-^{v_-})$$

so,

$$\Delta E_k = \Delta E_0 + \frac{RT}{(v_+ Z_+)F} \ln \left[\tilde{m}^v \gamma_\pm^v \left(v_+^{v_+} v_-^{v_-} \right) \right] \tag{18.24b}$$

The only residual problem is the evaluation of ΔE_0. This is usually done by measuring several points in the dilute region where the mean ionic activity coefficient can be obtained from Equation 18.18 approximated as

$$\ln \gamma_\pm^{dil} = - \frac{A_D |Z_+ Z_-| I^{1/2}}{1 + I^{1/2}}$$

The rewarding result is that the mean ionic activity coefficient values for different electrolytes obtained by the direct electrochemical measurements agree with the values obtained by measurement of the osmotic coefficient.

ELECTROCHEMICAL MEASUREMENT OF THE MEAN IONIC ACTIVITY COEFFICIENT WITH INDEPENDENT MEASUREMENTS FOR THE CATION AND THE ANION

Although, as discussed above, the mean ionic activity coefficient can be measured connecting both ISEs, for the cation and for the anion, directly to a voltmeter, it is also instructive to consider the case of measuring each voltage response separately against a single-junction reference electrode, as shown in Figure 18.4. The reference electrode is immersed in an internal standard solution that leaks into the sample solution, creating a liquid junction potential at the surface of contact between the two solutions. This contact between the sample solution and the reference solution is

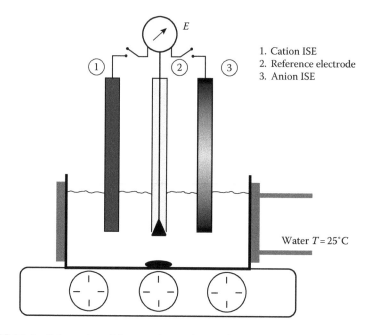

1. Cation ISE
2. Reference electrode
3. Anion ISE

Water $T = 25°C$

FIGURE 18.4   Schematics of the experimental setup for measurements of ionic activities.

necessary to close the electric circuit. In this case, the voltage response of the electrodes for an ion i takes the form

$$E_{i,k}^R = E_{i,0}^R + \frac{RT}{Z_i F} \ln(a_{i,k}) + E_{J,k} \qquad (18.25)$$

Thus, we write for the cation and for the anion

$$E_{+,k}^R = E_{+,0}^R + \frac{RT}{Z_+ F} \ln(a_{+,k}) + E_{J,k}$$

$$E_{-,k}^R = E_{-,0}^R - \frac{RT}{|Z_-| F} \ln(a_{-,k}) + E_{J,k}$$

We have used a superscript R to indicate that in this case, the voltage response is measured against a reference electrode and its value will be different from the value obtained in the previous experimental setup. The additional term, $E_{J,k}$, is due to the junction potential generated by the contact of the sample solution with the reference solution that happens at the tip of the reference electrode. This junction potential will have the same value for the responses of both ISEs connected to the same reference electrode in simultaneous measurements. Taking the difference of both independent readings, we obtain

$$\Delta E_k^R = \Delta E_0^R + \frac{RT}{F} \ln \left[a_{+,k}^{\frac{1}{Z_+}} \, a_{-,k}^{\frac{1}{|Z_-|}} \right] \qquad (18.24c)$$

Equation 18.24c is formally identical to Equation 18.24a and by calibration of the value of ΔE_0^R will yield values for the mean ionic activity coefficients of the electrolyte. Again, the rewarding result is that the mean ionic activity coefficient values for different electrolytes obtained by this approach agree with the values obtained by the direct electrochemical procedure and with the values obtained from measurements of the osmotic coefficient.

ACTIVITY OF A SECOND SOLUTE IN AN AQUEOUS ELECTROLYTE SOLUTION

The direct measurement of the mean ionic activity coefficient of an electrolyte using ISEs opens the possibility of determining the activity of a water-soluble solute dissolved in an electrolyte aqueous solution. According to Equation 7.11, for a solute A in an aqueous solution with an electrolyte E,

$$\left[\frac{\partial \mu_A}{\partial \tilde{m}_E}\right]_{T,\tilde{m}_A} = \left[\frac{\partial \mu_E}{\partial \tilde{m}_A}\right]_{T,\tilde{m}_E}$$

Considering

$$\mu_A = \mu_A^\theta + RT \ln\left(\tilde{m}_A \, \gamma_A\right)$$

and

$$\mu_E = \mu_E^\theta + RT \ln\left(\tilde{m}_E^\nu \, \gamma_\pm^\nu \, \nu_+^{\nu_+} \, \nu_-^{\nu_-}\right)$$

we obtain

$$\left[\frac{\partial \ln \gamma_A}{\partial \tilde{m}_E}\right]_{T,\tilde{m}_A} = \nu \left[\frac{\partial \ln \gamma_\pm}{\partial \tilde{m}_A}\right]_{T,\tilde{m}_E}$$

Thus, integrating between 0 and $\tilde{m}_E$, at constant m_A,

$$\ln\left[\frac{\gamma_A^{(2)}}{\gamma_A^{(1)}}\right] = \nu \int_0^{\tilde{m}_E} \left[\frac{\partial \ln \gamma_\pm}{\partial \tilde{m}_A}\right]_{T,\tilde{m}_E} d\tilde{m}_E \qquad (18.26)$$

In this expression, the superscripts (1) and (2) indicate the value of the activity coefficient of the solute A in the absence of electrolyte and in the presence of electrolyte, respectively, both at the same molality of solute A. The former value should be available from independent measurements, usually from the isopiestic method in the case of nonelectrolytes. In order to obtain the activity coefficient of the solute A in the presence of the electrolyte, measurements are made of the mean ionic activity coefficient of the electrolyte at fixed values of the molality of the electrolyte changing the concentration of the solute A. These results are then correlated by a function of the form

$$\ln\left[\frac{\gamma_\pm^{(2)}}{\gamma_\pm^{(1)}}\right] = \sum_{k=1} C_k\, m_A^k$$

Usually, no more than three or four terms are required in the right-hand side of this equation. Again here, the superscripts (1) and (2) indicate the value of the activity coefficient of the electrolyte in the absence of solute A and in the presence of solute A, respectively, both at the same molality of the electrolyte. The former value should be available from independent measurements. The values of the coefficients C_k are correlated as a function of the molality of the electrolyte.

$$C_k = \sum_{j=1} c_j\, m_E^j$$

The rest is simple algebra necessary to obtain the value of the right-hand side of Equation 18.26 and get values of $\gamma_A^{(2)}$.

AQUEOUS SOLUTIONS OF WEAK ELECTROLYTES

For weak electrolytes, only a fraction α of the electrolyte added to the solution will ionize so that the activity of the electrolyte that can be obtained from the activity of water, by means of the Gibbs–Duhem equation, is an effective value. In this case, with a degree of dissociation α, the molality of the ion i in solution would be $\nu_i \tilde{m}\alpha$ and its activity would be $\gamma_i \nu_i \tilde{m}\alpha$, while for the undissociated electrolyte in solution the activity would be $\gamma_E \tilde{m}(1-\alpha)$. The dissociation constant from Equation 18.3 is

$$K_T^{dis} = \frac{a_+^{\nu_+}\, a_-^{\nu_-}}{a_E}$$

Replacing and rearranging,

$$\frac{\alpha}{(1-\alpha)^{\frac{1}{\nu}}} = \frac{1}{\gamma_\pm}\left[\frac{\gamma_E\, K_T^{dis}}{\tilde{m}^{\nu-1}\, \nu_+^{\nu_+}\, \nu_-^{\nu_-}}\right]^{\frac{1}{\upsilon}}$$

A weak electrolyte has a weak dissociation, so the left-hand side is close to α. For a 1:1 weak electrolyte in dilute solution, $\gamma_E \approx 1$,

$$\alpha \approx \frac{1}{\gamma_\pm}\left[\frac{K_T^{dis}}{\tilde{m}}\right]^{\frac{1}{2}} \tag{18.27}$$

This approximation is interesting as a way of understanding the effect of additional ions in the dissociation of the weak electrolyte. If an electrolyte that does not contain the cation or the anion of the weak electrolyte is added to the solution, the ionic strength will increase, and thus the value of the mean ionic activity coefficient will decrease, resulting in an increase of the degree of dissociation α. If the added electrolyte contains either the cation or the anion of the weak electrolyte, the activity of the common ion will increase, and as the dissociation constant does not change, the activity of the undissociated electrolyte must increase; that is, the degree of dissociation decreases. This is the so-called common ion effect.

CHARGED ORGANIC MOLECULES

Amino acids are organic compounds having an amino group, $-NH_2$, and a carboxylic group, $-COOH$, attached to a hydrocarbon chain. Without going into the chemistry of amino acids, with the sole interest in their behavior in aqueous electrolyte solutions, we observe that most of the common amino acids at pH above 9.4 have the carboxyl group negatively charged, $-COO^-$, and at pH below 2.2, the amino group is positively charged, $-NH_3^+$. At a pH between 2.2 and 9.4, the amino acid has both a negative carboxyl group and a positive amino group. This state with zero net charge is known as the zwitterionic form, from the German word *zwitter*, meaning "hybrid." At the midpoint between the two limiting values of pH, the amino acid has a weak positive charge in the amino group and a weak negative charge in the hydroxyl group that closely compensate. This state is known as the isoelectric point, and the pH is designated as the pI of the amino acid. The interesting point is that at the isoelectric point the amino acid presents a minimum in its solubility in water. Amino acids associate in chains known as peptides, and peptides associate in larger molecules known as proteins, of which enzymes are a special kind presenting catalytic properties. In aqueous solutions, the presence of electrolytes, such as $(NH_4)_2SO_4$, for example, affects the solubility of proteins. In a dilute electrolyte solution, the protein is more soluble than in pure water, but if the concentration of the electrolyte is increased beyond a certain point, the protein is less soluble than in pure water. These behaviors are known as the salting in and salting out of the protein by changing the concentration of the electrolyte.

Different kinds of charged organic molecules are the so-called ionic liquids. These are organic salts that are in liquid phase at temperatures below the boiling point of water. This temperature limit is rather arbitrary, but it comes from the possibility of using these ionic liquids as solvents for green chemistry synthesis. Although ionic liquids are rather viscous, they have low volatility and high thermal stability properties that make them recyclable.

EXPERIMENT TO DETERMINE THE CHARGE OF A PROTEIN

The molecules of proteins behave as polyions in aqueous solutions and are too large to pass through some membranes, while other ions can. When one such polyelectrolyte is in an aqueous solution on one side of a membrane and some electrolyte formed by small ions, NaCl, for example, and water can freely pass through the membrane, we are in the presence of the Donnan equilibrium described in Chapter 10. Due to the condition of electroneutrality on both sides of the membrane and the inability of a polyion to migrate, ions are at different concentrations on each side of the membrane. This causes a difference in electric potential to arise between the two phases. As an illustration of the chemical equilibrium between ionic species separated by a permeable membrane, we discuss here an experiment designed for the determination of the charge of the sodium salt of an anionic protein. There are many kinds of membranes. Thus, the first step is to decide what kind of membrane is appropriate for the purpose. Some membranes are permeable only to the solvent; some are more permeable to cations than to anions or vice versa. It is known that biological membranes behave differently when they are part of a living tissue than when they are not. For this reason, the consideration of equilibrium in living biological systems may be a very poor assumption. As a general rule, it has been observed that single charged ions, such as Na^+ or Cl^-, permeate through membranes more easily than polycharged ions, such as Ba^{2+} or SO_4^{2-}; hence, for this experiment NaCl is chosen as the electrolyte. The membrane is chosen to be permeable to both the sodium and the chloride ions. Figure 18.5a presents the experimental setup in which the same weight of deionized water is charged on both sides of the membrane, a molality m_P of the protein is charged to the left side of the cell, and a molality m of sodium chloride is charged to the right side. After equilibration, a molality $m_{-,R}$ of Cl^- is measured by titration of the solution in the right-hand side of the membrane (Figure 18.5b).

The dissociation equilibrium of the sodium salt of the anionic protein P with a negative charge $|Z_P|$ is given by

$$PNa_{|Z_P|} = P^{-|Z_P|} + |Z_P|Na^+$$

The charge balance of the solution in the left side of the membrane where there is the polyion plus chloride and sodium ions is

$$|Z_P| m_P + m_{-,L} = m_{+,L} \qquad (18.28)$$

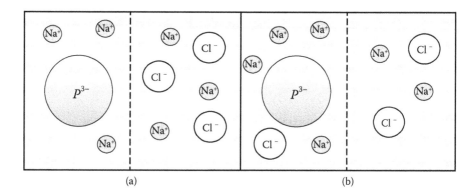

(a)   (b)

FIGURE 18.5 The system protein + electrolyte separated by an electrolyte-permeable membrane: (a) immediately after loading and (b) after the equilibrium is established.

or

$$|Z_P| = \frac{m_{+,L} - m_{-,L}}{m_P} \qquad (18.28a)$$

The molality of chloride ions on the right-hand side of the membrane after $m_{-,L}$ moles of chloride ion migrated to the left-hand side of the membrane is

$$m_{-,R} = m - m_{-,L}$$

Hence, the charge balance on the right-hand side of the membrane gives

$$m_{+,R} = m_{-,R} = m - m_{-,L} \qquad (18.29)$$

In order to obtain the molality of sodium in the left-hand side of the membrane, we apply the expression for the Donnan equilibrum,

$$\left[\frac{a_{+,L}}{a_{+,R}} \right]^{v_+} = \left[\frac{a_{-,R}}{a_{-,L}} \right]^{v_-}$$

For NaCl, $v_+ = v_- = 1$, so we write

$$\left[\frac{m_{+,L}\ \gamma_{+,L}}{m_{+,R}\ \gamma_{+,R}} \right] = \left[\frac{m_{-,R}\ \gamma_{-,R}}{m_{-,L}\ \gamma_{-,L}} \right]$$

Although in complete thermodynamic rigor the dimensionless molalities should be used for activities, in order to simplify the nomenclature, we are ignoring this requirement here. Thus, using the definition of the mean ionic activity coefficient,

$$m_{+,L} = \left[\frac{m_{+,R}\ m_{-,R}}{m_{-,L}} \right] \left[\frac{\gamma_{\pm,R}}{\gamma_{\pm,L}} \right]^2$$

Replacing in this expression the molalities of sodium and chloride in the right-hand side of the membrane given by Equation 18.29, we write

$$m_{+,L} = \frac{(m - m_{-,L})^2}{m_{-,L}} \left[\frac{\gamma_{\pm,R}}{\gamma_{\pm,L}} \right]^2 \qquad (18.30)$$

Thus, $|Z_P|$ is directly obtained from Equation 18.28a. The mean ionic activity coefficient of sodium chloride in the right-hand side of the membrane can be obtained using the Debye–Hückel expression at low concentrations of NaCl or more complex correlations for concentrated solutions. On the other hand, the mean ionic activity coefficient of sodium chloride in the protein side of the membrane is not easily evaluated because the charged protein increases the ionic strength of the solution on that side of the membrane. If we assume that the mean activity coefficients of NaCl have the same value on both sides of the membrane, from Equation 18.30,

$$m_{+,L}\, m_{-,L} = (m - m_{-,L})^2$$

Replacing back $m_{+,L}$ from Equation 18.28,

$$\left(|Z_P|m_P + m_{-,L}\right)m_{-,L} = (m - m_{-,L})^2$$

and expanding and rearranging,

$$m_{-,L} = \frac{m}{\left(2 + \dfrac{|Z_P|m_P}{m}\right)}$$

From this approximate expression, it can be seen that less than half of the NaCl added to the right-hand side permeated to the left-hand side. Thus, the contribution to the ionic strength of the charged protein on the left-hand side is compensated to some extent by the smaller concentration of NaCl, and the ratio of the activity coefficients in Equation 18.30 can be set to unity as a good approximation.

CONCLUSIONS

In this chapter, we have presented a general view and the main concepts necessary for the understanding of the behavior of aqueous electrolyte solutions. This is just the starting point for more detailed studies. One case of interest, for example, is the measurement of the activity of individual ions, which is discussed in Chapter 27.

REFERENCES

1. Robinson, R. A., and Stokes, R. H. 1959. *Electrolyte Solutions*, 2nd ed. London: Butterworths Scientific, Appendix 8.10, Table 1.
2. Hamer, W.J., and Wu, Y.-Ch., 1972. *J. Phys. Chem. Ref. Data.* 1:1047–1099.
3. Wilczek-Vera, G., and Vera, J. H. 2003. Peculiarities of the Thermodynamics of Electrolyte Solutions: A Critical Discussion. *Can.J.Chem.Eng.* 81:70–79.
4. Debye, P., and Hückel, E. 1923. Zur Theorie der Elektrolyte. I. Gefrierpunktserniedrigung und verwandte Erscheinungen. *Phys. Z.* 24: 185–206.

Section III

Applications

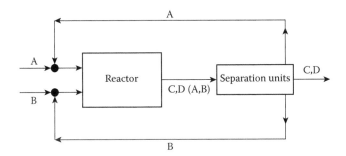

Going beyond the equilibrium conversion

19 The Thermodynamics of Chemical Reactions

In Section I, we reviewed the fundamentals, while in Section II, Chapters 5–17, we studied the behavior of mixtures. In this chapter, the first specifically considering an application of the thermodynamics of mixtures, attention is centered on a reacting system. As usual in thermodynamics, we do not consider the kinetics of the process but, as discussed in Chapters 8 and 12, we only pay attention to the final state of the system at equilibrium, after reactions have occurred.

STATEMENT OF THE PROBLEM AND BASIC DEFINITIONS

In order to clarify the applicability of chemical reaction equilibrium, let us state (as shown in Figure 19.1) that some reactions get so displaced toward the products that, for all practical purposes, there is no need to perform equilibrium calculations. This is the case, for example, of the combustion of many organic compounds to yield carbon dioxide and water. Considering that the main purpose of these reactions is usually the generation of heat, their study is presented in Chapter 21 dealing with heat effects.

The study of chemical reaction equilibrium is important when the final state consists of a mixture of both reactants and products in significant amounts. In this case, as at equilibrium, there is no net reaction, and close to equilibrium the rate of reaction tends to slow down; the design of reactors is based on kinetic rather than thermodynamic considerations. Moreover, although for this case the calculation of the composition of a reacting mixture at equilibrium gives the maximum possible conversion that can be expected at the conditions of operation, in principle it is possible to carry out a reaction in such a way that all the reactants fed to the system are completely converted into products.

Consider the case shown in Figure 19.2 in which reactants A and B enter a continuous reactor to give an output stream of products C and D plus unreacted A and B. The output stream is then sent to a separation system from which products C and D are obtained while the unreacted A and B compounds are recycled to the reactor. The net effect is that A and B enter the overall system and products C and D leave the overall system.

Thus, one may ask why worry about chemical equilibrium when even with a very low conversion it is possible to convert all the reactants to products. The answer is that the size of the reactor and the number and complexity of the separation units depend on the amount of chemicals that need to be recycled. The energy consumed also depends on how efficient the reactor is. The higher is the conversion that can be obtained in the reactor in a single pass, the better.

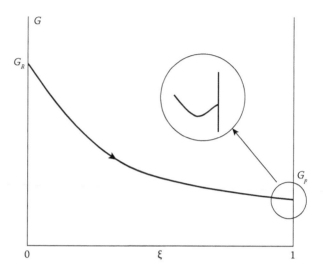

FIGURE 19.1 Gibbs function of a closed system in which a chemical reaction is almost complete.

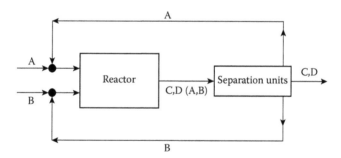

FIGURE 19.2 Diagram of a continuous reactor.

This maximum conversion is limited by the composition attainable at equilibrium. Thus, the study of chemical equilibrium is important as a way to establish the maximum possible *extent of reaction*, ξ, in a discontinuous batch reactor for a given feed composition at specified conditions of temperature and pressure. No matter how fast or how slow the kinetics of the reaction is, the value of the extent of reaction at equilibrium will determine the maximum conversion of each reactant in the reactor, that is, the maximum possible yield of products at the operating conditions. The *conversion* χ_i *of a reactant* i, for which the initial and final numbers of moles are n_{i0} and n_i, respectively, is defined as

$$\chi_i = \frac{n_{i0} - n_i}{n_{i0}} \tag{19.1}$$

Integrating Equation 8.9 between $(n_{i,0}, \xi = 0)$ and (n_i, ξ) and rearranging, we write

$$n_i = n_{i0} + v_i \xi \qquad (19.2)$$

and combining this result with Equation 19.1, we have

$$\chi_i = \frac{1}{n_{i0}}(-v_i)\xi \qquad (19.3)$$

where the *stoichiometric number*, v_i, is intrisically negative for a reactant and positive for a product.

Hence, the conversion of reactant i in a closed system at equilibrium is determined by the equilibrium value of the extent of reaction, ξ.

If the actual conversion obtained in such a closed system is much less than the equilibrium conversion, then kinetic factors are limiting the advancement of the reaction and the use of a catalyst should be considered. On the other hand, if the actual conversion is close to the equilibrium conversion, then thermodynamic factors are limiting and the attention should be directed to the separation and recycling of products.

In the case of a single reaction, for a given value of the extent of reaction and a knowledge of the initial number of moles of the chemical species, n_{i0}, the mole fractions of the compounds are obtained from Equation 19.2 as

$$y_i = \frac{n_{i0} + v_i \xi}{n_0 + \xi \Delta v} \qquad (19.4)$$

with

$$n_0 = \Sigma_k n_{k0} \qquad (19.5)$$

and, as discussed in Chapter 8, the change in the number of moles of the stoichiometric reaction,

$$\Delta v = \Sigma_k v_k = \Sigma_{products}\lambda_p - \Sigma_{reactants}\lambda_R \qquad (19.6)$$

where λ_k is the coefficient of compound k in the stoichiometric equation for the reaction. If there is more than one independent reaction, Equations 19.2 through 19.4 are directly generalized to

$$n_i = n_{i0} + \sum_r v_{i,r}\xi_r \qquad (19.2a)$$

$$\chi_i = \frac{1}{n_{i0}}\left(-\sum_r v_{i,r}\xi_r\right) \qquad (19.3a)$$

and again

$$y_i = \frac{n_{i0} + \sum_r v_{i,r}\xi_r}{n_0 + \sum_r \xi_r \Delta v_r} \tag{19.4a}$$

In Equations 19.4 and 19.4a, we have used the symbol y_i for the mole fraction, but obviously similar expressions apply for the mole fraction x_i of a compound in a liquid mixture.

While the reaction proceeds, one of the reactants may be totally consumed. This reactant will then be the *limiting reactant* fixing the extent of reaction at a value ξ_L. This can be calculated setting the number of moles of this reactant to zero in Equation 19.2. Thus,

$$\xi_L = -\frac{n_{L,0}}{v_L}$$

In some practical cases, one of the reactants is much more expensive than the others, and it is convenient to use it as the limiting reactant.

For those cases in which the attainment of chemical equilibrium is important, it is desirable to determine the value of the extent of reaction ξ at equilibrium conditions when the reactants are charged in their stoichiometric amounts to a batch reactor, that is, to a system that it is closed after the reactants are loaded.

DETERMINATION OF THE NUMBER OF INDEPENDENT REACTIONS

Equations 19.4 and 19.4a show that first step to get the equilibrium composition is to know how many *independent reactions* are possible, and then determine the value of the extent of reaction at equilibrium for each of the independent reactions. Although there are different ways of determining the number of possible independent reactions (one of them purely mathematical based on the rank of the matrix of stoichiometric numbers), we present here a method centered on the reactions of formation. This approach is more intuitive and connects well with the conceptual steps used for evaluation of the equilibrium constant.

In this method, the number of independent reactions is obtained by writing the stoichiometric equations of formation, from its constituent elements, of each chemical species considered to be present in the system. Then all the elements considered not to be present as elements, that is, in their elemental form, are eliminated in a systematic way. The final number of independent equations resulting from this process must be the same no matter the path chosen to eliminate the elements not present as elements. This procedure is illustrated in the examples that follow.

Each independent reaction has an equilibrium value of the extent of reaction, ξ_r. The solution of the equilibrium problem for any of the sets of independent reactions gives a unique set of values ξ_r, that is, a unique set of equilibrium concentrations.

The important point is that it is not necessary to know which are the independent reactions that happen in reality. If the equilibrium compositions are obtained for the set of reactions generated by the above procedure, these will be the equilibrium compositions of the compounds in the real system. When there is any question about the possible presence of a particular element in its elemental form in the equilibrium mixture, it is safe to assume that the particular element is present as element. If it were not present, the final result would show that at equilibrium, its "presence" is negligible and it could have been ignored from the start. The only price to pay for a wrong guess is the additional work of solving some extra equations. The same is true when it is not known what products will be obtained from multiple reactions. It is wise to include as many products as one could assume to be present, do the additional work, and from the results see which products could have been left out to begin with. As in most cases of applications, ideas are better clarified through examples.

Example 19.1: Catalytic Oxidation of Ammonia

In the catalytic oxidation of ammonia with air to produce nitric acid, the following chemical species are present in the reactor(s): NH_3, O_2, N_2, NO, NO_2, H_2O, and HNO_3. Determine the number of independent reactions for this system considering that (1) all chemical species participate in the reaction and (2) under the reaction conditions, N_2 can be considered inert.

Solution
Reactions of formation (marked as f) of compounds considered to be present:

f_1. $N_2 + 3H_2 = 2NH_3$
f_2. $N_2 + O_2 = 2NO$
f_3. $N_2 + 2O_2 = 2NO_2$
f_4. $2H_2 + O_2 = 2H_2O$
f_5. $H_2 + N_2 + 3O_2 = 2HNO_3$

Elements forming the compounds: N, H, O
Elements present as elements: N_2, O_2
Elements not present as elements: H_2

Thus, with reaction f_1, H_2 is eliminated from reactions f_4 and f_5, and reaction f_1 is then eliminated.
The final reactions are

1. $N_2 + O_2 = 2NO$
2. $N_2 + 2O_2 = 2NO_2$
3. $2N_2 + 6H_2O = 4NH_3 + 3O_2$
4. $2N_2 + 9O_2 + 2NH_3 = 6HNO_3$

As all elements not present in their elemental form have been eliminated, there are four independent reactions ($r = 4$). As an exercise, one can use equation f_4 to eliminate H_2 from f_1 and f_5 and see that although a different set of equations is obtained, again $r = 4$.

In practice, it is known that N_2 does not participate in the reactions. N_2 is one of the elements forming the compounds, but if it had not been added as an inert, it would not be produced in the reactor and it would not be present as an element in the final products. The situation of nitrogen in this case is similar to the situation of hydrogen. Clearly, if N_2 is replaced by argon in the mixture with oxygen, as a way to "dilute" the chemicals as N_2 does, nothing will change in the reactions. In this case, argon will not appear in the reactions of formation of the chemical compounds, and thus it would not be necessary to eliminate it from equations in which it is not present. Therefore, as N_2, being one of the elements forming the compounds, would not appear as an element in the final products, it is necessary to eliminate it from Equations 1–4 above. Eliminating N_2 using Equation 1, the final reactions in this case are

1. $O_2 + 2NO = 2NO_2$
2. $6H_2O + 4NO = 4NH_3 + 5O_2$
3. $7O_2 + 4NO + 2NH_3 = 6\ HNO_3$

The result then is $r = 3$.
According to the technical literature, the reactions that actually take place are

$$4NH_3 + 5O_2 = 6H_2O + 4NO \text{ (reaction 2 above, reversed)}$$
$$O_2 + 2NO = 2NO_2 \text{ (reaction 1 above)}$$

and

$$H_2O + 3NO_2 = NO + 2HNO_3$$

The third reaction is obtained by linear combination of reaction 3, with the two other.

Example 19.2: Production of 1,3-Butadiene

In the production of 1,3-butadiene by dehydrogenation of n-butane, it is expected to find the following chemical species: n-butane, hydrogen, 1,3-butadiene ($CH_2=CH-CH=CH_2$ or $1,3-C_4H_6$), and the side products: 1-butene ($CH_2=CH-CH_2-CH_3$ or $1-C_4H_8$), 2-butene ($CH_3-CH=CH-CH_3$ or $2-C_4H_8$), 1,2-butadiene ($CH_2=C=CH-CH_3$ or $1,2-C_4H_6$), 1-butyne ($CH\equiv C-CH_2-CH_3$ or $1-C_4H_6$), and 2-butyne ($CH_3-C\equiv C-CH_3$ or $2-C_4H_6$). Determine the number of independent reactions to be considered when (1) only the above compounds are present at equilibrium and (2) N_2 is added as an inert to minimize side reactions.

Solution
Reactions of formation of compounds considered to be present:

f_1. $4C + 5H_2 = C_4H_{10}$
f_2. $4C + 3H_2 = 1,3-C_4H_6$
f_3. $4C + 4H_2 = 1-C_4H_8$
f_4. $4C + 4H_2 = 2-C_4H_8$
f_5. $4C + 3H_2 = 1,2-C_4H_6$

f_6. $4C + 3H_2 = 1\text{-}C_4H_6$
f_7. $4C + 3H_2 = 2\text{-}C_4H_6$

Elements forming the compounds: C, H
Element present as element: H_2
Element not present as element: C

Use equation f_1 above to eliminate C from all the rest:

1. $C_4H_{10} = 1,3\text{-}C_4H_6 + 2H_2$
2. $C_4H_{10} = 1\text{-}C_4H_8 + H_2$
3. $C_4H_{10} = 2\text{-}C_4H_6 + H_2$
4. $C_4H_{10} = 1,2\text{-}C_4H_6 + 2H_2$
5. $C_4H_{10} = 1\text{-}C_4H_6 + 2H_2$
6. $C_4H_{10} = 2\text{-}C_4H_6 + 2H_2$

As no other elimination is required, $r = 6$.
Since N_2 is an inert and it does not participate in the reactions, $r = 6$.

Example 19.3: Isomerization of 1,3-Butadiene

In the isomerization of 1,3-butadiene to obtain 1,2-butadiene, it is expected that we will also find 1-butyne and 2-butyne in the products. Determine the number of independent reactions to be considered.

Solution
Reactions of formation:

f_1. $4C + 3H_2 = 1,3\text{-}C_4H_6$
f_2. $4C + 3H_2 = 1,2\text{-}C_4H_6$
f_3. $4C + 3H_2 = 1\text{-}C_4H_6$
f_4. $4C + 3H_2 = 2\text{-}C_4H_6$

Elements forming the compounds: C, H
No elements present as elements

Thus, with reaction f_1 eliminate C and H_2 in all other equations. This eliminates one reaction; then, $r = 3$.

THERMODYNAMIC EQUILIBRIUM CONSTANT IN TERMS OF THE EQUILIBRIUM COMPOSITIONS

To simplify matters, we consider here the case of a single reaction. As discussed in Chapter 12, once the stoichiometry of the reaction is decided, the thermodynamic equilibrium constant, K_T, is a function of temperature only, and it is related to the activities of the compounds by

$$K_T \equiv \frac{\prod_k a_{Pk,\theta}^{\lambda_{Pk}}}{\prod_j a_{Rj,\theta}^{\lambda_{Rj}}} \tag{12.15}$$

The dependence of the value of K_T on the stoichiometry of the equation, as written, is clearly shown by the presence of the stoichiometric coefficients in Equation 12.15. Thus, giving a numerical value of the *Thermodynamic equilibrium constant*, without specifying the stoichiometry of the equation, is meaningless. On the other hand, once the stoichiometric equation has been specified and the same stoichiometric coefficients are consistently used in all calculations, the compositions at equilibrium are independent of how the stoichiometric equation was written in the first place. This fact is better demonstrated with one simple example.

Consider a typical reaction,

$$\lambda_A A + \lambda_B B \leftrightarrow \lambda_C C + \lambda_D D$$

for which the thermodynamic equilibrium constant takes the form

$$K_T \equiv \frac{a_C^{\lambda_C} a_D^{\lambda_D}}{a_A^{\lambda_A} a_B^{\lambda_B}}$$

According to Equation 12.16, the value of the thermodynamic equilibrium constant in this case will be obtained from the Gibbs energies of the compounds in their standard states as

$$\ln K_T = \frac{[(\lambda_A G_A^0 + \lambda_B G_B^0) - (\lambda_C G_C^0 + \lambda_D G_D^0)]}{RT}$$

An independent thinker may decide to work with the chemical equation multiplied by a factor F, which can even be a negative number, thus reversing the sense of the reaction. For the reaction written as

$$(F\lambda_A)A + (F\lambda_B)B \leftrightarrow (F\lambda_C)C + (F\lambda_D)D$$

the equilibrium constant takes the form

$$K_T^* \equiv \frac{a_C^{F\lambda_C} a_D^{F\lambda_D}}{a_A^{F\lambda_A} a_B^{F\lambda_B}} = \left(\frac{a_C^{\lambda_C} a_D^{\lambda_D}}{a_A^{\lambda_A} a_B^{\lambda_B}} \right)^F$$

where the asterisk, used as a superscript on the left-hand side, indicates that the numerical value of this thermodynamical equilibrium constant will be different from the previous one. In fact, the value of this new thermodynamic equilibrium constant will be obtained as

$$\ln K_T^* = \frac{\left[\left(F\lambda_A G_A^0 + F\lambda_B G_B^0 \right) - \left(F\lambda_C G_C^0 + F\lambda_D G_D^0 \right) \right]}{RT}$$

$$= \frac{\left[\left(\lambda_A G_A^0 + \lambda_B G_B^0 \right) - \left(\lambda_C G_C^0 + \lambda_D G_D^0 \right) \right] F}{RT}$$

or

$$\frac{1}{F} \ln K_T^* = \ln K_T$$

So, this new numerical value will give

$$K_T^* = [K_T]^F = \left(\frac{a_C^{\lambda_C} a_D^{\lambda_D}}{a_A^{\lambda_A} a_B^{\lambda_B}} \right)^F$$

As shown by this result, a change in the way of writing the stoichiometric equation does not change the relation between the activities, that is, the relation between the equilibrium compositions. Hence, once the stoichiometric equation has been decided, it is necessary to keep the same stoichiometric coefficients for all calculations in order to make sense of the numerical value of the thermodynamic equilibrium constant.

Another general point that deserves mentioning is that the sign of the standard Gibbs energy change only indicates that if it is negative, the equilibrium constant is larger than unity, and if it is positive, the equilibrium constant is smaller than unity for the equation as written. The larger the value of the equilibrium constant, the more the equilibrium composition is displaced toward the products. The sign of the standard Gibbs energy change should not be interpreted as meaning that the reaction as written is spontaneous or not. As this value is obtained considering the reactants and the products in their standard state, and not the conditions of the reaction, any conclusion regarding the spontaneity of the reaction is unreliable. In addition, even when the conditions could be favorable for a spontaneous reaction, the reaction may not happen due to the need of activation energy to initiate it. An example for this is the reaction of oxygen with the stoichiometric amount of hydrogen to form water. The standard Gibbs energy change to form liquid water is –237.14 kJ/mol of water, and to form water vapor is –228.61 kJ/mol of water. This reaction with an extremely favorable equilibrium constant will not proceed unless initiated by a spark or another means. Once started, however, it will proceed with explosive speed.

A reaction with an unfavorable equilibrium constant can be displaced toward the products by adding an excess of one of the reactants that it is either easy to separate and recover or inexpensive. Moreover, even a reaction with an unfavorable value of the equilibrium constant can be forced to proceed toward completion by continuously withdrawing one of the products of the reaction.

SOLVAY PROCESS

A good example of a successful procedure to carry out a difficult reaction is the Solvay process to produce soda ash, Na_2CO_3, starting from brine, $NaCl_{(aq)}$, and lime, $CaCO_3$.

The desired reaction is

$$2NaCl + CaCO_3 \rightarrow Na_2CO_3 + CaCl_2, \quad \Delta G^0 = +96.9 \text{ (kJ/mol } Na_2CO_3),$$

which has a positive standard Gibbs energy change. Common salt, NaCl, and lime are inexpensive raw materials, while soda ash is a valued commodity with multiple industrial applications and calcium chloride, $CaCl_2$, is used as a road salt.

In 1791, Leblanc proposed a method to carry out the reaction with the help of sulfuric acid. The Leblanc process was replaced by the more economical Solvay process, proposed in 1861. Variations of the Solvay scheme have been proposed, but they do not change the basic feature of the process presented in Figure 19.3, together with the reactions in each unit.

This process promotes the reaction using ammonia, NH_3. The key steps happen in the ammonization unit, in which the brine is mixed with ammonia, becoming alkaline, and in the Solvay unit, where the alkaline brine is contacted with carbon dioxide gas, CO_2, to form sodium bicarbonate, $NaHCO_3$. In an alkaline environment, the sodium bicarbonate is less soluble than NaCl, and it is withdrawn from the unit as a precipitate. The precipitate is finally calcined to give sodium carbonate.

The standard process needs a small makeup of ammonia to compensate for losses. If ammonia is easily available, it can be fed directly to the process and the unit of ammonia recovery can be eliminated. In this case, the by-products would be ammonium chloride and calcium hydroxide, both having useful applications. Moreover, if the carbon dioxide is obtained from flue gas of combustion, the whole feed of calcium carbonate can be eliminated and the only by-product will be ammonium chloride. In any case, this process shows that a reaction with a very unfavorable equilibrium constant can be carried almost to completion.

Once the value of thermodynamics in the study of chemical reactions is properly understood, the next step is to relate the thermodynamic equilibrium constant to the

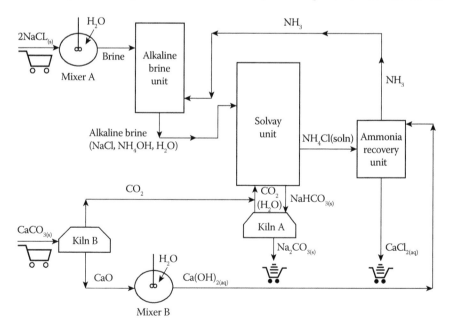

FIGURE 19.3   Solvay process.

composition of the reacting mixture at equilibrium. For this, we recall that the activity of a compound is given as

$$a_{i,\theta} \equiv \frac{f_i}{f_i^{(\theta)}} \tag{12.13}$$

REACTIONS IN THE GAS PHASE

For a reaction in the gas phase, the fugacity of compound i is given by

$$f_i = y_i \, \phi_i \, P \tag{13.1}$$

and the standard state is the pure compound i as ideal gas at one unit of pressure, normally taken as 1 bar. Thus, $f_i^{(\theta)} = 1$ (unit of pressure used to measure the pressure P), and from Equation 12.13, remembering that by dividing by the unit of pressure the symbol P represents a dimensionless quantity, we write

$$a_{i,\theta} = y_i \, \phi_i \, P$$

Hence, Equation 12.15 is rewritten as

$$K_T = \frac{\Pi_k y_{Pk,\theta}^{\lambda_{Pk}} \; \Pi_k \phi_{Pk,\theta}^{\lambda_{Pk}}}{\Pi_j y_{Rj,\theta}^{\lambda_{Rj}} \; \Pi_j \phi_{Rj,\theta}^{\lambda_{Rj}}} \, P^{\Delta v} \tag{19.7}$$

with Δv given by Equation 19.6. By analogy with Equation 12.15, this expression is written in shorthand notation as

$$K_T = K_y \, K_\phi \, P^{\Delta v} \tag{19.7a}$$

We recall here that the thermodynamic equilibrium constant, K_T, is only a function of temperature, and its value can be obtained from the first principles. On the other hand, our interest is to evaluate K_y and then, with the help of Equation 19.4, evaluate ξ. Thus, Equation 19.7a is conveniently rewritten as

$$\ln K_y = \ln K_T - \ln K_\phi - \Delta v \ln P \tag{19.7b}$$

As a good approximation at moderately high pressures, we can consider that the gas mixture behaves as an ideal mixture. In this case, the fugacity coefficients of the compounds in the mixture are equal to their values for the pure compounds at the same temperature and pressure. If this is the case, K_ϕ and all of the right-hand side of Equation 19.7b are independent of composition. Furthermore, at low pressure we can consider that the gas mixture behaves as an ideal gas, in which case the fugacity coefficients are all equal to unity and Equation 19.7b reduces to

$$\ln K_y^{id} = \ln K_T - \Delta v \ln P \tag{19.8}$$

where the superscript *id* has been used as a reminder that this form considers an ideal gas phase. This simplified form separates the effect of temperature and pressure, showing more clearly that if the number of moles increases by the reaction, an increase in pressure will decrease the value of K_y; that is, it will favor the displacement of the reaction toward the reactants, and vice versa. This is a form of the Le Chatelier principle.

At very high pressure, when the assumption of ideal gas mixture is not valid, an iterative procedure may be required to correct for the composition dependence of K_ϕ.

REACTIONS IN THE LIQUID PHASE

Solutions of electrolytes in water require special consideration, and its discussion was covered in Chapter 18. For nonelectrolyte liquids and solids, the standard state is usually taken as the pure compound in its natural state at 1 bar and 298.15 K. For solutes, sometimes molality is used as a measure of composition and the standard state is the solute in a hypothetical ideal solution, in Henry's sense, at unit molality and 298.15 K. In the case of using mole fraction as a measure of composition, the most common case for mixtures of nonelectrolytes, the activity of compound i is directly given by Equation 15.13,

$$a_{i,\theta} = x_i \, \gamma_{i,\theta} \qquad (15.13)$$

and Equation 12.15 takes the simple form,

$$K_T = \frac{\prod_k x_{Pk}^{\lambda_{Pk}} \prod_k \gamma_{Pk,\theta}^{\lambda_{Pk}}}{\prod_j x_{Rj}^{\lambda_{Rj}} \prod_j \gamma_{Rj,\theta}^{\lambda_{Rj}}} = K_x K_\gamma \qquad (19.9)$$

where we have used the same shorthand notation used in Equation 19.7a for reactions in the gas phase. Thus, for reactions in liquid phase we write

$$\ln K_x = \ln K_T - \ln K_\gamma \qquad (19.9a)$$

Even when in appearance this form is simpler than Equation 19.7b, in practice its use may get complicated by a variation in the value of K_γ, which may even be a function of the initial composition of the liquid mixture.

This observation is better illustrated with one exaggerated example. Consider a mixture in which acetic acid and ethyl alcohol are the reactants and water and ethyl acetate are the products:

(Acetic acid) + (Ethyl alcohol) $\Leftrightarrow$ (Ethyl acetate) + (Water)

While the thermodynamic equilibrium constant is only a function of temperature, the value of K_γ will be different if the initial mixture is formed with a drop of ethyl acetate in a large amount of water or with a drop of acid in a large amount of alcohol. In the first case, the activity coefficient of water (one of the products) will be close to unity, while the activity coefficients of ethyl acetate, acid, and alcohol will be close to their values at infinite dilution in water. In the second case, the activity coefficient of alcohol (one of the reactants) will be close to unity, while the activity coefficients of water, acetic acid, and ethyl acetate will be close to their values at infinite dilution in ethyl alcohol. Although this case is exaggerated, it shows that the value of K_γ depends on the composition of the feed. In addition, the value of K_γ cannot be accurately calculated due to the scarcity of data on activity coefficients in reacting mixtures. In the simple example considered here, it would be required to know the values of the activity coefficients of all four participants in the reaction as a function of composition of the quaternary mixture. One practical way of proceeding is to measure the compositions at equilibrium under typical conditions of operation, and knowing the value of the thermodynamic equilibrium constant K_T, back-calculate K_γ. In this case, as the activity coefficients are normally a weak function of temperature, for small variations in the initial composition of the mixture, the value of K_γ can be assumed to be constant. An example of the application of this idea is presented later in the chapter.

REACTIONS WITH COMPOUNDS IN THE SOLID PHASE

In some cases, a compound (elemental or nonelemental) is present in solid state at equilibrium with other compounds, either solid, liquid, or gas. In these cases of heterogeneous reactions, Equation 12.15 is used with the activities expressed in different forms for different compounds.

For the compounds in the vapor phase, the activities are obtained from Equation 13.1, as

$$a_{i,\theta} = y_i \, \phi_i P$$

For the compounds in the liquid phase, the activities are obtained from Equation 15.13 as

$$a_{i,\theta} = x_i \, \gamma_{i,\theta}$$

For those compounds that are pure solids, Equation 12.13 is used as

$$a_{i,\theta} = \frac{f_i}{f_i^{(\theta)}(1bar,T)} = \exp\frac{v_i^s(P-1)}{RT}$$

where P must be in bars. The molar volumes of solids are very small when compared with the ratio $RT/(P-1)$; thus, for pure solids, as a good approximation,

$$a_{i,\theta} = 1$$

For this reason, in many situations the activities of pure solid compounds do not show in the right-hand side of Equation 12.15. However, if the pure solid is not an element in its natural (elemental) state at 1 bar and 298.15 K, the value of the Gibbs energy of formation of the solid must be included in the calculation of K_T, as discussed below. On the other hand, if the pure solid is an element in its natural elemental form, by convention, the Gibbs energy of formation is taken as zero.

CALCULATION OF THE THERMODYNAMIC EQUILIBRIUM CONSTANT

In any case, for reactions in a gas phase, in a liquid phase, or in the presence of a solid phase, it is necessary to get the value of the thermodynamic equilibrium constant, K_T, at the temperature, T, of the system. This is done in two steps. In order to use standard thermodynamic tables, first its value is obtained at the temperature $T_0 = 298.15$ K, and then the value at the temperature, T, of the system is calculated.

The first step is to write the standard reaction to be considered. In the standard reaction, the stoichiometric amounts of reactants are converted into the stoichiometric amounts of products. We then assume that each of the compounds participating in the stoichiometric reaction is formed from a reaction of its elements in their natural elemental state at 298.15 K and 1 bar. These are imaginary reactions that are used only for tabulation purposes. The Gibbs energies and the enthalpies of the elements in their standard states are arbitrarily set equal to zero. *Lange's Handbook of Chemistry* [1] gives values of the standard Gibbs energies of formation, $\Delta G^0_{f,i,T_0}$, and the standard enthalpies of formation, $\Delta H^0_{f,i,T_0}$, of different pure compounds, i. These values correspond, respectively, to the standard chemical potential and the standard enthalpy of the compound in the nomenclature used in Chapter 12.

From Equation 12.14, the standard Gibbs energy change for the stoichiometric reaction at 298.15 K is then

$$\Delta G^\theta_{T_0} \equiv \sum_k^c v_k \Delta G^0_{f,k,T_0} \equiv \sum_k^c v_k \mu^\theta_k \tag{19.10}$$

and from Equation 12.18, the standard enthalpy change for the stoichiometric reaction at 298.15 K is

$$\Delta H^\theta_{T_0} \equiv \sum_k^c v_k \Delta H^0_{f,k,T_0} \equiv \sum_k^c v_k h^\theta_k \tag{19.11}$$

From Equation 12.16, the value of the thermodynamic equilibrium constant at $T_0 = 298.15$ K can be obtained directly as

$$\ln K_{T_0} = -\frac{\Delta G^\theta_{T_0}}{RT_0} \tag{19.12}$$

The value of the thermodynamic equilibrium constant at a temperature T different from T_0 is then obtained by integration of Equation 12.17 as

$$\ln K_T = \ln K_{T_0} + \int_{T_0}^{T} \frac{\Delta H^\theta_T}{RT^2} dT \tag{19.13}$$

Again, here Equation 19.11 gives the standard enthalpy change of reaction at $T_0 = 298.15$ K and the value of the standard enthalpy change at a temperature T is required. From Equation 5.5, we write

$$\Delta C_P^\theta = \left(\frac{\partial \Delta H_T^\theta}{\partial T} \right)_{P,n}$$

with

$$\Delta C_P^\theta \equiv \sum_k^c \nu_k c_{P,k} \tag{19.14}$$

Thus, by integration we obtain

$$\Delta H_T^\theta = \Delta H_{T_0}^\theta + \int_{T_0}^{T} (\Delta C_P^\theta) dT \tag{19.15}$$

Equation 19.11, using the heats of formation of the compounds at 298.15 K, gives the value of the first term of the right-hand side of Equation 19.15, while the integral is evaluated using information on the specific heats, $c_{P,k}$, of the compounds. Different correlations for specific heats use different series as a function of temperature. For generality, we use here the following series in which particular correlations may set to zero some of the coefficients:

$$c_{P,k} = c_{0,k} + c_{1,k}\, T + c_{2,k}\, T^2 + c_{3,k}\, T^3 + c_{4,k}\, T^{-2} \tag{19.16}$$

Thus, from Equation 19.14,

$$\Delta C_P^\theta = \Delta C_0 + \Delta C_1 T + \Delta C_2 T^2 + \Delta C_3 T^3 + \Delta C_4 T^{-2}$$

with

$$\Delta C_0 \equiv \sum_k^c \nu_k c_{0,k}$$

and similar expressions for ΔC_1, ΔC_2, ΔC_3, and ΔC_4. Although the algebra is trivial, for the purposes of discussion we give here the final forms obtained after integration. From Equation 19.15,

$$\Delta H_T^\theta = \Delta H_{T_0}^\theta + (\Delta C_0)(T - T_0) + \frac{(\Delta C_1)}{2}\left(T^2 - T_0^2\right) + \frac{(\Delta C_2)}{3}\left(T^3 - T_0^3\right)$$

$$+ \frac{(\Delta C_3)}{4}\left(T^4 - T_0^4\right) - (\Delta C_4)\left(\frac{1}{T} - \frac{1}{T_0}\right) \tag{19.17}$$

From Equation 19.13, then,

$$\ln K_T = \ln K_{T_0} - \frac{\Delta H_0}{R}\left(\frac{1}{T} - \frac{1}{T_0}\right) + \frac{\Delta C_0}{R}\ln\frac{T}{T_0} + \frac{\Delta C_1}{2R}(T - T_0)$$

$$+ \frac{\Delta C_2}{6R}\left(T^2 - T_0^2\right) + \frac{\Delta C_3}{12R}\left(T^3 - T_0^3\right) + \frac{\Delta C_4}{2R}\left(\frac{1}{T^2} - \frac{1}{T_0^2}\right)$$

$$(19.18)$$

with

$$\Delta H_0 = \Delta H_{T_0}^\theta - \Delta C_0 T_0 - \frac{\Delta C_1}{2}T_0^2 - \frac{\Delta C_2}{3}T_0^3 - \frac{\Delta C_3}{4}T_0^4 + \frac{\Delta C_4}{T_0} \qquad (19.19)$$

where $\Delta H_{T_0}^\theta$ is given by Equation 19.11.

Two simplifications of the above equations are common. If the specific heats of all compounds are considered to be constant, that is, independent of temperature, and as a good approximation they are evaluated at their mean value in the temperature range from T_0 to T, then Equation 19.18 takes the form

$$\ln K_T = \ln K_{T_0} - \frac{\Delta H_0}{R}\left(\frac{1}{T} - \frac{1}{T_0}\right) + \frac{\Delta \overline{C}}{R}\ln\frac{T}{T_0}$$

with

$$\Delta H_0 = \Delta H_{T_0}^\theta - \Delta \overline{C}T_0$$

Thus, combining both results,

$$\ln K_T = \ln K_{T_0} - \frac{\Delta H_{T_0}^\theta}{R}\left(\frac{1}{T} - \frac{1}{T_0}\right) + \frac{\Delta \overline{C}}{R}\left(\frac{T_0}{T} - 1 + \ln\frac{T}{T_0}\right)$$

The last term of the right-hand side of this equation normally is negligible in comparison with the other two terms, so the approximated form is simply written as

$$\ln K_T = \ln K_{T_0} - \frac{\Delta H_{T_0}^\theta}{R}\left(\frac{1}{T} - \frac{1}{T_0}\right)$$

For an exothermic reaction, for example, $\Delta H_{T_0}^\theta \leq 0$, a plot of ($\ln K_T$) versus ($1/T$) will show a straight line with positive slope, where the high temperature range is to the left of the diagram. At higher temperatures, the value of the equilibrium constant decreases for an exothermic reaction, indicating that the reaction displaces toward the reactants. The reverse is valid for an endothermic reaction. This result, like the approximation shown for Equation 19.7b, is an example of the Le Chatelier principle. The reaction displaces its equilibrium in a sense that tends to counteract the effect of a change in temperature or pressure.

EXAMPLES OF EVALUATION OF THE THERMODYNAMIC EQUILIBRIUM CONSTANT

The numerical value of the thermodynamic equilibrium constant depends on the way that the stoichiometric equation is written. Thus, giving a value of the equilibrium constant without specifying the stoichiometric equation is meaningless. On the other hand, once the stoichiometry has been specified, the same stoichiometric coefficients must be used for all calculations. If this is done, the equilibrium compositions obtained are independent of how the stoichiometric equation was written.

In order to evaluate the equilibrium constant at $T_0 = 298.15$ K, we need the value of $\Delta G^\theta_{T_0}$ and use Equation 19.11. Complete tables of standard heats and Gibbs energy changes for the reactions of formation of pure compounds at 298.15 K and 1 bar are available in the literature [1].

As for the determination of the number of independent reactions in a system, the reactions of formation of the compounds present in the equilibrium mixture are the basic pieces of information needed to study the chemical equilibria between different species. In order to evaluate K_T at a temperature different from 298.15 K, as indicated by Equations 19.18 and 19.19, we need information on the specific heats of the compounds and on the standard heats of reaction $\Delta H^\theta_{T_0}$, which again, according to Equation 19.11, is obtained from tables for the standard heat of formation.

The standard heats and standard Gibbs energy of formation of the elements in their natural (elemental) states at 1 bar and 298.15 K are conventionally assigned a zero value. This convention is particularly important since it provides a basis to express the enthalpy and Gibbs energy of all compounds on the same basis.

Example of a Heterogeneous Reaction

Calculate the standard enthalpy change and the standard Gibbs energy change at 1 bar and 298.15 K, for the following reaction:

$$Mg(OH)_{2(s)} + 2HCl_{(g)} = MgCl_{2(s)} + 2H_2O_{(g)}$$

Note that it is important to specify per mole of which compound the value of $\Delta H^\theta_{T_0}$ or $\Delta G^\theta_{T_0}$ is given. As different compounds have different stoichiometric coefficients, once the stoichiometric equation is specified, it is necessary to keep it unchanged to make sense of numerical values of $\Delta H^\theta_{T_0}$ and $\Delta G^\theta_{T_0}$.

We write first the reactions of formation of 1 mole of all nonelemental species starting from the constituent elements in their standard states. From *Lange's Handbook of Chemistry* [1], we get the values of $\Delta H^0_{f,T_0}$ and $\Delta G^0_{f,T_0}$ (see Table 19.1).

The overall reaction R is obtained by adding reaction f_3 plus twice reaction f_4 and subtracting reaction f_1 plus twice reaction f_2. After performing this operation, all elements cancel out and we recover the overall reaction: $R = f_3 + 2f_4 - (f_1 + 2f_2)$. The same operation should be carried out with the values of $(\Delta H^B_{f,T_0})$ in order to get the overall enthalpy change of the reaction.

$$\Delta H^\theta_{T_0} = \left[-641.3 + 2(-241.826)\right] - \left[(-924.7) - 2(-92.31)\right] = -16.13 \, kJ/mol \; MgCl_2$$

TABLE 19.1
Molar Enthalpies and Molar Gibbs Energies for the Reactions of Formation

No.	Reactions of Formation	$\Delta H_{f,T_0}^0$ [1] kJ/mol	$\Delta G_{f,T_0}^0$ [1] kJ/mol
f_1	$O_{2(g)} + H_{2(g)} + Mg_{(s)} = Mg(OH)_{2(s)}$	−924.7	−833.7
f_2	$\frac{1}{2} H_{2(g)} + \frac{1}{2} Cl_{2(g)} = HCl_{(g)}$	−92.31	−95.30
f_3	$Cl_{2(g)} + Mg_{(s)} = MgCl_{2(s)}$	−641.3	−591.8
f_4	$H_{2(g)} + \frac{1}{2} O_{2(g)} = H_2O_{(g)}$	−241.826	−228.61

Source: Data taken from Lange, N.A., and Speight, J.G. *Lange's Handbook of Chemistry*. New York: McGraw-Hill, 2005.

Thus, the reaction is exothermic. Similarly, for the standard Gibbs energy change of reaction,

$$\Delta G_{T_0}^\theta = \left[(-591.8) + 2(-228.61)\right] - \left[(-833.7) - 2(-95.3)\right] = -24.72 \text{kJ/mol MgCl}_2$$

In principle, the standard reaction has a favorable equilibrium. Now we can calculate the value of the thermodynamic equilibrium constant at $T_0 = 298.15$ K for the reaction.

From Equation 19.12,

$$\ln K_{T_0} = -\frac{(-24.72 \cdot 10^3)}{(8.314)(298.15)} \left[\frac{J/_{mol}}{\left(J/_{(mol \cdot K)} \right) K} \right] = +9.97_2 \left[\text{dimensionless} \right]$$

Thus,

$$K_{T_0} = 2.14_3 \cdot 10^4$$

The very large value of K_T indicates that at equilibrium the reaction is almost totally displaced toward the products (see Figure 19.1).

Example of a Reaction in the Gas Phase

The Haber process produces ammonia in a batch reactor operating at 100 bar and 450°C with a 50% excess over the stoichiometric proportion of N_2 with respect to H_2.

As a first step toward understanding the necessity of operating at the above conditions, we calculate the value of the thermodynamic equilibrium constant at 25°C and 450°C for the stoichiometric reaction written as

$$N_2 + 3H_2 = 2NH_3$$

For this calculation, we distinguish useful from useless information. The thermodynamic equilibrium constant is only a function of temperature. The pressure and the proportion of the chemical species are totally immaterial. From *Lange's Handbook of Chemistry* [1], for the formation of 1 mole of ammonia,

$$\Delta H^0_{f,T_0} = -45.94 \left(\frac{kJ}{mol\ NH_3} \right)$$

$$\Delta G^0_{f,T_0} = -16.4 \left(\frac{kJ}{mol\ NH_3} \right)$$

Thus, for the equation as written, considering a zero value for the elements in their elemental form,

$$\Delta H^\theta_{T_0} = -91.88 \left(\frac{kJ}{mol\ N_2} \right)$$

$$\Delta G^\theta_{T_0} = -32.8 \left(\frac{kJ}{mol\ N_2} \right)$$

From Equation 19.12,

$$\ln K_{T_0} = -\frac{-32.8 \cdot 10^3}{8.314 \cdot 298.15} = 13.23$$

$$K_{T_0} = 5.58 \cdot 10^5$$

This value is so high that, from the thermodynamics point of view, it would seem that the reaction should be complete at room temperature. Unfortunately, a very stable triple bond in the nitrogen molecule requires a use of a metal catalyst in the Haber process. The reaction will not proceed without the catalyst, and this catalyst only works at temperatures in the 400°C–500°C range.

This explains why we should operate at 450°C. Let us then calculate K_T at 450°C = 723.15 K:

$$\ln K_T = 13.23 - \frac{\left(-91.88 \cdot 10^3 \right)}{1.9878.314} \left(\frac{1}{723.15} - \frac{1}{298.15} \right) = -8.55$$

$$K_T = 1.93 \cdot 10^{-4}$$

Thus, at this temperature the value of the thermodynamic equilibrium constant decreased by nine orders of magnitude. As the catalyst requires this high temperature, some other way is needed to improve the conversion of the reaction. At a high pressure, according to the principle of Le Chatelier, the reaction tends to displace from 4 moles of reactants to 2 moles of product. The value of the thermodynamic equilibrium constant does not change with pressure, but the all-important value of K_y will change. Thus, now we need to calculate the value of K_y at 100 bar.

Using the approximation of Equation 19.7b, that sets $K_\phi = 1$, it is immaterial that the reactor is operating with a 50% excess of N_2 over the stoichiometric proportion. For the reaction, as written, $\Delta v = -2$; thus, at 100 bar, as a first approximation,

$$\ln K_y^* = -8.53 - (-2)\ln(100) = 0.677$$

and

$$K_y^* = 1.97$$

The increase of pressure increases the value of K_y by four orders of magnitude, but it also makes the whole process very expensive. Note that the value of P used in the above equation must be in bars. If P had been given in any other units, it would have been mandatory to convert it to bars. (Why?)

To evaluate K_ϕ, we recall the convention introduced in Equation 19.7:

$$K_\phi = \frac{\phi_{NH_3}^2}{\phi_{N_2}\phi_{H_2}^3}$$

As a first approximation, we can assume an ideal mixture of real gases, and consider $\phi_k = \phi_k^0$. In this case, the initial composition of the mixture has no effect on the value of K_y. At very high pressure, the use of equations of state with mixing rules may be required to calculate the values of the fugacity coefficients. In such a case, an iterative method can be needed and the initial composition of the mixture will slightly affect the value of K_ϕ.

Example of a Reaction in Liquid Phase

For the following reaction at equilibrium at 298.15 K,

(Acetic acid [A]) + (Ethanol [E]) = (Ethyl acetate [EA]) + (Water [W])

the value of K_T can be obtained using data from Tables 1.56 and 2.53 of *Lange's Handbook of Chemistry* [1], and the values of $\Delta G_{f,T_0}^0$ for A, E, EA, and W are −390.2, −174.8, −332.7, and −237.14 kJ/mol, respectively. Thus,

$$\Delta G_{T_0}^\theta = \left[(-237.14)+(-332.7)\right]-\left[(-390.2)+(-174.8)\right] = -4.8 \text{kJ/mol}$$

In this case, it is not necessary to specify per mole of which compound $\Delta G_{T_0}^\theta$ is given.

$$\ln K_{T_0} = \frac{-(-4.8)\cdot 10^3}{8.314 \cdot 298.15} = 1.952$$

$$K_{T_0} = 7.05$$

In practical work, it is sometimes useful to define a "constant" K_C in terms of concentrations by

$$K_c \equiv \frac{c_C^{\lambda_C} c_D^{\lambda_D}}{c_A^{\lambda_A} c_B^{\lambda_B}}$$

This so-called constant, which is not really a constant, can be directly related to the thermodynamic equilibrium constant by considering that the mole fraction of compound i can be written as

$$x_i = \frac{n_i}{\Sigma n_k} = \frac{n_i}{v}\frac{v}{\Sigma n_k} = c_i(\rho_m)^{-1}$$

where ρ_m is the molar density of the mixture at equilibrium. Thus,

$$\ln K_x = \ln K_c - \Delta v \ln \rho_m$$

and from Equation 19.9a, $\ln K_c$ is directly related to K_T. Notably for reactions without a change in the number of moles, that is, reactions for which $\Delta v = 0$, as in the case of the reaction of acetic acid with ethanol, $K_c = K_x$.

EXPRESSION OF K_X AS A FUNCTION OF THE CONVERSION OF ONE OF THE REACTANTS

It is of interest to relate K_C to the conversion of one of the reactants. For the general reaction,

$$\lambda_A A + \lambda_B B \leftrightarrow \lambda_C C + \lambda_D D$$

defining the conversion in terms of reactant B, from Equation 19.1,

$$n_{B0} - n_B = \chi_B n_{B0}$$

or

$$n_B = n_{B0}(1 - \chi_B) = n_{B0} - \chi_B n_{B0}$$

For reactant A,

$$n_{A0} - n_A = \frac{\lambda_A}{\lambda_B}(n_{B0} - n_B)$$

or

$$n_A = n_{A0} - \frac{\lambda_A}{\lambda_B}\chi_B n_{B0}$$

For products C and D,

$$n_C = n_{C0} + \frac{\lambda_C}{\lambda_B} \chi_B n_{B0}$$

$$n_D = n_{D0} + \frac{\lambda_D}{\lambda_B} \chi_B n_{B0}$$

The total number of moles at equilibrium then is

$$n_T = n_0 + \chi_B n_{B0} \left(\frac{\lambda_C}{\lambda_B} + \frac{\lambda_D}{\lambda_B} - \frac{\lambda_A}{\lambda_B} - 1 \right) = n_0 + \frac{\chi_B n_{B0}}{\lambda_B} \Delta v \quad \text{with} \quad n_0 = \sum_k n_{0k}$$

Therefore,

$$K_x = \frac{\left[n_{C0} + \frac{\lambda_C}{\lambda_B} \chi_B n_{B0} \right]^{\lambda_C} \left[n_{D0} + \frac{\lambda_D}{\lambda_B} \chi_B n_{B0} \right]^{\lambda_D}}{\left[n_{A0} - \frac{\lambda_A}{\lambda_B} \chi_B n_{B0} \right]^{\lambda_A} \left[n_{B0} - \chi_B n_{B0} \right]^{\lambda_B}} \left[n_0 + \frac{n_{B0}}{\lambda_B} \chi_B \Delta v \right]^{-\Delta v}$$

For the reaction of acetic acid with ethanol, assuming that initially there are no products present and remembering that $\Delta v = 0$,

$$K_x = \frac{\chi_B^2}{[r - \chi_B][1 - \chi_B]}$$

or

$$(K_x - 1)\chi_B^2 - K_x (r + 1)\chi_B + K_x r = 0$$

where

$$r = n_{A,0}/n_{B,0}$$

A laboratory experiment at 298.15 K using $r = 1$ gave a conversion of 2/3; thus,

$$K_x = K_c = \frac{(2/3)^2}{[1 - 2/3][1 - 2/3]} = 4$$

As a result, from Equation 19.9 we obtain $K_\gamma = K_\gamma/K_x = 1.76$. More importantly, assuming K_x is constant, that is, K_γ is constant, it is possible to estimate the conversion at other values of r. Table 19.2 compares calculated values with experimental values of the conversion measured in the lab for different values of r.

This exercise based on real data demonstrates that with the help of the thermodynamic theory, one can get reasonable estimates in practical situations.

TABLE 19.2
Comparison of Calculated Conversion Values χ_B with the Experimental Ones

$r = n_{A,0}/n_{B,0}$	$\chi_{B, observed}$	$\chi_{B, calculated}$
0.05	0.050	0.049
0.08	0.078	0.078
0.18	0.171	0.171
0.28	0.226	0.258
0.33	0.293	0.298
0.50	0.414	0.423
1.00	0.665	0.667
2.24	0.876	0.864
8.00	0.966	0.967

CHEMICAL EQUILIBRIUM WITH MORE THAN ONE INDEPENDENT REACTION

As indicated by Equations 19.2a through 19.4a, when there is more than one independent reaction, for each independent reaction, there is an extent of reaction ξ_r. The number of equations for equilibrium constants and the number of unknown extents of reaction are equal to the number of independent reactions, and any method to solve a set of nonlinear equations can be used to solve the problem. The technique required to solve the set of nonlinear equations can be quite sophisticated, and the discussion of such techniques is beyond the scope of this text. Here, we will only mention that as an alternative to solving the set of equations of the equilibrium constants, it is possible to directly minimize the Gibbs energy of the system constrained by the material balances of the elements. While the number of moles of each compound and the total number of moles of the system change when a chemical reaction occurs, the amount of each element present in the system is conserved. A mathematical technique to solve a constrained minimum makes use of the Lagrange multipliers.

REFERENCE

1. Lange, N. A., and Speight, J. G. 2005. *Lange's Handbook of Chemistry*. New York: McGraw-Hill.

20 The Thermodynamics of Equilibrium-Based Separation Processes

Separation processes usually represent a major component in the cost of chemical products, as most industries need to purify chemicals to be fed to reactors and separate the reaction products obtained. Some industries, such as petroleum refining, are mostly a sequence of separation steps.

Separation processes can be classified into two broad categories: equilibrium-stage processes and rate-controlled processes. Only the first of these categories is in the realm of thermodynamics.

Although equilibrium is seldom attained, thermodynamics is of help in the design of equilibrium-stage processes. In these processes, a phase containing the compound to be separated is contacted with a second phase which has a much lower concentration of that compound. The two phases are left to equilibrate; the second phase is removed, and the desired compound is recovered from it. Sometimes, the first phase is contacted again with another amount of (fresh) second phase. This sequence of contacts, in different equilibrium stages, is the basis of the design of separation processes discussed in this chapter. The key problem is to evaluate the relative proportions of the compounds when the phases contacted are in equilibrium.

DISTRIBUTION COEFFICIENT AND SELECTIVITY

Figure 20.1 depicts a typical separation unit in which two phases, designated generically by α and β, are in direct contact at a pressure, P, and a temperature, T, common to both phases.

Phase α can be a liquid phase and phase β a vapor phase, both α and β can be liquid phases, or one of the phases can be solid and the other either liquid or gas. Our interest is to determine the relative amounts of the compounds in each phase in the system at equilibrium. Mole fractions are normally used as a measure of composition. In the particular case of vapor–liquid equilibrium, the symbols x_i and y_i are used for the mole fractions of compound i in the liquid phase and in the vapor phase, respectively. For liquid–liquid equilibrium, the two liquid phases are designated as phases L_1 and L_2. The distribution coefficient K_i of a compound i between the two phases α and β is defined as

$$K_i^{\beta\alpha} \equiv \frac{x_i^{\beta}}{x_i^{\alpha}} \tag{20.1}$$

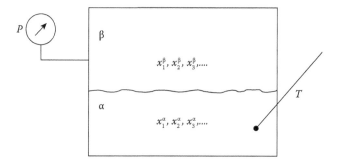

FIGURE 20.1   Two-phase system unit.

In vapor–liquid equilibrium, the distribution coefficient is called the K-factor, the superscripts α and β are eliminated, and we write

$$K_i \equiv \frac{y_i}{x_i} \qquad (20.1a)$$

A value of K_i larger than unity, $K_i > 1$, indicates that compound i tends to concentrate in phase β. In vapor–liquid equilibrium nomenclature, for "light" compounds, that is, compounds that tend to concentrate in the vapor phase, the K-factor is larger than unity, and for "heavy" compounds, that is, compounds that tend to concentrate in the liquid phase, the K-factor is less than unity. In the separation of a multicomponent mixture, the comparison of the values of the distribution coefficients for two compounds i and j shows which of the components is selectively taken out from phase α by phase β. The selectivity α_{ij} is defined as

$$\alpha_{ij} = \frac{K_i^{\beta\alpha}}{K_j^{\beta\alpha}} \qquad (20.2)$$

In vapor–liquid equilibrium, the selectivity is called relative volatility, and it is written as

$$\alpha_{ij} = \frac{K_i}{K_j} = \frac{\left(y_i/x_i\right)}{\left(y_j/x_j\right)} \qquad (20.2a)$$

The thermodynamic treatment of a phase equilibrium problem always starts by equating the fugacities, or the chemical potentials, of each compound in all phases in which the compound is present. The values of the distribution coefficient and the selectivity are then obtained solving the resulting system of equations. For a vapor–liquid equilibrium of compounds that are present in significant amounts in both phases, using Equation 16.2a we write

$$K_i \equiv \frac{y_i}{x_i} = \gamma_{i,0} \frac{(P_i^s)'}{P} \qquad (20.3)$$

At moderate pressures, the corrections to the vapor pressure can be safely ignored and we write

$$K_i \approx \gamma_{i,0} \frac{P_i^s}{P} \tag{20.3a}$$

For mixtures of compounds of the same homologous series, the assumption of ideal solution in the Lewis sense is a good approximation, so we write

$$K_i^{id} = \frac{P_i^s}{P}$$

or

$$\ln K_i^{id} = \ln P_i^s - \ln P$$

For the case of the vapor–liquid equilibrium of a light compound (j) and a heavy compound (i), say methane dissolved in a hexane, the light compound is predominantly in the vapor phase and we may use Henry's convention for the solute. Thus, using Equation 13.1 for the vapor phase and Equation 15.9 for the solute j in the liquid solvent i, we write

$$K_j = \gamma_{j,H} \frac{H_{j,i}}{\phi_j P}$$

Considering the case of an ideal solution in Henry's sense, at moderate pressure, we write

$$\ln K_j^{id} = \ln H_{j,i} - \ln P$$

Hence, for both the solvent i and the solute j, a plot of the logarithm of the distribution coefficient versus the logarithm of the pressure will give a straight line with a slope of -1 in the diluted region. A typical plot of this kind, not drawn to scale, is shown in Figure 20.2 with $T_2 > T_1$.

At $P = 1$ (one unit of pressure), the value of $\ln H_{j,i}$, in which $H_{j,i}$ has the same units as P, can be found in the scale of $\ln K_j$. At low pressure, the logarithms of the distribution coefficients for the light and heavy compounds are a linear function of the logarithm of the pressure. At higher pressures, the curves bend due to nonidealities, and the two curves join at the mixture critical point.

For the case of two liquid mixtures α and β in equilibrium, using Equation 16.5 we write

$$K_i \equiv \frac{x_i^\beta}{x_i^\alpha} = \frac{\gamma_{i,0}^\alpha}{\gamma_{i,0}^\beta} \tag{20.4}$$

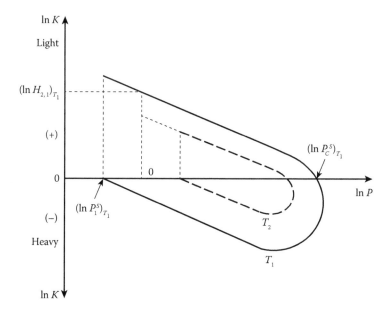

FIGURE 20.2   Logarithm of the distribution coefficient as a function of pressure.

and

$$\alpha_{ij} = \frac{\left(\gamma_{i,0}^{\alpha}/\gamma_{i,0}^{\beta}\right)}{\left(\gamma_{j,0}^{\alpha}/\gamma_{j,0}^{\beta}\right)}$$

If both mixtures were ideal in Lewis's sense, the composition of both phases would be the same and there would not be any phase separation. As discussed in Chapter 17, mixtures separate into two liquid phases only in the case of strong positive deviations from ideality in Lewis's sense. If this is the case, as the mole fractions of compound i are inversely proportional to the values of its activity coefficients in both phases, the phase in which the activity coefficient is smaller will have a larger mole fraction of compound i. If two sparsely miscible liquid pure compounds are put in contact, one of the compounds will be almost pure in one liquid phase and at high dilution in the other. Say that the compounds are water (1) and benzene (2), the aqueous phase is phase α and the organic phase is phase β. In Lewis's convention, the activity coefficient of compound 1 in phase α will be close to unity and in phase β it will be a value much larger than unity. In order to work with activity coefficients of the order of unity, the ideal behavior in Henry's sense can be used for the compound at high dilution. In this convention, at low concentration its activity coefficient will be close to unity. For component 2, the reverse will hold. Thus, considering ideal behavior in Lewis's sense for the solvent and in Henry's sense for the solute, we write

$$x_1^{\alpha}(P_1^s)' = x_1^{\beta}H_{1,\beta}$$

$$K_1^{id} \equiv \frac{x_1^\beta}{x_1^\alpha} = \frac{\left(P_1^s\right)'}{H_{1,\beta}}$$

Similarly for compound 2,

$$K_2^{id} = \frac{x_2^\alpha}{x_2^\beta} = \frac{\left(P_2^s\right)'}{H_{2,\alpha}}$$

The value of x_1^β is the solubility of water in the organic phase, and the value of x_2^α is the solubility of benzene in the aqueous phase. If, in addition, the two liquid phases α and β are at equilibrium with a vapor phase,

$$y_1 \phi_1^0 P = x_i^\alpha \left(P_1^s\right)' = x_1^\beta H_{1,\beta}$$

$$y_2 \phi_2^0 P = x_2^\alpha H_{2,\alpha} = x_2^\beta \left(P_2^s\right)'$$

In this case, one can define two different K-factors for each compound depending on the phase chosen. At moderate conditions, we write

$$y_1 P = x_1^\beta H_{1,\beta}$$

$$y_2 P = x_2^\beta \left(P_2^s\right)'$$

Adding and rearranging, we obtain

$$x_1^\beta = \frac{P - \left(P_2^s\right)'}{H_{1,\beta} - \left(P_2^s\right)'}$$

A similar relation can be written for x_2^α. These equations can be used to estimate Henry's constant if the solubilities are small. Normally, the solubility is measured at atmospheric pressure and at the temperature required for a practical application. Thus, it is necessary to correct the value of Henry's constant for its use at pressures of interest. For simplicity, consider that compound 2 is at high dilution in solvent 1. From Equation 15.8, in Henry's convention,

$$f_2^{dil} = x_2 H_{2,1}$$

or

$$\ln f_2^{dil} = \ln x_2 + \ln H_{2,1}$$

Recalling Equation 14.15, we write for the case of compound 2 at high dilution,

$$\left(\frac{\partial \ln H_{2,1}}{\partial P}\right)_{T,n} = \left(\frac{\partial \ln f_2^{dil}}{\partial P}\right)_{T,n} = \frac{\overline{v}_2^{\infty}}{RT}$$

and integrating between 1 atm and the pressure P, we obtain

$$\ln H_{2,1}(P) = \ln H_{2,1}(1 \text{ atm}) + \frac{\overline{v}_2^{\infty}(P-1)}{RT} \qquad (20.5)$$

This result gives a good approximation for most practical applications.

One last case of interest is the one of a compound at high dilution in two immiscible solvents in contact. This is, for example, the case of a small amount of liquid butanol added to a two-liquid-phase water–benzene system. Calling the solute, butanol, component 1, and the aqueous and organic phases α and β, respectively, we write

$$x_1^{\alpha} H_{1,\alpha} = x_1^{\beta} H_{1,\beta}$$

$$K^{id} = \frac{x_1^{\beta}}{x_1^{\alpha}} = \frac{H_{1,\alpha}}{H_{1,\beta}}$$

Due to the simplifications made, in this case K_i^{id} is a function of temperature only. When nonidealities of the phases become important, the K-factor becomes a function of the composition of the phases. For completeness, we mention here that in practical liquid–liquid equilibrium work, sometimes an empirical distribution coefficient is used in terms of concentration units (mol/dm³),

$$K_{i,\text{practical}} = \frac{c_i^{\beta}}{c_i^{\alpha}}$$

This empirical term includes the ratio of the molar densities of the phases in the value of K.

For organic mixtures, Berg rules, discussed in Chapter 17, are of help in the design of separation processes. Mixtures of organic liquids with water can sometimes be separated by the addition of salts. This is the case, for example, of a mixture of methanol and water when K_2CO_3 is added as shown schematically in Figure 20.3.

The effect is usually referred to as "salting out" the mixture. Similar unusual effects are obtained by the addition of water-soluble polymers to water. For a polymer with a hydrocarbon skeleton to be water soluble, it must have active groups able to combine with water. The hydrocarbon "tail" and the water will keep away

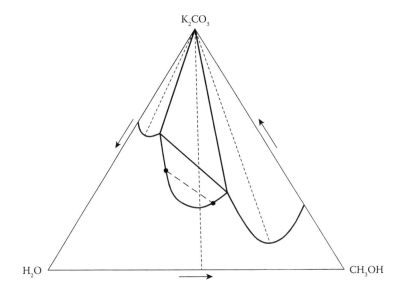

FIGURE 20.3   Ternary system of methanol–water–K_2CO_3.

from each other as much as they can. In a practical situation, the hydrophobic hydrocarbon tail will go to the surface and tail-out, while the active hydrophilic group will face the water body. This configuration will obviously affect the surface properties of water (surface tension), and such a compound is properly called a surfactant. At larger concentrations of the surfactant, the surface will get too crowded and eventually, at a so-called critical micelle concentration (CMC), the surfactant molecules will form "drops" inside the body of water. These drops, called micelles, will have all the hydrocarbon tails facing inside the drop, with the active groups at the surface of the micelle facing the water. The nonpolar, hydrocarbon microphase thus formed can dissolve organic soluble materials. At a high micellar concentration of an aqueous phase in contact with an organic phase, the structure can reverse to give a continuous organic phase with water drops surrounded by the active groups. These so-called reverse micelles can be used to "trap" water-soluble solutes in organic phases for their eventual separation. The formation of micelles and reverse micelles is a clear example of local composition in mixtures and has major potential for the development of separation techniques in bioprocesses.

AZEOTROPIC AND SOLVENT DISTILLATION

The study of vapor–liquid equilibrium is of special importance because it is estimated that about 90% of the industrial separations are done by distillation. In this case, the separating agent is heat. A distillation column consists of a reboiler and either a packed or a plate column on top of the reboiler where there is partial condensation returning the heavy compound by gravity down to the reboiler and collecting the light compound at the top of the column. As the separation is never complete, the top product is condensed and a part of it is refluxed to the column. The fresh feed normally enters the column at an intermediate height. The details of the design of distillation columns are presented in specialized texts. From the point of view of applied thermodynamics, we are only interested in the fundamentals for solving the equilibrium equations. Before going into detail, we mention here two cases of distillation inspired in an understanding of thermodynamics. A general rule for separation processes is to avoid adding a compound that is not present in the original mixture. Introducing additional compounds would require additional separations. Another general rule for separation processes is to leave the most difficult separation for the end. Thus, in many cases a separation process ends up with the separation of a binary mixture with close boiling point compounds, and in extreme cases, with the problem of separating an azeotropic mixture. For these cases of mixtures difficult to separate by conventional methods, thermodynamics suggests that the addition of a new compound may be advantageous. A classic example is the separation of ethanol–water mixtures, which presents a maximum pressure, minimum boiling point azeotrope containing 89 mol%, that is 96 wt%, ethanol. Thus, if an aqueous mixture containing 10 mol% ethanol is distilled, the product obtained at the top of the column would be the azeotropic mixture and the bottom product would be pure water. It is not possible to obtain anhydrous ethanol in a simple distillation column. To obtain anhydrous ethanol, it is necessary to add a third compound to form a minimum boiling point azeotrope with water having a boiling point that is lower than the boiling point of ethanol. Considering the modified Berg classification of liquids and their mixtures, benzene, toluene, and cyclohexane are good candidates to form a heteroazeotrope with water that, upon condensation of the vapor, will separate into two liquid phases. Cyclohexane is normally preferred. Thus, in a first column, the ethanol–water azeotrope is separated from the excess water and sent as a feed to a second column, to which cyclohexane is fed close to the top of the column in the amount required to give the vapor phase composition of the azeotropic mixture water–cyclohexane. In this second column, anhydrous ethanol is obtained at the bottom and the azeotropic mixture of water–cyclohexane is obtained as the top product, condensed, and separated into an aqueous phase and a cyclohexane phase. The cyclohexane, together with fresh cyclohexane to make up for any losses, is recycled close to the top of the second column. The aqueous phase containing a small contamination of cyclohexane is sent as a feed to a third column, where the cyclohexane is obtained at the top and the water is obtained at the bottom.

A different application of thermodynamics involving the addition of a compound not previously present in the mixture is the case of solvent distillation, also called extractive distillation. In this case, a compound that is a good solvent for the heavier component in a binary mixture difficult to separate is added at the top of the distillation column to "wash" that component down. Normally, the solvent is a heavier compound of the same homologous series of the heavy component to be "washed down." As indicated by Berg's classification, compounds of the same homologous series form solutions presenting close to ideal behavior, and thus they are good solvents. There are cases, however, in which the two compounds forming the mixture of difficult separation are of a similar homologous series. In this case, a compound of a different homologous series that it is a particularly good solvent for one of the compounds in the initial mixture can be used. A typical example of this case of solvent distillation is the use of furfural, a polar compound, to separate 1-butene, an unsaturated hydrocarbon, from isobutane, a saturated hydrocarbon. Furfural and 1-butene form an electron–donor acceptor complex. In solvent distillation, the separating agent is fed to the top of the column, while in azeotropic distillation, it is fed by the middle of the column. In the latter case, columns have a larger diameter in their lower section. These physical aspects of the equipment clearly reflect the different thermodynamic requirements for proper use of the separating agent.

VAPOR–LIQUID EQUILIBRIUM

In this section, we discuss in some detail a few examples of vapor–liquid equilibrium calculations that are of practical interest. For the vapor phase, invariably the fugacity of each compound i is expressed using Equation 13.1. For the fugacity of a compound in the liquid phase, we may use Equation 12.5 and write Equation 12.10 as

$$x_i \phi^L_{i(T,P,x)} P = y_i \phi_{i(T,P,y)} P$$

or (20.6)

$$x_i \phi^L_{i(T,P,x)} = y_i \phi_{i(T,P,y)}$$

This is the so-called ϕ approach, in which the fugacities of the compounds are expressed in terms of the fugacity coefficient in each phase. For the vapor phase, we use simply the symbol ϕ_i, while for the liquid phase, we use ϕ^L_i. In general, all fugacity coefficients of the vapor phase must be calculated from a single equation of state (EOS) and all fugacity coefficients of the liquid phase must be calculated from a single EOS. The equations of state do not need to be the same for both phases, although usually the same (cubic) EOS is used. For the vapor phase at low pressures, for example, we may find it convenient to set all values $\phi_i = 1$ (ideal gas behavior), while for the liquid phase, even the assumption of an ideal mixture, $\phi^L_i = \phi^{0,L}_i$, could be a poor one.

For the compounds in the liquid phase, we may express the fugacities in terms of activity coefficients and write Equation 16.2 as

$$x_i \gamma_{i,0(T,P,x)} (P^s_i)^\# = y_i \phi_{i(T,y)} P \tag{20.7}$$

For generality, in Equation 20.7 we preferred to keep the vapor phase correction to ideality separate from the liquid phase correction. From Equations 16.2 and 16.3 then,

$$(P^s_i)^\# = \phi^0_i \exp\left[\frac{v^{0L}_i (P - P^s_i)}{RT}\right] P^s_i \tag{20.8}$$

In this treatment, called the $\gamma - \phi$ approach, all activity coefficients for the compounds in the liquid phase should originate from a single expression for the excess Gibbs energy of the mixture, and all fugacity coefficients for the compounds in the vapor phase from a single EOS. Analytical solution of groups (ASOG) [6,7] is a particularly useful method to obtain activity coefficients for vapor–liquid equilibrium calculations with $\gamma - \phi$. This method is discussed in Chapter 26 and in Appendix D.

Alternatively, for the compounds in the liquid phase, we can use Equation 15.9 if the compound is diluted and write

$$x_i \gamma_{i,H(T,P,x)} H_{i,r} = y_i \phi_{i(T,P,y)} P \tag{20.9}$$

BUBBLE AND DEW POINT CALCULATIONS

For a "dew pressure" problem, all the values of the vapor phase mole fractions y_i ($i = 1$, 2, ..., c) and the temperature T are known. The values of the liquid phase mole fractions x_i ($i = 1, 2, ..., c$) and the pressure P are to be calculated. Similarly for a "bubble temperature" problem, all compositions of the liquid phase and pressure are known and the values of the vapor phase mole fractions and temperature are to be calculated. The unknown variable, pressure or temperature, is the one defining the problem to be solved, while the use of "dew" or "bubble" indicates the phase in which the compositions are known. Because in both kind of problems the composition of one of the phases and either the pressure or the temperature are known, we can present a generalized treatment. Let us call α the phase for which all compositions X_i^α are known. This can be either the liquid (bubble point calculation) or the vapor (dew point calculation) phase. Let us designate by ξ_A the known intensive variable, T or P, and by ξ_B the unknown variable (P or T). Thus, the known variables are $X_i^\alpha (i = 1, 2, ..., c)$ and ξ_A, and the unknown variables are $X_i^\beta (i = 1, 2, ..., c)$ and ξ_B. In addition, as Equations 20.6, 20.7, and 20.9 are symmetrical, we generalize them under the form

$$X_i^\alpha \Phi_{i(\xi_A, \xi_B, X^\alpha)}^\alpha = X_i^\beta \Phi_{i(\xi_A, \xi_B, X^\beta)}^\beta \tag{20.10}$$

The terms included in Equation 20.10, for each type of bubble or dew point calculation, are presented in Table 20.1.

The basic idea behind the calculation scheme is shown in Flowchart 20.1.

In step 1, the known variables are read. In step 2, the missing value (T or P) is estimated, and then in step 3, preliminary values of all Φ_i^α are obtained. In the first pass by step 4, the unknown phase β is considered to behave like an ideal mixture so that $\Phi_i^\beta = \Phi_i^{id}$ at T and P. This is a very poor approximation for dew point calculations and probably a not so poor approximation for bubble point calculations at low pressure. In any case, for the first pass this approximation is totally justified. In subsequent passes, the values of Φ_i^β are calculated with the new value of ξ_B obtained in step 13 and the last values X_i^β obtained in step 11. The unknown compositions X_i^β are calculated in step 7 from the general form of Equation 20.10.

TABLE 20.1
Summary of Different Bubble and Dew Calculations

Problem	Known (X_i^α, ξ_A)	Unknown (X_i^β, ξ_B)	Approach	Φ_i^α	Φ_i^β
Bubble P	x_i, T	y_i, P	ϕ	ϕ_i^L	ϕ_i
			$\gamma - \phi$	$\gamma_i \left(P_i^s \right)^{\#}$	$\phi_i P$
Bubble T	x_i, P	y_i, T	ϕ	ϕ_i^L	ϕ_i
			$\gamma - \phi$	$\gamma_i \left(P_i^s \right)^{\#}$	$\phi_i P$
Dew P	y_i, T	x_i, P	ϕ	ϕ_i	ϕ_i^L
			$\gamma - \phi$	$\phi_{i,P}$	$\gamma_i \left(P_i^s \right)^{\#}$
Dew T	y_i, P	x_i, T	ϕ	ϕ_i	ϕ_i^L
			$\gamma - \phi$	$\phi_i P$	$\gamma_i \left(P_i^s \right)^{\#}$

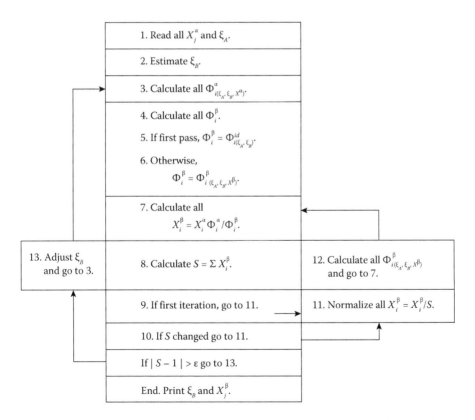

1. Read all X_j^α and ξ_A.

2. Estimate ξ_B.

3. Calculate all $\Phi^\alpha_{i(\xi_A, \xi_B, X^\alpha)}$.

4. Calculate all Φ_i^β.

5. If first pass, $\Phi_i^\beta = \Phi^{id}_{i(\xi_A, \xi_B)}$.

6. Otherwise,

$$\Phi_i^\beta = \Phi^\beta_{i(\xi_A, \xi_B, X^\beta)}.$$

7. Calculate all

$$X_i^\beta = X_i^\alpha \Phi_i^\alpha / \Phi_i^\beta.$$

13. Adjust ξ_B and go to 3.

8. Calculate $S = \Sigma X_i^\beta$.

12. Calculate all $\Phi^\beta_{i(\xi_A, \xi_B, X^\beta)}$ and go to 7.

9. If first iteration, go to 11.

11. Normalize all $X_i^\beta = X_i^\beta / S$.

10. If S changed go to 11.

If $|S - 1| > \varepsilon$ go to 13.

End. Print ξ_B and X_j^β.

FLOWCHART 20.1   General scheme for vapor–liquid calculation for a tolerance $\left| \Sigma x_i^\beta - 1 \right| \le \varepsilon.$

Since, based on previous assumptions, the values $X_1^\beta, X_2^\beta, X_3^\beta$, and so forth, are calculated independently, they will not add to unity as they should. For this reason, their sum is calculated in step 8, and if it is the first iteration, the values of Φ_i^β need to be recalculated. In step 11, the X_i^β values are normalized, and in step 12, the value of Φ_i^β is calculated. The normalization step is very important because the functions to compute Φ_i^β, either by an EOS or a g^E function, are designed to work with normalized values of composition. Steps 3–10 are then repeated until no further change in S is observed. Then the value of S is compared with unity, and if it differs by less than a previously specified error ε, the result is accepted. If the error is larger than ε, the variable ξ_B is adjusted in step 13, and the process is repeated. The missing information in Flowchart 20.1 is how to estimate ξ_B (P or T), and how to adjust its value in consecutive iterations. For the initial estimates, we assume that both phases behave ideally. From Equation 16.4, assuming ideal gas phases,

$$P = \sum x_i \gamma_i P_{i(T)}^s \tag{20.11}$$

and assuming ideal gas and liquid phases,

$$P = \frac{1}{\sum \left(\dfrac{y_i}{P_{i(T)}^s} \right)} \tag{20.12}$$

Equation 20.11 is useful for bubble point calculations, and Equation 20.12 for dew point calculations. When the temperature is known, the pressure is directly estimated. When the pressure is known, the estimation of temperature requires trial and error. However, since only an estimate is required, it is enough to consider a weighted average ($T = \sum x_i T_b$ or $T = \sum y_i T_b$) of the boiling temperatures of the components at the given pressure.

From Antoine's equation, Equation 5.17b, the boiling temperature of a pure compound can be calculated without iterations:

$$t_b = [B/(A - \log_{10} P)] - C \tag{5.17c}$$

Thus, a rough estimate of ξ_B (T or P) can be made in step 2.

For the adjustment of ξ_B in step 13, the following rules of thumb are used for the first adjustment:

1. If $\sum y_i > 1$, the estimated P was too low or the estimated T was too high. Thus, in a bubble point calculation, $P = 1.05P$ or $T = 0.95T$ is used for the second iteration of the outside loop if $S > 1$. Conversely, if $S < 1$, $P = 0.95P$ or $T = 1.05T$ is used for the second iteration. Before continuing, it is necessary to make $S_{OLD} = S$ and $\xi_{OLD} = \xi$.

2. If $\Sigma x_i > 1$, the estimated P was too high or the estimated T was too low. Thus, for the second iteration in a dew point calculation, $P = 0.95P$ or $T = 1.05T$ is used if $S > 1$. Conversely, $P = 1.05P$ or $T = 0.95T$ is used for the second iteration if $S < 1$. Before continuing, make $S_{OLD} = S$ and $\xi_{OLD} = \xi$.

Starting from the second iteration, the adjustment can be made using Newton's method in incremental form. This method uses the information of the two previous iterations. For ξ_{OLD}, the value of the sum is S_{OLD}, and for ξ, it is S. For ξ_{NEW}, it is desired to have the sum equal 1. Thus,

$$\frac{\xi_{NEW} - \xi}{(1 - S)} = \frac{\xi - \xi_{OLD}}{S - S_{OLD}}$$

or

$$\xi_{NEW} = \xi + \frac{\xi - \xi_{OLD}}{S - S_{OLD}}(1 - S)$$

Although there are many alternative and accelerating methods, the above general guidelines are enough to solve most vapor–liquid equilibrium problems of practical interest. There is one particular case, however, that leads itself to a much simpler method of solution. This is the common case of a bubble point calculation at low pressure and subcritical conditions for all compounds. In this case, using the $\gamma - \phi$ approach, Equation 20.7 reduces to

$$x_i \gamma_{i(T,P,x)} P_i^s = y_i \phi_{i(T,P)}^o P \qquad (20.13)$$

Since the x values and either T or P are known, the calculation of each y_i can be made once P or T is specified. In the general terms of Flowchart 20.1, Φ_B is not a function of X_B. In this case, from step 6 the calculation goes to step 9, eliminating steps 7, 8, 11, and 12. Exactly the same can be said in the ϕ approach, if it is used for this case.

FLASH CALCULATIONS

For flash calculations, the approach is different. In this case, the feed flow rate, F, the temperature, T, the pressure, P, and the overall composition of the mixture, z_i, are known. The mixture is separated into a vapor with flow rate V and composition y_i, and a liquid stream with flow rate L and composition x_i.

There are many approaches to solve this problem. C. J. King (1971) [1] recommends the approach of Rachford and Rice (1952) [2]. This approach is simple, and it is particularly suited for Newton's method of numerical solution. Before carrying out the solution of a flash problem, it is advisable to verify that the mixture of composition $z_i^{a\dagger}$ is indeed in the two-phase region at the conditions of P and T. This is done by performing one bubble pressure and one dew pressure calculation at the desired temperature. In order to satisfy the flash condition, the value of pressure P should be between the bubble pressure and the dew pressure of the mixture. Once this is verified, the calculation can proceed as follows. The material balance (see Appendix A) gives

$$F = L + V$$

and for each compound i,

$$z_i F = x_i L + y_i V$$

In addition, from the equilibrium condition,

$$K_i = \frac{y_i}{x_i}$$

Combining the first two equations to eliminate L and the resulting equation with the third to eliminate y_i, after rearrangement we obtain

$$x_i = \frac{z_i}{1 + (V/F)(K_i - 1)} \tag{20.14}$$

and then

$$y_i = K_i x_i = \frac{K_i z_i}{1 + (V/F)(K_i - 1)} \tag{20.15}$$

The function to be zeroed according to Rachford and Rice is

$$M(V/F) = \Sigma y_i - \Sigma x_i$$

Combining with Equations 20.14 and 20.15,

$$M(V/F) = \sum \frac{z_i (K_i - 1)}{1 + (V/F)(K_i - 1)} \tag{20.16}$$

The variable of iteration is (V/F). Its value must always be between 0 and 1, and the flash temperature should be between the bubble and the dew temperatures of the mixture. The problem is solved when $M(V/F) = 0$, within some predetermined

---

$\dagger$ Note that the symbol z_i is used here for the composition of the feed. It should not be confused with the compressibility factor.

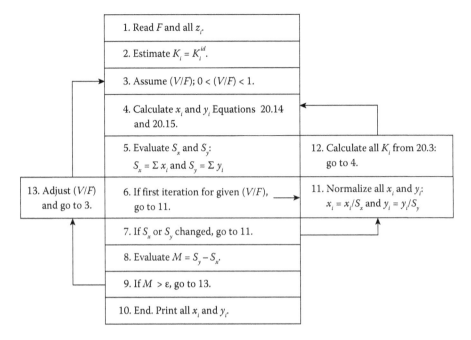

FLOWCHART 20.2   Iteration scheme for flash calculations.

accuracy ε. A scheme of solution is presented in Flowchart 20.2. It is assumed here that a decision has been previously made to use the $\gamma - \phi$ approach with a given EOS for the vapor phase and a given excess Gibbs energy function for the liquid phase. The adjustment of (V/F) can be made by the incremental form of Newton's method as

$$\frac{(V/F)_{NEW} - (V/F)}{O - M} = \frac{(V/F) - (V/F)_{OLD}}{M - M_{OLD}}$$

or

$$(V/F)_{NEW} = (V/F) - M\frac{(V/F) - (V/F)_{OLD}}{M - M_{OLD}}$$

The new value of (V/F) must be between 0 and 1. In this particular case, the Newton method can be applied in its differential form, if desired, due to the fact that from Equation 20.16 it is possible to evaluate $dM/d(V/F)$ under the assumption that all values of K_i are constant.

LIQUID–LIQUID EQUILIBRIUM

Most treatments of liquid–liquid equilibrium employed in practice make use of activity coefficients to express the fugacities of the compounds in the liquid phase. Although the model for the excess Gibbs energy need not be the same in both liquid phases, almost invariably a single model is used for both phases. Depending on the case, either the Lewis or the Henry convention can be used for the normalization of the activity coefficients. Only in special cases is Henry's convention preferred. When Lewis's convention is used for both phases, Equation 20.4 shows that the distribution coefficient for a compound is equal to the ratio of its activity coefficients in each of the phases. Thus, the model used to calculate the activity coefficients is quite important. The most commonly used models are the nonrandom two-liquid equation (NRTL) [3] equation and the universal quasi-chemical (UNIQUAC) [4] equation. The Wilson [5] equation, which is so flexible for vapor–liquid equilibrium calculations, cannot be used for liquid–liquid calculations. The reason for this is that, as discussed in Chapter 17, this equation is unable to satisfy the condition for liquid phase instability.

At the present stage of development, the calculation of liquid–liquid equilibrium for a ternary system, starting with parameters for the excess Gibbs energy functions of the three binaries, is at best approximate. The usual case is having a known overall composition of a ternary mixture and wanting to determine the compositions of the two liquid phases in equilibrium, if any two such phases exist. Formally, this case is similar to a flash problem, so a calculation scheme like the one shown in Flowchart 20.2 could be used, except that K_i is given by Equation 20.4 rather than by Equation 20.3. Thus, the estimate of K_i for step 2 needs to be done with a crude guess of the distribution of the third compound into the two partially miscible liquid phases. A more sound thermodynamic approach, seldom used in practice, is the following. Consider 1 mole of liquid mixture containing n_A and n_B moles of two partially miscible compounds A and B and n_C moles of solute C that distributes into the two liquid phases at the temperature of interest. After stirring the mixture and letting it equilibrate, the phase rich in compound A will have α moles and the phase rich in compound B will have $(1 - \alpha)$ moles. The overall mole fractions z_i of the three compounds in the feed are known; hence, by material balance,

$$z_i = n_i^\alpha + n_i^\beta = \alpha x_i^\alpha + (1 - \alpha) x_i^\beta$$

with

$$x_i^\alpha = \frac{n_i^\alpha}{\alpha}$$

and

$$\alpha = \sum n_k^\alpha$$

Thus, if n_A^α, n_B^α, and n_C^α could be evaluated, the number of moles in both phases and the mole fractions in both phases could be obtained. In addition, we know that n_i^α

must be positive and it cannot be larger than z_i, so we can transfer the problem to the evaluation of three unbound variables ζ_i defined by

$$n_i^\alpha = \frac{z_i}{1 + |\zeta_i|}$$

The first estimate of the variables ζ_i is obtained from knowledge of the solubility of B in A, plus a reasonable estimate of the distribution of C between the phases. Their final values are obtained by minimization of the Gibbs energy of the system at equilibrium.

$$G = \sum n_i^\alpha \mu_i^\alpha + \sum n_i^\beta \mu_i^\beta \quad \text{with } \mu_i^\alpha = \mu_i^\theta + RT \ln\left(x_i^\alpha \gamma_i^\alpha\right) \text{ and}$$

$$\mu_i^\beta = \mu_i^\theta + RT \ln\left(x_i^\beta \gamma_i^\beta\right)$$

Rearranging, we write

$$\tilde{G} = \frac{G - \sum z_i \mu_i^\theta}{RT} = \sum n_i^\alpha \ln\left(x_i^\alpha \gamma_i^\alpha\right) + \sum n_i^\beta \ln\left(x_i^\beta \gamma_i^\beta\right)$$

Thus, when $\tilde{G}$ is at its minimum, G also reaches its minimum and all compositions can be determined. Although the method has a sound thermodynamic base, its successful application depends strongly on the validity of the excess Gibbs energy model used to evaluate the activity coefficients of the three compounds in the two phases. In practice, most of the liquid–liquid equilibrium problems found in liquid extraction processes are solved with empirical diagrams drawn from laboratory data, but the analytical method is sometimes used as a first approximation.

SOLID–LIQUID EQUILIBRIUM

There are two simple cases of interest for a binary system. Either the solvent is solid in equilibrium with a solution, as in the case of an iceberg in an aqueous solution of salt like the ocean, or the solute is solid, as in a saturated aqueous solution of salt. It is also possible for the solid phase to be a solid solution or a hydrate (solvate), but we restrict the discussion here to a pure compound solid phase.

Let us designate by x_i the mole fraction in the liquid phase of the compound (solvent or solute) that is pure in a solid phase. At equilibrium, we can write

$$f_i^{0,s} = x_i \gamma_i f_i^{0,L} \tag{20.17}$$

where $f_i^{0,L}$ is the fugacity of compound i in a hypothetical pure subcooled liquid at the temperature T of the system. Hence,

$$\ln x_i = \ln \frac{f_i^{0,S}}{f_i^{0,L}} - \ln \gamma_i \tag{20.18}$$

TABLE 20.2

Steps for the Calculations of Δh_k and Δs_k

Step k	Δh_k	Δs_k
1. Subcooled liquid at T and P is heated to the melting point T_m at P.	$C_p^L(T_m - T)$	$C_p^L \ln \dfrac{T_m}{T}$
2. Liquid at T_m and P is solidified; Δh_m is the "heat of fusion."	$-\Delta h_m$	$-\dfrac{\Delta h_m}{T_m}$
3. Solid is cooled from T_m to T.	$C^S p(T - T_m)$	$C_p^S \ln \dfrac{T_m}{T}$

The first term on the right-hand side corresponds to the dimensionless molar Gibbs energy change of the pure compound i when passing from the state of subcooled liquid to solid at temperature T:

$$\frac{\Delta g_i}{RT} = \frac{\Delta h_i - T \Delta s_i}{RT} = \ln \frac{f_i^{0,S}}{f_i^{0,L}} \tag{20.19}$$

The enthalpy and entropy changes are accounted for in the three steps shown in Table 20.2. Hence, the total changes are

$$\Delta h = \Delta h_1 + \Delta h_2 + \Delta h_3 = \left(C_p^L - C_p^S\right)(T_m - T) - \Delta h_m$$

$$\Delta s = \Delta s_1 + \Delta s_2 + \Delta s_3 = \left(C_p^L - C_p^S\right)\ln \frac{T_m}{T} - \frac{\Delta h_m}{T_m}$$

Calling $\Delta C_p = C_p^L - C_p^S$, we write

$$\ln \frac{f_i^{0,S}}{f_i^{0,L}} = -\frac{\Delta h_m}{R}\left(\frac{1}{T} - \frac{1}{T_m}\right) + \frac{\Delta C_p}{R}\left(\frac{T_m}{T} - 1 - \ln \frac{T_m}{T}\right) \tag{20.20}$$

Combining Equations 20.18 and 20.20, we write

$$\ln x_i = -\frac{\Delta h_m}{RT_m}\left(\frac{T_m}{T} - 1\right) + \frac{\Delta C_p}{R}\left(\frac{T_m}{T} - 1 - \ln \frac{T_m}{T}\right) - \ln \gamma_i \tag{20.21}$$

As a first approximation, the effect of the ΔC_p term can be neglected and we write

$$\ln x_i = -\frac{\Delta h_m}{RT_m}\left(\frac{T_m}{T} - 1\right) - \ln \gamma_i \tag{20.22}$$

As an additional approximation, we can assume ideal solution in the liquid phase, $\gamma_i = 1$, and write

$$\ln x_i^{id} = -\frac{\Delta h_m}{RT_m}\left(\frac{T_m}{T}-1\right) \tag{20.23}$$

When the solvent is present as a solid, T_m is the melting point, or freezing point, of the pure solvent. In this case, Δh_m refers to the heat of fusion of the pure solvent, and Equation 20.23 is used to calculate the "freezing point depression" due to the presence of a solute.

Alternatively, when the solute is present as a pure solid, both T_m and Δh_m refer to the solute and Equation 20.23 is used to calculate the "ideal solubility." More realistic values of the solubility can be obtained by using some g^E model to calculate γ_i. Sometimes, Equation 20.23 is written in a different form using

$$\frac{\Delta h_m}{RT_m} = \frac{\Delta S_m}{R}$$

$$\ln x_i^{id} = \frac{\Delta S_m}{R}\left(1-\frac{T_m}{T}\right) \tag{20.24}$$

PHASE EQUILIBRIUM THROUGH MEMBRANES

With the increasing interest in bioseparations, old concepts are recovering importance in chemical engineering thermodynamics. In this section, we briefly consider the basic principles of osmosis. The osmotic phenomena occur when a solution of a nonvolatile solute in a solvent is in contact with pure solvent through a membrane only permeable to the solvent. This can be the case of an aqueous solution of sugar, a nonionic solute, in contact with water through a semipermeable membrane like collodion, or of an aqueous solution of sodium chloride, an ionic solute, in contact with water through a semipermeable membrane such as a cell wall. The physical effect caused by such a contact is an overpressure on the side of the solution with respect to the side of the pure solvent. This overpressure is called the osmotic pressure, π. The osmotic pressure can be measured by means of an experiment such as the one shown in Figure 20.4.

The osmotic pressure, at a given temperature, is a function of the concentration of the solution. The thermodynamic treatment of the osmotic phenomena is simple. At equilibrium, the fugacity of the solvent in the solution at T, $(\pi + P)$ and x must be equal to the fugacity of the pure solvent (water) at T and P.

Hence,

$$f_W(T,P+\pi,x) = f_W^0(T,P)$$

$$x_W \gamma_W f_W^0(T,P+\pi) = f_W^0(T,P)$$

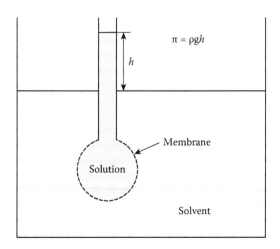

FIGURE 20.4   Osmotic equilibrium.

$$\ln\left(x_W\gamma_W\right)+\ln\frac{f_W^0\left(T,P+\pi\right)}{f_W^o\left(T,P\right)}=0 \tag{20.25}$$

The second term of Equation 20.25 refers to the pure solvent and is directly obtained from Equation 14.17 as

$$\ln\frac{f_W^0\left(T,P+\pi\right)}{f_W^0\left(T,P\right)}=\frac{v_W^0}{RT}\pi \tag{20.26}$$

where v_W^0 is the molar volume of pure liquid water assumed to be incompressible. Thus,

$$\pi=\frac{RT}{v_W^0}\left(-\ln a_W\right) \tag{20.27}$$

where we have used the definition of activity coefficient

$$a_W\equiv x_W\gamma_W$$

If we consider the special case of a very dilute solution, that is, almost pure water on the solution side of the membrane, $\gamma_W\approx 1$ and $a_W\approx x_W$ and

$$-\ln a_W\approx-\ln x_W\approx-\ln(1-x_s)\approx x_s$$

where $x_s\approx n_s/n_W$ is the mole fraction of solute in the very dilute solution. Thus,

$$\pi^{\text{dilute}}=\frac{n_sRT}{\left(n_Wv_W^0\right)}=n_s\frac{RT}{V_W^0} \tag{20.28}$$

or with

$$n_s/V_W^0 = \tilde{C}_s$$

$$\pi^{\text{dilute}} = \tilde{C}_s RT$$

Equation 20.28 is the well-known van't Hoff equation for dilute solutions. For concentrated solutions, $(-\ln a_W)$ will differ from (n_s/n_W) and Equation 20.27 is used to define the osmotic coefficient φ.

$$(-\ln a_W) \equiv \frac{n_s}{n_W}\varphi \tag{20.29}$$

Thus,

$$\pi = \frac{n_s RT}{V_W^0}\varphi \tag{20.30}$$

RECAPITULATION OF CONCEPTS

The solution to a phase equilibrium problem always starts from the equality of the fugacity for each compound in the different phases, in which it is present at equilibrium, at a given temperature T and pressure P. How do we express the fugacity in each case is a matter of convenience. Equations of state are almost invariably used for the vapor phase. Simple equations of state, such as the virial pressure expansion truncated after the second virial coefficient, are particularly useful for low-pressure, moderate-temperature vapor–liquid equilibrium calculations with the $\gamma - \phi$ approach. In this case, for the liquid phase, the fugacity of each compound is expressed in terms of activity coefficients derived from a single expression for the excess Gibbs energy of the mixture. If the experimental data for activity coefficients are missing, they can be predicted using a group contribution method, like ASOG [6,7].

For high-pressure calculations, and even for low-pressure calculations, the use of cubic equations of state has become popular in the last decades. They are discussed in Chapter 25, and the details for application of the Peng–Robinson–Stryjek–Vera (PRSV) EOS [8,9] are given in Appendix C. Other more complex equations of state have also been used.

For liquid–liquid equilibria and for solid–liquid equilibria, the fugacities are almost invariably expressed in terms of activity coefficients.

The techniques to solve phase equilibrium problems change somewhat from problem to problem, but they are all based on the solution of the material balances and the equality of the fugacities at equilibrium.

REFERENCES

1. King, C. J. 1971. *Separation Processes*. New York: McGraw-Hill.
2. Rachford, H. H., and Rice, J. D., Jr. 1952. Procedure for use of electronic digital computers in calculating flash vaporization hydrocarbon equilibrium. *J. Petrol. Technol.* 4(10): 19–20.
3. Renon, H., and Prausnitz, J. M. 1969. Derivation of the three-parameter Wilson equation for the excess Gibbs energy of liquid mixtures. *AIChE J.* 15(5): 785–787.
4. Abrams, D. S., and Prausnitz, J. M. 1975. Statistical thermodynamics of liquid mixtures. New expression for the excess Gibbs energy of partly or completely miscible systems. *AIChE J.* 21(1): 116–128.
5. Wilson, G. M. 1964. Vapor-liquid equilibrium. XI. A new expression for the excess free energy of mixing. *J. Am. Chem. Soc.* 86(2): 127–130.
6. Kojima, K., and Tochigi, K. 1979. *Prediction of Vapor-Liquid Equilibria by the ASOG Method*. Tokyo: Kodansha-Elsevier.
7. Tochigi, K., and Gmehling, J. 2011. Determination of ASOG parameters—Extension and revision. *J. Chem. Eng. J. Jpn.* 44: 304–306.
8. Stryjek, R., and Vera, J. H. 1986. PRSV: An improved Peng-Robinson equation of state for pure compounds and mixtures. *Can. J. Chem. Eng.* 64(2): 323–333.
9. Stryjek, R., and Vera, J. H. 1986. PRSV2: A cubic equation of state for accurate vapor-liquid equilibria calculations. *Can. J. Chem. Eng.* 64: 820–826.

21 Heat Effects

HEAT EFFECTS IN PURE COMPOUNDS

Engineers have prepared tables of thermodynamic properties for most compounds of practical interest. In these tables, the values of the molar or specific volume, v, enthalpy, h, and entropy, s, are given as a function of temperature and pressure. To use these tables, one must first determine the physical state of the compound at the known values of T and P. The key piece of information is the value of the vapor pressure, P^s (also called saturation pressure), at the temperature of interest. This value is normally reported in the first column of the tables, together with the values of v^s, h^s, and s^s for both the liquid and the vapor. If the pressure of interest is higher than the saturation pressure, the compound is in the state of compressed liquid. In this case, as for all practical purposes, liquids are incompressible, the difference in pressure is ignored, and the values of the thermodynamic properties of the liquid compound are considered to be equal to those it has as saturated liquid. Only for work at high pressure may corrections be needed. In the case that the pressure of interest is lower than the saturation pressure, the compound is in the gas phase and its properties are those reported for that pressure. Usually, interpolation between two values of pressure from the tables is required. Sometimes, it is also necessary to interpolate between two values of temperature before proceeding with the interpolation of pressure. In technical language, a compound in the gas phase is referred to as a superheated vapor, meaning that its temperature is higher than the saturation temperature corresponding to its pressure.

Figure 21.1 depicts a schematic of an isotherm in a pressure–enthalpy (P-h) diagram.

In the special case that the pressure corresponds exactly with the vapor pressure at the temperature of interest, the compound can be in one of the following states: as saturated liquid, as a system of saturated liquid in contact with saturated vapor, as wet saturated vapor, or as dry saturated vapor. The wet saturated vapor is a saturated vapor phase containing small droplets of saturated liquid. All these cases, generically known as a "saturated mixture," will have different values of the molar or specific volume, enthalpy, or entropy. In the case of using saturated steam as a heating medium, it is necessary to determine in which of the above cases is the system. For heating purposes, saturated steam is used instead of saturated (boiling) water or superheated steam. This is because the heat of condensation of steam is orders of magnitude larger than the heat that can be obtained from cooling a single-phase fluid.

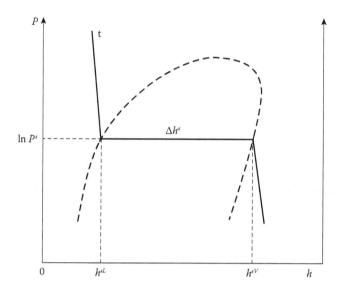

FIGURE 21.1 A schematic pressure–enthalpy diagram for a liquid–vapor system.

In order to determine the content of dry saturated steam in a saturated mixture, engineers have designed a device called adiabatic throttling calorimeter (Figure 21.2).

Knowing the temperature of the saturated steam to be used for heating, the saturation pressure and the enthalpies of the saturated liquid and the saturated vapor phases, h^{sL} and h^{sV}, are readily obtained from tables. If the saturated mixture

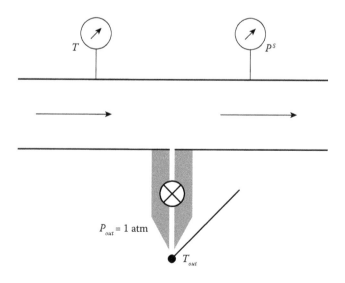

FIGURE 21.2 Adiabatic throttling calorimeter.

contains a mass fraction χ of dry saturated steam, the specific enthalpy of the saturated mixture is

$$h_m^s = (1-\chi)h^{sL} + \chi h^{sV}$$

$$= h^{sL} + \chi(h^{sV} - h^{sL})$$

$$= h^{sL} + \chi \Delta h^s$$

or

$$\chi = \frac{h_m^s - h^{sL}}{\Delta h^s} \qquad (21.1)$$

The mass fraction χ of dry saturated steam is called the *quality of the steam*. The value of h_m^s is obtained by purging a small stream of saturated mixture from the steam pipe into the atmosphere. As shown in Figure 21.2, this is done through an insulated pipe having a high-pressure valve. The temperature of the steam at the exit of the valve is measured after a short time of purging. When steady state is reached and the temperature t_{out} stabilizes, h_{out} can be determined knowing P_{atm} and t_{out}. From the first law for an open flow system (Chapter 3), neglecting kinetic and potential energy changes, as there is no heat exchanged with the surroundings and no shaft work involved, from Equation 3.14 we get

$$\Delta h_{pipe}^{in} = 0$$

or

$$h_{out} = h_m^s$$

Thus, the quality of the steam, χ, is evaluated from Equation 21.1, and it is usually expressed in percent.

Example 21.1: Determination of Steam Quality

The manometer installed in a steam pipe indicates a pressure of 14 bar at the point where the saturated steam mixture passes the entrance of an adiabatic throttling calorimeter. After opening the throttling valve of the calorimeter, the temperature of the superheated steam leaving the calorimeter is 400 K. The outside pressure is 1 atm (1 atm = 1.013 bar). Determine the quality of the steam in the pipe.

Table 2-305 of Perry's handbook, pages 2-413 to 2-414 [1], provides information presented in Table 21.1. As the tabulated values of pressure are in megapascals, we need to convert the pressure data to proper units:

14 [bar] = 1.4 [MPa] and 1 [atm] = 0.1 [MPa]

For the value of h_{out} at 400 K and 0.1 MPa, we read 49.189 [kJ/mol].
Thus,

$$\chi = \frac{h_m^s - h^{sL}}{\Delta h^s} = \frac{49.189 - 14.942}{50.238 - 14.942} = 0.97$$

The quality of steam is 97%.

TABLE 21.1
Vapor and Liquid Enthalpies of the
Saturated Steam Mixture

Pressure (MPa)	h^{sV} (kJ/mol)	h^{sL} (kJ/mol)
1.171	50.134	14.293
1.455	50.263	15.098
Interpolating		
1.4	50.238	14.942

Source: Green, D. W., and Perry, R. H., *Perry's Chemical Engineers' Handbook*, 8th ed., McGraw-Hill Professional Publishing, Blacklick, OH, 2007, Table 2-305, pp. 2-413 to 2-414.

HEAT EFFECTS IN MIXTURES

Except in the rare case of ideal solutions, each time that different chemical species are mixed, there are heat effects. If the mixing process is adiabatic, the temperature of the final mixture will be different from the mixed mean temperature, which we could calculate from knowledge of the masses, the temperatures, and the specific heats of the pure compounds' initial states. If the mixing process is carried out under isothermal conditions in a thermostatic bath, heat will be exchanged with the surroundings.

As it is always done in thermodynamics, we classify first the main cases encountered in practice and then systematize the experimental information in order to be able to treat new situations. There are two broad categories of mixing processes: (1) mixing processes in which there is no chemical reaction between the species mixed and (2) mixing processes in which chemical reactions occur. In this second category, two heat effects are superimposed: the heat effect due to the chemical reactions and the heat effect due to the purely physical interactions of mixing. The heat effects due to reactions are much larger than the heat effects due to purely physical interactions, so when a reaction occurs, the latter can be ignored. We refer to the first case as the study of heat of mixing and the second case as the study of heat of reaction. For the cases of mixing of the compounds, either in an open flow system without shaft work, Equation 3.14, or in a closed system at constant pressure, Equation 3.17, we write:

$$q = \Delta h \tag{21.2}$$

In the rare cases when a mixing process occurs in a closed system at constant volume, there is no work done, and from Equation 2.3, the heat effect is equal to the change in internal energy, not the change in enthalpy. However, for liquid mixtures the difference between internal energy and enthalpy is totally negligible, and frequently we retain the form of Equation 21.2. We recall that the difference in enthalpy is always taken as the final enthalpy minus the initial enthalpy, and that q is positive when heat enters the system. Conventionally then, heat evolved is indicated with a negative sign.

HEAT OF MIXING, Δh_M

The definition of the heat of mixing requires that the final mixture and all the pure compounds before mixing be at the same temperature and pressure. We recall that the effect of pressure is negligible in the case of liquids. For the molar or specific heat of mixing at a constant temperature T, we write

$$q = \Delta h_m = h - \sum_i x_i h_i^0 \qquad (21.3)$$

where h is the enthalpy of the mixture at temperature T and pressure P, and h_i^0 are the enthalpies of the pure compounds at the same conditions of temperature and pressure of the mixture.

In the literature, sometimes the heat of mixing is referred to as the excess enthalpy of the mixture. In fact, as it was discussed in Chapter 6, an ideal mixture is formed without heat effects, so for an ideal mixture in Lewis's sense, the second term of the right-hand side of Equation 21.3 corresponds to the enthalpy of the ideal mixture, and then by analogy with Equation 16.11b we write

$$q = \Delta h_m = h - h^{id} = h^E \qquad (21.4)$$

When forming a real mixture, starting with all pure compounds at the same temperature, the temperature of the mixture will have a different value. Considering that the definition of the heats of mixing, Equation 21.3, requires that the final mixture be at the same temperature as the initial temperature of the pure compounds before mixing, we can think of the process as occurring in two steps. First, we mix adiabatically the pure compounds, all at an initial common temperature T. In this step, there is no heat exchanged with the surroundings and the mixture will have an enthalpy h_{ad}. From Equation 21.3, with $q = 0$,

$$h_{ad} = \sum_i x_i h_i^0$$

Thus, in an adiabatic mixing the mixture has the same enthalpy as the sum of the input enthalpies of the pure compounds, but unless it is an ideal mixture, its temperature will be different from the temperature of the pure compounds before mixing. In a second step then, the mixture is heated or cooled to bring its temperature back to T. In this step, the system exchanges with the surroundings an amount of heat q, per mole or per unit mass of mixture.

$$q = h - h_{ad}$$

Thus, for this second step the amount of heat exchanged between the system and the surroundings is exactly that given by Equation 21.3. When the enthalpy of the mixture is larger than the sum of the enthalpies of the pure compounds at the same conditions of temperature and pressure, the temperature of the system after the adiabatic step is lower than the initial temperature of the pure compounds, that is, $\Delta h_m > 0$, then $q > 0$. Thus, heat must be supplied to the system to keep the temperature constant. In this case, we say that the process is endothermic. The opposite case is referred to as an exothermic mixing process.

For the mixing of compounds in the gas phase, the heat of mixing is usually negligibly small and can be ignored for most practical situations. In fact, for any ideal mixture in Lewis's sense, gas, liquid, or solid, the heat of mixing is exactly zero. Thus, although perfectly general, most of the treatment that follows is intended for *liquid* mixtures.

To fix ideas, let us consider the enthalpy of an ideal binary liquid mixture, in Lewis's sense,

$$h^{id,0} = x_1 h_1^0 + x_2 h_2^0 = h_2^0 + (h_1^0 - h_2^0)x_1$$

In this case, the enthalpy of the mixture is a linear function of composition, and on a plot of h versus x_1, it will be represented as a straight line joining the points ($x_1 = 0$, $h = h_2$) and ($x_1 = 1$, $h = h_1$). In this case, the mixture is formed without heat effects. On the other hand, Equation 21.3 tells us that for an endothermic case, $\Delta h_m > 0$, $q > 0$, and the curve of h versus x goes above the ideal straight line, as shown in Figure 21.3, and heat must be supplied to the system to keep the temperature constant. This is the most common case for organic mixtures, and it corresponds with positive deviations from ideality in Lewis's sense. The more dissimilar the compounds are, the larger are

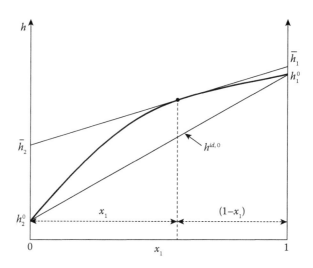

FIGURE 21.3   Enthalpy of binary liquid mixture for the endothermic case.

the positive deviations and the heat required to keep the temperature constant. Only a few binary mixtures presenting association between the two dissimilar compounds show exothermic behavior corresponding to negative deviations from ideality in Lewis's sense. Alcohol–amine systems or the system acetone–chloroform, for example, are cases in which the curve of h versus x falls below the ideal-mixture straight line, as well as the aqueous solutions of strong acids (e.g., H_2SO_4) and hydroxides (e.g., NaOH).

For completeness, in Figure 21.3 we have also shown the locations of the partial molar enthalpies $\bar{h}_1$ and $\bar{h}_2$ for a point at composition x_1. These values are found at the intersection of the tangent to the curve h–x, at the given composition x_1, with the y-axis at $x_1 = 1$ and $x_1 = 0$, respectively.

In fact, by analogy to Equation 6.17 we have

$$\left(\frac{\partial h}{\partial x_1}\right)_{T,P} = \frac{\bar{h}_1 - h}{1 - x_1} = \frac{h - \bar{h}_2}{x_1} \tag{21.5}$$

Thus, the tangent to the curve at a point x_1 fixes the position of both partial molar enthalpies at this composition.

Heats of mixing also provide the experimental information necessary to evaluate the temperature dependence of the parameters used in models for the excess Gibbs energy of binary systems. These models use the Lewis convention of ideality, and in this case, as shown by Equation 21.3, the heat of mixing Δh_m is equal to the excess enthalpy h^E. Adding Equation 16.13b over all compounds, we write

$$\Delta h_m = -RT^2 \left(\frac{\partial \left(g^E/RT\right)}{\partial T}\right)_{P,n} \tag{21.6}$$

Conversely, knowledge of (g^E/RT) as a function of temperature makes it possible to estimate values of Δh_m.

ENTHALPY–CONCENTRATION DIAGRAMS

The measurement of heats of mixing is performed in calorimeters. With this experimental information available and the help of Equation 21.3, it is possible to construct diagrams or tables of enthalpy versus composition (concentration) for a mixture of interest. Although for most cases of practical interest this information is handled directly by computer programs, for the treatment of new situations it is important to understand the basic principles used for the construction of such diagrams or tables. For this purpose, Equation 21.3 is better written as

$$h = \Delta h_m + \sum_i x_i h_i^0 \tag{21.3a}$$

This form of Equation 21.3 shows that while Δh_m is experimentally measurable, the value of h depends on the arbitrarily chosen reference states used to set the enthalpy to zero for the pure compounds. For water, to be in agreement with the reference state used in the steam tables, the enthalpy is set at zero for saturated liquid water at 273.15 K or 32.018°F. In fact, the pure compound as saturated liquid at 273.15 K is a common reference state for compounds that are liquids at this temperature. For solutes such as NaCl and NaOH, it is conventional to consider the partial molar enthalpy of the solute at infinite dilution in the solvent, water, for example, as zero at 273.15 K. This choice, which may seem odd at first, has its basis on Equation 21.3 written as

$$\Delta h_m = \sum_i^c x_i \left(\overline{h}_i - h_i^0 \right) \tag{21.3b}$$

For a binary system then,

$$\Delta h_m = x_1 \left(\overline{h}_1 - h_1^0 \right) + x_2 \left(\overline{h}_2 - h_2^0 \right)$$

The heat of mixing per mole of solute is given by

$$\tilde{\Delta h}_S \left[\frac{J}{\text{mol of solute}} \right] \equiv \frac{\Delta h_m \left[\frac{J}{\text{mol of solution}} \right]}{x_2 \left[\frac{\text{mol of solute}}{\text{mol of solution}} \right]}$$

For solid solutes, it is convenient to express the concentration of solute as molality, m. Then,

$$\tilde{\Delta h}_S \left[\frac{J}{\text{mol of solute}} \right] \equiv \frac{\Delta h_m \left[\frac{J}{\text{kg of solvent}} \right]}{m_2 \left[\frac{\text{mol of solute}}{\text{kg of solvent}} \right]}$$

At high dilution of a solute 2 in a solvent 1, x_2 is small and x_1 is virtually unity. In this case, $\overline{h}_1 = h_1^0$ and the first term of the right-hand side of Equation 21.3b cancels out. In addition, $\overline{h}_2 = h_2^\infty$. Thus, arbitrarily setting $\overline{h}_2^\infty = 0$, we write

$$\tilde{\Delta h}_S^\infty = \lim_{x_2 \to 0} \frac{\Delta h_m}{x_2} = -h_2^0 \tag{21.7}$$

The left-hand side of Equation 21.7 is the heat of solution, that is, the heat of mixing per mole of solute at high solute dilution.[†] This is a measurable quantity. Hence, the convention of arbitrarily setting $\bar{h}_2^\infty = 0$ fixes the value of the enthalpy of the pure solute at the temperature of the system, indicating once more that the enthalpy values are relative to the zero fixed for their scale. As an example of practical tables for enthalpy, Tables 2-182 and 2-183 (pages 2-203 to 2-206) of Perry's handbook [1] give heats of solution per mole of solute for inorganic and organic compounds. These values are reported as heat evolved, and thus they are negative values of the heats of solution. The values of Table 2-182 of Perry's handbook [1] are reported either at infinite dilution, indicating number of moles of water (e.g., 400 and 1000) necessary to dissolve 1 mole of solute, or at "aq.," indicating an aqueous solution of unspecified high dilution. This is justified because the heat of solution per mole of solute is quite constant from infinite dilution up to 20 moles of solvent per mole of solute. A practical consequence of this fact is that, as also suggested by Equation 21.7, for diluted solutions we can use the approximation

$$\Delta h_m^{dil} \approx \left(\Delta \tilde{h}_S^\infty\right) x_2 \tag{21.8}$$

At higher concentrations of solute, errors can be large and experimental values are required. In some cases, when heats of mixing or heats of solution are not available, it is possible to find information on heats of formation of aqueous solutions. These refer to the hypothetical formation of an aqueous mixture starting from the elements forming the solute and a certain number of molecules of water. As an example, Table 2-178 of Perry's handbook, page 2-203 [1], provides these data. The use of this information is better illustrated with Example 21.2.

Example 21.2: Calculation of a Heat of Dilution from Heats of Formation

For BaBr$_2$, Table 2-182 of Perry's handbook [1], page 2-203, reports that the heat evolved when forming an infinitely dilute solution in water is +5.3 kcal/mol of solute. Considering that Table 2-182 gives heat evolved, we write $\Delta h_s^\infty = -5.3\,(\text{kcal/mol})$. Verify this value using information on heats of formation.

Solution
Table 2-178 of Perry's handbook [1], page 2-186, reports the heats of formation of BaBr$_2$(c) and of BaBr$_2$(aq.,400) as –180.38 and –185.67 kcal/mol, respectively. These values refer to the following formation equations:

1. $Ba + Br_2 + (400)H_2O = BaBr_2(aq, 400)$ $\Delta h_f = -185.67\,\text{kcal/mol}$

2. $Ba + Br_2 = BaBr_2(c)$ $\Delta h_f = -180.38\,\text{kcal/mol}$

[†] One may ask how it is possible to have 1 mole of solute at infinite dilution and how to measure the heat effect in this case. A plot of the heat of solution against the concentration of the salt shows a curve that starts from the point (0,0). This curve is closely a straight line in the diluted region, and its slope is the value of the heat of solute per mole of solute at infinite dilution. The heat of solution should not be confused with the heat of dilution that is the heat effect when more solvent is added to an existing solution.

Subtracting Equation 2 from Equation 1, we get

$$BaBr_2(c) + (400)H_2O = BaBr_2 \text{ (aq,400)} \quad \Delta h = (-185.67) - (-180.38)$$
$$\Delta h = -5.29 \text{ (kcal/mol)}$$

This value compares well with $\Delta h_s^\infty = -5.3 \text{ (kcal/mol)}$ obtained from Table 2-182 of Perry's handbook [1].

Thus, either from experimental measurements or from estimates, we can have values of Δh_m as a function of composition. Subsequently, using Equation 21.2, we can prepare diagrams or tables of enthalpy versus composition.

Typical examples are the enthalpy–concentration diagrams for aqueous solutions of NaOH and H_2SO_4 presented in Figures 2-29 and 2-31 of Perry's handbook [1], pages 2-403 and 2-409. Similarly, as mentioned in Example 21.2, Table 2-182, page 2-203, of Perry's handbook [1] gives values of the heat evolved per mole of solute when forming dilute solutions. The conceptually interesting point is the relation between the enthalpy–concentration diagrams and the information given in this table. Again, in this case, the relation between these two pieces of information is better illustrated by an example.

Example 21.3: Calculation of the Enthalpy of a Solution

Figure 2-29, page 2-403, of Perry's handbook [1] contains the information presented in Table 21.2.

Compare these data with values obtained using Equation 21.3a:

$$h = \Delta h_m + \sum_i x_i \, h_i^0$$

Solution at 18°C
Finding the Enthalpies of Pure Compounds, h_i^0

- For NaOH: Temperatures of 64°F and 248°F correspond to 18°C and 120°C. From Table 2-182, page 2-203, of Perry's handbook [1], the heat evolved per mole of solute when forming an infinitely dilute solution of NaOH in water at 18°C is 10.18 (kcal/mol). Thus, the heat of solution per mole of solute at infinite dilution is the negative of this value:

$$\Delta \tilde{h}_s^\infty = -10.18 \text{ (kcal/mol of solute)}$$

  Therefore, according to Equation 21.7, the enthalpy of NaOH (solid) at 18°C is arbitrarily fixed at $h_2^0 = 10.18$ (kcal/mol) of NaOH or (using the factor 1800 [(Btu/lbmol)/(kcal/mol)])

TABLE 21.2

Enthalpy Data for Aqueous Solution of NaOH

Temperature	Enthalpy of the Solution, h (Btu/lb of solution)	
t (°F)	$X_{NaOH} = 0.05$	$X_{NaOH} = 0.10$
64	35	32
248	209	198

Source: Green, D. W., and Perry, R. H., *Perry's Chemical Engineers' Handbook*, 8th ed., McGraw-Hill Professional Publishing, Blacklick, OH, 2007, Figure 2-29, p. 2-403.

Note: The resolution of the plot allows for enthalpy readings within ±3 units on the last place. X_{NaOH} is the weight fraction of NaOH in the solution.

$$h_2^0 = 10.18(\text{kcal/mol}) \cdot 1800 \left[\frac{\text{Btu}/_{\text{lbmol}}}{\text{kcal}/_{\text{mol}}} \right] = 18,324 \left(\text{Btu}/_{\text{lbmol}} \right) \text{of NaOH}$$

or

$$h_2^0 = 18,324 \left(\text{Btu}/_{\text{lbmol}} \right) \frac{1(\text{lbmol})}{40(\text{lb})} = 458.1 \left(\text{Btu}/_{\text{lb}} \right) \text{of NaOH}$$

- For H_2O: For liquid water, Table 2-305 of Perry's handbook, page 2-413 [1], gives the information presented below in kilojoules per mole. As 18°C is equal to 291.15 K, we need to perform the interpolation to get the value of enthalpy at the required temperature (see Table 21.3). Thus,

$$h_1^0 = 1.3609(\text{kJ/mol}) \cdot 429.92 \left[\frac{\text{Btu}/_{\text{lbmol}}}{\text{kJ}/_{\text{mol}}} \right] = 585 \left(\text{Btu}/_{\text{lbmol}} \right) \text{or } h_1^0$$

$$= 32.5 \left(\text{Btu}/_{\text{lb}} \right) \text{of water}$$

Finding Δh_m
From Equation 21.8,

$$\Delta h_m^{dil} \approx \left(\Delta \tilde{h}_S^\infty \right) x_2$$

TABLE 21.3
Enthalpy of Liquid Water at Different Temperatures

Temperature (K)	h^{st} (kJ/mol)
290	1.2742
300	2.0279
Interpolating	
291.15	1.3609

Source: Green, D. W., and Perry, R. H., *Perry's Chemical Engineers' Handbook*, 8th ed., McGraw-Hill Professional Publishing, Blacklick, OH, 2007, Table 2-305, p. 2-413.

This approximation is valid when there are more than 20 moles of solvent per mole of solute.

For a more concentrated solution, at a weight fraction $X_2 = 0.10$, in 100 lb of mixture there are 10 lb of NaOH and 90 lb of H_2O. Remembering that $M_2 = 40.0$ and $M_1 = 18.0$, this is equivalent to 0.25 lb-mol of NaOH and 5.0 lb-mol of H_2O. Thus, the ratio of moles of water per mole of solute is 20, which falls in a safe region for application of Equation 21.8. In addition, from Equation 21.7,

$$\Delta \tilde{h}_S^\infty = -h_2^0$$

At a weight fraction $X_2 = 0.10$,

$$\Delta h_m = (-18324) \cdot \left(0.25 / 5.25\right) = -872.6 \left(\frac{Btu}{lbmol\ mixture}\right)$$

$$\Delta h_m = (-872.6) \cdot \left(5.25 / 100\right) = -45.8 \left(\frac{Btu}{lb\ mixture}\right)$$

From Equation 21.3a,
$h = (-45.8) + (0.90)(32.5) + (0.10)(458.1)$
$h = 29$ (Btu/lb)
This value should be compared with $h = 32$ (Btu/lb) as reported in Table 21.2.
At a weight fraction $X_2 = 0.05$, similar calculations give

$$\Delta h_m = -22.9 \left(\frac{Btu}{lb\ mixture}\right)$$

and
$h = 31$ (Btu/lb)
which compare well with $h = 35$ (Btu/lb) presented in Table 21.2.

Solution At 120°C
In order to estimate h at 120°C, we assume that Δh_m does not change with temperature, but that the enthalpies of the pure compounds need to be corrected for temperature.

- For H_2O: Table 2-305 of Perry's handbook, page 2-413 [1], gives for liquid water the information presented below in kilojoules per mole. As 120°C is equal to 393.15 K, we need to perform the interpolation to get the value of enthalpy at the required temperature (see Table 21.4). Thus,

$$h_1^0 = 3903 \left(\frac{Btu}{lb\ mol} \right) \text{ or } h_1^0 = 216.84 \left(\frac{Btu}{lb} \right) \text{ of water}$$

- For NaOH: For NaOH, no values of the heat of solution per mole of solute at infinite dilution are reported at other temperatures. On the other hand, values for C_p for NaOH are reported in [2]. The trend line in the temperature range of 298 K–400 K has the form

$$C_p \left(\frac{cal}{deg \cdot mol} \right) = 6 \cdot 10^{-5} t^2 + 0.0041 t + 14.122 \text{ with } R^2 = 1$$

where temperature t is in degrees Celsius. As the precision of the information available does not justify an integration, considering that at 18°C, $C_p = 0.355$ cal/(g deg) = 0.640 Btu/(lb·deg), and at 120°C, $C_p = 0.387$ cal/(g deg) = 0.697 Btu/(lb·deg), we assume $\bar{C}_p \approx 0.668$ Btu/(lb·deg) as an average value.

TABLE 21.4
Enthalpies of Saturated Liquid Water at Different Temperatures

Temperature (K)	h^{sl} (kJ/mol)
390	8.8354
400	9.6013
Interpolating	
393.15	9.0767

Source: Green, D. W., and Perry, R. H., *Perry's Chemical Engineers' Handbook*, 8th ed., McGraw-Hill Professional Publishing, Blacklick, OH, 2007, Table 2-305, p. 2-413.

$$\Delta h_2^0 = C_p\,\Delta T = 0.668\cdot(120-18) = 68.1 \text{ Btu/lb NaOH}$$

Then, at 120°C,

$$h_2^0 = 458.1 + 68.1 = 526.2\left(\text{Btu/lb NaOH}\right)$$

Thus, using Equation 21.3a,
At a weight fraction X_2 = 0.05: h = 206 Btu/lb versus h = 209 Btu/lb (Table 21.2)
At a weight fraction X_2 = 0.10: h = 195 Btu/lb versus h = 198 Btu/lb (Table 21.2)
These results compare well with those given in Table 21.2 that were obtained from the enthalpy–concentration diagram and show the consistency of different sources of information.

USE OF ENTHALPY–CONCENTRATION INFORMATION

Enthalpy–concentration diagrams or tables are useful tools in chemical engineering calculations. Here, we discuss two important cases: (1) mixing two solutions of different concentrations and at different temperatures and (2) concentrating a solution of a nonvolatile solute by evaporation of the solvent.

MIXING OF TWO SOLUTIONS

The general scheme of the mixing process is shown in Figure 21.4. This case also includes the dilution of a mixture by the addition of pure solvent, that is, X_s = 0. The symbols F, S, P (feed, solvent, product) denote the total amounts (during a given time interval) of the streams added; X_F, X_S, X_P denote the mole fractions or weight fractions of the solute in each stream; and t_F, t_S, t_P are the corresponding temperatures. We can write two equations for the material balances and one equation for the energy balance (neglecting the work of stirring and heat losses).

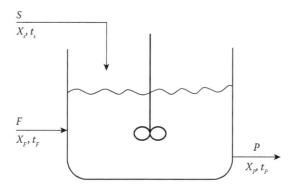

FIGURE 21.4   General scheme of the mixing process.

The material balances (see Appendix A) are

$$P = F + S \tag{21.9}$$

$$PX_P = FX_F + SX_S \tag{21.10}$$

and the energy balance is

$$Q = Ph_p - (Fh_F + Sh_s) \tag{21.11}$$

In most practical cases, the material balances can be solved separately and the results are then used to solve the energy balance. Table 21.5 shows some common situations. Other cases can be similarly solved.

From the point of view of the energy balance, two cases are of special interest. The first is the case of an adiabatic mixing process for which we know all the variables of the material balances, t_F and t_S, and we want to obtain t_p^{ad}. With (t_F, X_F) and (t_S, X_S), we determine h_F and h_S, respectively, from an enthalpy–concentration diagram. From Equation 21.11, with $Q^{ad} = 0$, we obtain

$$h_p^{ad} = (F\,h_F + S\,h_S)/P \tag{21.12}$$

Then, with h_p^{ad} and X_p we obtain t_p^{ad} from the enthalpy–concentration diagram.

The second case of interest is one in which we want to determine the heat Q that needs to be exchanged with the surroundings for a desired value of t_p. Here, we assume that first we form the mixture adiabatically, and then we exchange heat to change the temperature of the mixture from t_p^{ad} to t_p. Thus, with (t_p, X_p) we read h_p from the enthalpy–concentration diagram and

$$Q = P\left(h_p - h_p^{ad}\right) \tag{21.13}$$

In fact, substituting Equation 21.12 into Equation 21.13, we recover Equation 21.11. Eliminating F, S, and P from Equations 21.9, 21.10, and 21.12, it is possible to show that h_p^{ad} can be directly read in an enthalpy–concentration diagram from the

---

TABLE 21.5
Different Cases for Solving the Material and Energy Balances of Mixing

Known	To Be Evaluated		
F, X_F, S, X_S	$P = F + S$	$X_P = (FX_F + SX_S)/P$	
F, X_F, P, X_P	$S = P - F$	$X_s = (PX_P - FX_F)/S$	
F, X_F, X_S, X_P	$P = F(X_F - X_S)/(X_P - X_S)$	$S = P - F$	
X_F, X_s, X_P, P	$F = P(X_P - X_S)/(X_F - X_S)$	$S = P - F$	

intersection of the straight line joining the points (X_F, t_F), (X_S, t_S) with the vertical line at X_p.

Example 21.4: Mixing of Diluted and Concentrated H_2SO_4

Calculate the heat, in kilojoules per hour, that needs to be added or withdrawn from a continuous mixer receiving 4500 kg/h of dilute sulfuric acid, 5% by weight and 200°F, and 8000 kg/h of concentrated sulfuric acid, 90% by weight and 70°F, if the desired temperature of the resulting mixture is 100°F.

Solution
Let us identify the dilute acid with the stream S, $S = 4500$ kg/h, $X_S = 0.05$, $t_s = 200°F$, and the concentrated acid with the stream F, $F = 8000$ (kg/h), $X_F = 0.90$, $t_f = 70°F$. For the product P, $t_p = 100°F$. From the material balances,

$$P = F + S = 12,500 \text{ (kg/h)}$$

and

$$X_P = \frac{FX_F + SX_S}{P} = \frac{8000 \cdot 0.90 + 4500 \cdot 0.05}{12,500} = 0.59$$

From Figure 2-31, page 2-409, of Perry's handbook [1],
    $X_F = 0.9$, $t_f = 70°F$; $h_f = -60$(Btu/lb) $= -140$ (kJ/kg)
    $X_S = 0.05$, $t_s = 200°F$; $h_s = +145$ (Btu/lb) $= +337$ (kJ/kg)
    $X_P = 0.59$, $t_p = 100°F$; $h_p = -100$ (Btu/lb) $= -233$ (kJ/kg)
    Then, from Equation 21.11,
    $Q = -3309 \cdot 10^3$ (kJ/h)
Thus, heat needs to be withdrawn from the system. Note the fact that the values of X are on a weight basis and thus include the vapor in Figure 2-31 of Perry's handbook [1]. This is of no concern below the line of boiling points at 1 atm.

CONCENTRATION OF A SOLUTION OF A NONVOLATILE SOLUTE BY EVAPORATION OF SOLVENT

The general scheme of the evaporation process is shown in Figure 21.5.

In this case, the stream S containing pure solvent ($X_S = 0$) leaves the system. Usually, the temperatures of the streams S and P are the same, $t_S = t_P$. Since the boiling temperature of the concentrated solution P is higher than the boiling point of the solvent at the same pressure, the assumption of thermal equilibrium, $t_S = t_P$, means that the solvent leaves the system as superheated vapor. In this case, the material balances take the form

$$F = S + P \tag{21.14}$$

$$F X_F = P X_P \tag{21.15}$$

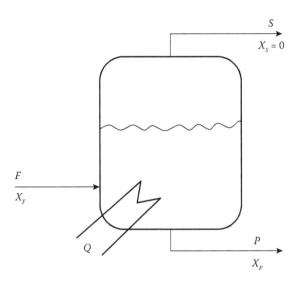

FIGURE 21.5   General scheme of the evaporation process.

and the energy balance can be written as

$$Q = (Sh_S + Ph_p) - Fh_F \qquad (21.16)$$

Example 21.5: Dilution of a Concentrated NaCl Solution

Calculate the heat required to concentrate 12,000 kg/h of a dilute solution of sodium hydroxide from 5% to 65% by weight of solute. The diluted solution enters the process at 70°F, and the concentrated solution leaves the evaporator at 260°F. The evaporator operates at atmospheric pressure.

Solution
In this case, F = 12,000 kg/h, X_F = 0.05, t_F = 70°F; $t_p = t_s$ = 260°F, X_P = 0.65.
    From the material balances,

$$P = F\, X_F/X_P = (12,000)(0.05)/(0.65) = 923 \text{ (kg/h)}$$

$$S = F - P = 12,000 - 923 = 11,077 \text{ (kg/h)}$$

From Figure 2-29, page 2-403, of Perry's handbook [1],

$$X_F = 0.05, t_F = 70°F; h_F = 30 \text{ (Btu/lb)} = 70 \text{ (kJ/kg)}$$

$$X_P = 0.65, t_p = 260°F; h_p = 355 \text{ (Btu/lb)} = 826 \text{ (kJ/kg)}$$

$$X_S = 0, t_s = 260°F; h_S = 230 \text{ (Btu/lb)} = 535 \text{ (kJ/kg)}$$

From Equation 21.16,

$$Q = [(11,077)(535)+(923)(826)] - (12,000)(70)$$

$$Q = 5849 \ 10^3 \ (kJ/h)$$

Thus, as expected, it is necessary to add heat to the system.

HEAT EFFECTS IN REACTIVE SYSTEMS

Heats of reaction are much larger than heats of mixing. Thus, when chemical reactions occur, the small contribution to the overall heat effect due to the mixing (or demixing) of compounds can be totally ignored.

To calculate heat effects in reacting mixtures, we make use of the standard heat of reaction discussed in Chapter 19. This standard heat of reaction at any temperature T, ΔH_T^θ can be calculated from Equations 19.11 and 19.15 or 19.17. We recall that this value of ΔH_T^θ corresponds to the enthalpy change of the standard reaction; that is, when the reactants enter in stoichiometric proportions, all at the same temperature T, the reaction is complete and the stoichiometric amounts of products leave at the same temperature T. In practical cases, the reactants are rarely added in stoichiometric proportions or added all at the same temperature. Frequently, "inert" compounds, that is, compounds that do not participate in the reactions, are present in one or more streams. The reaction itself is usually not complete or at equilibrium. Finally, there can be more than one reaction in the system. To fix ideas, Figure 21.6 illustrates the case of a reactor in which an excess of reactant A is fed at a temperature t_A to react with an expensive reactant B, diluted with an inert compound I, at a temperature t_B.

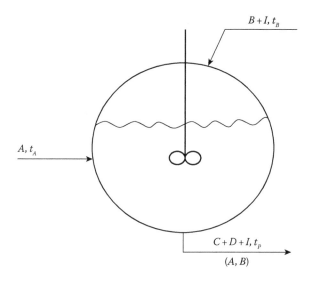

FIGURE 21.6   Schematics of a reactor. A, reactant present in excess; B, expensive reactant; I, inert dilutant for B; C and D, reaction products.

The products C and D, plus unreacted amounts of A and B and the inert I, leave the reactor at a temperature t_P. The study of heat effects in reacting systems always starts by writing the equations for the material balances and the equation for the energy balance and solving first, if at all possible, the material balances. In some situations, it may be necessary to solve simultaneously the material and energy balances. In the discussion that follows, we concentrate on how to solve the energy balance equation assuming the extent of reaction to be known.

For the energy balance, we start from Equation 21.1, written either for a certain lapse of time or for a certain amount of a given stream. The value of Δh is usually calculated assuming four separate steps:

1. The amounts of materials in the feed that actually react are taken from their temperature at the entrance conditions to a common temperature T conveniently chosen.
2. The reactions are carried out at the temperature T, to the extent that they actually occur in the system.
3. The products of reaction are taken from T to the temperature t_P of the product stream.
4. The unreacted reactants plus inerts are taken from their temperature at the feed to t_P. Each of these steps is described in detail below.

For simplicity, the discussion is presented in terms of flows, F (mol/h). Other bases may also be used.

Step 1: We first arbitrarily select a temperature at which the reactions will be carried out. In most cases, required data are tabulated at 25°C; thus, by choosing the reaction temperature as $T = T_0 = 298.15$ K, the calculations for step 2 are greatly simplified. In any case, designating by T the chosen temperature, the enthalpy change for step 1 is

$$\Delta h_1 = \sum_i F_i^R \overline{C_{P_i}} (T - T_i) \tag{21.17}$$

where the sum is taken over all the reactants in the amounts F_i^R that actually react, each at its temperature T_i.

Step 2: In this step, we calculate the standard heat of reaction for each reaction that has taken place, being careful to express the result of each reaction per mole of a particular compound k, arbitrarily chosen but clearly specified. This standard heat of reaction is calculated with Equation 19.11 if $T = T_0$ or with Equation 19.15 if $T \neq T_0$. In any case, for step 2,

$$\Delta h_2 = \sum_r \left[F_k \Delta H_T^\theta (\text{per mol of k}) \right]_r \tag{21.18}$$

where F_k denotes the moles of the compound k, chosen as reference, that are actually participating in the particular reaction r and the sum is carried out over all reactions.

The base for F_k (per hour, per 100 lbs of product, etc.) must be the same as the base used for F_i in Equation 21.17.

Step 3: In this step, all the products of the reactions, designated generically by the index j, are taken from the temperature T to the temperature T_P of the product stream:

$$\Delta h_3 = \sum_j F_j \overline{C_{P_j}} (T_P - T) \tag{21.19}$$

Step 4: Finally, all the unreacted reactants plus inerts entering into the feed, in the amounts F_i^I, are taken from their respective temperatures at entrance T_i to T_P.

$$\Delta h_4 = \sum_i F_i^I \overline{C_{P_i}} (T_P - T_i) \tag{21.20}$$

Note that the value of C_{Pi} for step 1 is the mean heat capacity in temperature range $(T_i - T)$, while in step 4 it is its value in the range $(T_P - T_i)$. These two values are not necessarily the same.

Finally, Equation 21.1 can be rewritten as

$$q = \Delta h_1 + \Delta h_2 + \Delta h_3 + \Delta h_4 \tag{21.21}$$

One type of reaction for which the heat effects are of special importance is the combustion of organic compounds. These reactions are so important that their standard heats of reaction at $T_0 = 298.15$ K are available in tabulations. For example, values for the combustion of carbon, carbon monoxide, and other inorganic compounds are reported in Table 2-178, pages 2-186 to 2-194, of Perry's handbook [1]. Table 2-179, pages 2-195 to 2-200, contains the same information for hydrogen and organic compounds. It is important to observe that the values reported in this table are the negative values of the heat of combustion, that is, $\Delta H_{c,T_0}$, and that it is necessary to decide if the product of combustion is liquid water or gaseous water. In general, for an organic compound of the form $C_nH_{m(g)}$, the heat of combustion to produce liquid water refers to the reaction

$$\text{A. } C_nH_{m(g)} + \left(n + \frac{m}{4}\right)O_{2(g)} \rightarrow nCO_{2(g)} + \frac{m}{2}H_2O_{(l)} \quad \Delta H_{c,C_nH_{m(g)},T_0}$$

The combustion of $C_{(s)}$ or $H_{2(g)}$ are just special cases of this reaction and correspond to the reactions of formation of $CO_{2(g)}$ or $H_2O_{(l)}$, respectively:

$$\text{B. } C_{(s)} + O_{2(g)} \rightarrow CO_{2(g)} \qquad \Delta H^0_{f,CO_{2(g)},T_0} = \Delta H^0_{c,C_{(s)},T_0}$$

$$\text{C. } H_{2(g)} + \frac{1}{2}O_{2(g)} \rightarrow H_2O_{(l)} \qquad \Delta H^0_{f,H_2O_{(l)},T_0} = \Delta H^0_{c,H_{2(g)},T_0}$$

We recall that, by convention, the heats of formation of the elements in their elemental standard state at $T_0 = 298.15$ K are considered to be equal to zero.

Adding reaction B multiplied by n to reaction C multiplied by $(m/2)$ and subtracting reaction A, we obtain

$$nC_{(s)} + \frac{m}{2}H_{2(g)} \rightarrow C_nH_{m(g)} \qquad \Delta H^0_{f,C_nH_{m(g)}}$$

Thus,

$$\Delta H^0_{f,C_nH_{m(g)},T_0} = n\Delta H^0_{f,CO_{2(g)},T_0} + \frac{m}{2}\Delta H^0_{f,H_2O_{(l)},T_0} - \Delta H^0_{c,C_nH_{m(g)},T_0} \qquad (21.22)$$

For the case of the standard heat of formation of a liquid compound $C_nH_{m(l)}$, the last term of the right-hand side of Equation 21.22 must be $\Delta H^0_{c,C_nH_{m(l)},T_0}$. Similarly, when one chooses to use the value of the heat of combustion of the compound C_nH_m, either liquid or gas, referred to the formation of water gas instead of liquid water, the value of $\Delta H^0_{f,H_2O_{(l)},T_0}$ in Equation 21.22 must be replaced by $\Delta H^0_{f,H_2O_{(g)},T_0}$.

Table 2-179 of Perry's handbook [1] contains values of ideal gas enthalpy of formation and the heats of combustions for organic compounds with gaseous water as one of the products. Both of these values can be used if one takes into account the enthalpies of vaporization.

If the compound of interest is not included in either of these two tables, it is possible in most cases to use a group contribution method to obtain the value of its heat of formation. Values of the group contributions for the method of Domalski–Hearing are given in Table 2-343, pages 2-479 to 2-484, of Perry's handbook [1]. Other group methods are also available.

As a final remark, it is worth mentioning here that for reactions involving only organic compounds for which the heats of combustion are available, it is possible to skip the step of calculating the heat of formation of each compound in order to evaluate the standard heat of reaction $\Delta H^{\theta}_{T_0}$. According to Equation 19.11, the standard heat of reaction or standard enthalpy change of reaction is evaluated as

$$\Delta H^{\theta}_{T_0} \equiv \sum_{k}^{c} v_k \, \Delta H^0_{f,k,T_0} \qquad (19.11)$$

We recall that in Equation 19.11, v_k is positive for products and negative for reactants. The standard heats of combustion refer to the breaking down of an organic compound into CO_2 and H_2O by the action of oxygen, while the standard heat of formation refers to the formation of the compound starting from the elements in their standard states. In addition, as the atoms of each element forming the organic molecules on both sides of the stoichiometric equation are balanced, the terms including the heat of formation of CO_2 and of H_2O cancel out. Thus,

$$\Delta H^{\theta}_{T_0} \equiv -\sum_{k}^{c} v_k \Delta H^0_{c,k,T_0} \qquad (21.23)$$

The sum on the right-hand side of Equation 21.23 includes the values of the standard heat of combustion of all compounds participating in the reaction. Thus, if $H_{2(g)}$ or $C_{(s)}$, as such, participate in the reaction, their heats of combustion must also be included. In addition, the heats of combustion must all refer to the formation of either liquid water or water gas; in no case should the information be mixed.

Example 21.6: Calculation of the Heat of Formation

Calculate the value of the heat of formation of propylene by the following methods: (a) using heats of combustion with $H_2O_{(g)}$ as a product, (b) using heats of combustion with $H_2O_{(l)}$ as a product, and (c) using Domalski–Hearing's group method.

Solution

1. From Table 2-179, Perry's handbook [1], the reported heat of combustion of propylene is with $H_2O_{(g)}$ as a product. Thus,

$$\Delta H^0_{c,C_3H_{6(g)}} = -1926.2 \, kJ/mol$$

$$\Delta H^0_{c,H_{2(g)}} = \Delta H_{f,H_2O_{(g)}} = -241.826 \, kJ/mol$$

$$\Delta H_{c,C_{(s)}} = \Delta H_{f,CO_{2(g)}} = -393.514 \, kJ/mol$$

Using Equation 21.21 with $n = 3$, $m = 6$,

$$\Delta H_{f,C_3H_6\,(g)} = (3)(-393.514) + \left(\frac{6}{2}\right)(-241.826) - (-1926.2)$$

$$= 20.18 \, kJ/mol$$

The calculated value compares well to the one reported in Table 2-179 [1] of 20.23 kJ/mol.

2. From [3], we obtain a value of heat of combustion with $H_2O_{(l)}$ as a product.

$$\Delta H^0_{c,C_3H_6(g)} = -2057.8 \, kJ/mol$$

$$\Delta H^0_{c,H_2(g)} = \Delta H_{f,H_2O(l)} = -68.3174 \, kJ/mol$$

$$\Delta H^0_{f,C_3H_6(g)} = (3)(-393.514) + \left(\frac{6}{2}\right)(-285.84) - (-2057.8)$$

$$= 19.74 \, kJ/mol$$

3. Domalski and Hearing group contribution method [4].
   To apply this method, it is necessary to understand its nomenclature. The groups are denoted by the atom other than hydrogen joined with

TABLE 21.6
Calculation of ΔH_f^0 Propene from the Domalski and Hearing Group Contribution Method

Groups	ΔH_f^0 (kJ/mol)
C_d-(2H)	26.32*
C_d-(H)(C)	36.32
C-(3H)(C)	-42.26
$CH_2 = CH-CH_3$	20.38

Note: Observe (and correct) error of the sign in Perry's Table 2-343 [1]. The correct sign can be found in the original source [4]. In the sixth edition of Perry's Table 3-335, for the group contribution method of Verma and Doraiswamy, there is an error of sign for the $-CH_3$ group.

a line to the other atoms to which it is joined in the molecule. For example, C–(3H)(C) means "atom of carbon joined to another atom of carbon and to three hydrogen atoms," that is, the group $-CH_3$. When an atom of carbon has a double bond to another atom of carbon, it is denoted by C_d and only the other two connections are specified. For example, C_d–(H)(C) means that the other two connections are carbon atoms (joined to others by single bonds) and an atom of hydrogen. On the other hand, C_d–(2H) means that a carbon atom is connected to a double bond and two hydrogen atoms. Table 21.6 reports the appropriate values from Table 2-343 of Perry's handbook [1]. For propylene, the structure is CH_2=CH–CH_3. The value obtained is 20.38 kJ/mol.

Example 21.7: Calculation of Standard Heats of Reaction

Obtain the values of the standard heat of reaction at 298.15 K for the following reactions:

1. (Propylene) + (Chlorine) = (Allylchloride) + (Hydrochloric acid)

$$CH_2 = CH - CH_3 + Cl_2 \rightarrow CH_2 = CH - CH_2Cl + HCl$$

2. (Propylene) + (Chlorine) = 1,2-Dichloropropane

$$CH_2 = CH - CH_3 + Cl_2 \rightarrow CH_2 Cl - CHCl - CH_3$$

Solution
In order to use Equation 19.11 or 21.22, we need the values of the heats of formation or combustion for allyl chloride and 1,2-dichloropropane, and these values are not available in Table 2-179 of Perry's handbook [1].

In order to use the Domalski–Hearing group contribution method, we need the values of the contributions of all groups, and in Table 2-343 [1], the contribution for the group Cl is not included. Benson's method, as reported in Table 3.4

in [5], contains enough information to estimate the standard heat of formation for 1,2-dichloropropane and allyl chloride (see Tables 21.7 and 21.8, respectively) in the gaseous state.

$$\Delta H^0_{f,gas} \text{ (1,2-dichloropropane)} = -73.21 \text{ kJ/mol}$$

This is a first approximation to the result, because at 25°C, 1,2-dichloropropane is in a liquid state. The Benson's method, on the other hand, provides estimates for ideal gas properties. Therefore, it is necessary to take into account the heat of vaporization ΔH_{vap} [6]. Thus,

$$\Delta H^0_{f,liq.} \text{ (1,2-dichloropropane)} = -173.21 - 35.26 = -208.47 \text{ kJ/mol}$$

$$\Delta H^0_{f,gas} \text{ (allyl chloride)} = -6.49 \text{ kJ/mol}$$

As above, at 25°C allyl chloride is also a liquid. Thus, an analogous correction has to be made:

$$\Delta H^0_f \text{ (allyl chloride)} = -6.49 - 29.04 = -35.53 \text{ kJ/mol}$$

TABLE 21.7

Calculation of ΔH^0_f of 1,2-Dichloropropane (CH_2Cl–$CHCl$–CH_3) from the Benson Group Contribution Method

Groups	ΔH^0_f (kJ/mol)
CH_2Cl–C	−69.07
CHCl–(2C)	−61.95
CH_3–(C)	−42.19
CH_2Cl–CHCl–CH_3	−173.21

TABLE 21.8

Calculation of ΔH^0_f of Allyl Chloride (CH_2Cl–CH =CH_2) from the Benson Group Contribution Method

Groups	ΔH^0_f (kJ/mol)
CH_2Cl–(=C)	−68.65
=CH–(C)	35.96
=CH_2	26.20
CH_2Cl–CHCl–CH_3	−6.49

From Table 2-178 of Perry's handbook, for HCl,

$$\Delta H^0_{f,HCl} = -22.063\,\text{kcal/mol} = -92.31\,\text{kJ/mol}$$

and for propylene, from Example 21.6,

$$\Delta H_{f,C_3H_6(g)} = 20.23\,\text{kJ/mol}$$

For reaction 1, from Equation 19.11, in the gaseous state approximation, we get

$$\Delta H^\theta_{T_0,gas} = (-92.31) - 6.49 - 20.23 = -119.37\,\text{kJ/mol}$$

and

$$\Delta H^\theta_{T_0,liq.} = (-92.31) - 35.53 - 20.23 = -148.07\,\text{kJ/mol}$$

taking into account the liquid state.
For reaction 2, correspondingly we obtain the following results:

$$\Delta H^\theta_{T_0,gas} = (-173.21) - 20.23 = -193.44\,\text{kJ/mol}$$

$$\Delta H^\theta_{T_0,liq.} = (-208.47) - 20.23 = -228.70\,\text{kJ/mol}$$

Example 21.8: Heat Effects in Reacting Mixtures

A gas mixture containing 80 mol% propane and 20 mol% butane is burned with 50% excess air. Assuming that the combustion is complete, calculate the heat per kilogram of gas that is transferred to a boiler if the initial gas mixture and the air enter the burner at 20°C and the products of combustion enter the stack at 180°C. The boiler heating chamber is at atmospheric pressure.

Solution
Material Balance
The line of reasoning followed below is quite standard in combustion problems, and for this reason, it is explained in detail. In this case, we take as a basis of calculation 1 kg of gas entering the burner. The molecular weights of propane and butane are, respectively, 44.09 and 58.12. One mole of mixture weighs [(0.8) (44.09) + (0.2)(58.12)] = 46.90 g, and thus in 1000g (l kg) we have 21.32 moles of mixture, that is, 17.06 moles of propane and 4.26 moles of butane. The combustion reactions are

$$C_3H_8 + 5O_2 = 3CO_{2(g)} + 4H_2O_{(g)}$$

$$C_4H_{10} + \frac{13}{2}O_2 = 4CO_{2(g)} + 5H_2O_{(g)}$$

The complete burning of 17.06 moles of propane consumes 85.30 moles of O_2 and produces 51.18 moles of CO_2 and 68.24 moles of H_2O. Similarly, for 4.26 moles of butane, the respective amounts are 27.72, 17.06, and 21.32 moles. Thus, the total production of CO_2 and H_2O is 68.24 and 89.56 moles, respectively. The stoichiometric requirement of O_2 is 113.02 moles. Using 50% excess air means feeding 169.52 moles of O_2, from which 113.02 react and 56.51 pass the burner unreacted. The amount of nitrogen fed to the system with the air is (0.79/0.21) (169.52) = 637.76 moles. These results are presented in Table 21.9.

Energy Balance
We choose to carry out the reaction at $T_0 = 298.15$ K and follow the four steps given by Equations 21.16 through 21.20. For steps 1, 3, and 4, we need average values of the specific heats of the compounds. Hence, with the help of information from Section C, page A-35, of reference [5], Table 21.10 can be prepared. On the right-hand side of Table 21.10, we have included the mean values of c_P^* at an average temperature for steps 1, 3, and 4.

TABLE 21.9
Data for Example 21.8

		(mol/kg fuel gas)			
Compound	Feed Stream	Consumed	Produced	Unreacted	Product Stream
$C_3H_8(g)$	17.06	17.06	0	0	0
$C_4H_{10}(g)$	4.26	4.26	0	0	0
$O_2(g)$	169.52	113.02	0	56.51	56.51
$N_2(g)$	637.76	0	0	637.76	637.76
$CO_2(g)$	0	0	68.24	0	68.24
$H_2O(g)$	0	0	89.56	0	89.56

TABLE 21.10
Data Used in the Energy Balance of Example 21.8

$$c_P^*/R = a + bT + cT^2 + dT^3 + eT^4, T(K)$$

						(cal/mol)		
						$\left(\bar{c}_P^*\right)_1$	$\left(\bar{c}_P^*\right)_3$	$\left(\bar{c}_P^*\right)_4$
Compound	a	$b \cdot 10^3$	$c \cdot 10^5$	$d \cdot 10^8$	$e \cdot 10^{11}$	20°C–25°C	25°C–180°C	20°C–80°C
C_3H_8	3.847	5.131	6.011	−7.893	3.079	17.50		
C_4H_{10}	5.547	5.536	8.057	−10.571	4.134	23.47		
O_2	3.63	−1.794	0.658	−0.601	0.179	6.96		
N_2	3.539	−0.261	0.007	0.157	−0.099			7.17
CO_2	3.259	1.356	1.502	−2.374	1.056		9.58	7.00
H_2O	4.395	−4.186	1.405	−1.564	0.632		8.17	

Step 1: Preheating Reactants from 20°C to 25°C

$$\Delta H_1 = [(17.06)(17.50) + (4.26)(23.33) + (112.99)(7.01)](298.15 - 293.15)$$
$$= +6.0 \text{ kcal/kg fuel}$$

Step 2: Combustion
From Table 2-179 of Perry's handbook [1], with $H_2O_{(g)}$ as a product,

$$\Delta H^0_{c,C_3H_8} = -2.04311 \cdot 10^9 \text{ J/kmol} = -488.315 \left(\text{kcal/mol } C_3H_8\right)$$

$$\Delta H^0_{c,C_4H_{10}} = -2.65732 \cdot 10^9 \text{ J/kmol} = -635.115 \left(\text{kcal/mol } C_3H_8\right)$$

$$\Delta H_2 = (17.06)(-488.315) + (4.26)(-635.115) = -11{,}039 \text{ kcal/kg fuel}$$

Step 3: Heating Products from 25°C to 180°C

$$\Delta H_3 = [(68.24)(9.48) + (89.56)(8.19)](453.15 - 298.15) = +214.0 \text{ kcal/kg fuel}$$

Step 4: Heating Nonreacting Gases from 20°C to 180°C

$$\Delta H_4 = [56.51)(7.17) + (637.73)(7.00)](453 - 293) = +778.9 \text{ kcal/kg fuel}$$

Thus, the total energy produced per kilogram of fuel, using Equation 21.20, is

$$Q = 11{,}039 + 6 + 214 + 779 = 10{,}254 \text{ kcal/kg fuel}$$

Example 21.9: Production of Allyl Chloride in an Adiabatic Continuous Stirred Tank Reactor

For the production of allyl chloride from propylene and chlorine, J. M. Smith [9] has shown that the best yields are obtained with an adiabatic continuous stirred tank reactor. In a given run of a pilot plant reactor with a feed of 0.170(lb-mol/h) of Cl_2 and 0.680 (lb-mol/h) of propylene, 70% of the Cl_2 feed is converted to allyl chloride and 14% is converted to the undesired by-product 1,2-dichloropropane. The small conversion to other by-products can be ignored. The reactants are initially at 20°C, and after being preheated separately, they enter the adiabatic reactor at 200°C. Calculate (1) the total heat effect of the process, in (kcal/h) and (2) the temperature of the gas products leaving the reactor.

Solution
Material Balance
The reactions are those given in Example 21.7. In 1 h of operation, 0.170 lb-moles of Cl_2 and 0.680 lb-moles of propylene, C_3H_8, are fed to the reactor of the chlorine feed, (0.70)(0.170) = 0.119 lb-moles react to produce 0.119 lb-mol of allyl chloride and 0.119 lb-mol of hydrochloric acid, and (0.14)(0.170) = 0.024 lb-mol react to produce 0.024 lb-mol of 1,2-dichloropropane. The unreacted chlorine, 0.027 lb-mol, leaves with the product stream. For each lb-mol of chlorine reacted, in either reaction, 1 lb-mol of propylene is consumed. Thus, (0.119 + 0.024) = 0.143 lb-moles of propylene react and (0.680 − 0.143) = 0.537 lb-moles go to the product stream unreacted. These results are summarized in Table 21.11.

TABLE 21.11

Data for Material Balance for Example 21.9

Compound	Feed Stream	Consumed	Produced	Inerts and Unreacted	Product Stream
			F (lb-mol/h)		
C_3H_6	0.680	0.143	0	0.537	0.537
Cl_2	0.170	0.143	0	0.027	0.027
C_3H_5Cl	0	0	0.199	0	0.199
HCl	0	0	0.199	0	0.199
$C_3H_6Cl_2$	0	0	0.024	0	0.024

Energy Balance

As most of the process happens at fairly high temperatures, we start with the approximation of the gaseous state.

In the preheating stage, the reactants are taken from 293.2 K to 473.2 K. In the adiabatic reactor, we carry out the reaction at 298.2 K, and then we take all products of the reaction from $T_0 = 298.2$ K to T_p (unknown), and also the unreacted reactants from 473 K to T_p. Thus, for the preheating stage we need c_p for C_3H_6 and Cl_2 for the range 293 K–473 K. For the same compounds, we need c_p for the range 298.2 to 473 K and for the range 473 K to T_p. For the products of reaction we need c_p for the range 298.2 K to T_p. From Example 21.7, we know that both reactions are exothermic, so the temperature T_p is higher than 473 K. We can guess a temperature and use approximated values of c_p. Subsequently, using any optimizing technique, Solver in Excel, for example, we calculate the required value of T_p. Say that we assume $T_p = 800$ K. From Section C, page A-35 of [5], we get the information needed to prepare Table 21.12.

Preliminary Step: Preheating Reactants from 20°C to 200°C

Converting the pound-moles to moles, 453.6 (mol/lb-mol), and using the average values of $\left(\overline{c_P^*}\right)$,

$$\Delta H_0 = \Delta H_{preheating} = [(0.680)(18.48) + (0.170)(8.32)] \, (473 - 293)(0.4536) = 1141.8$$
$$\text{(kcal/h)}$$

Since the reactor is adiabatic, the total heat effect of the process is $Q = 1142$ (kcal/h); this heat enters the system ($Q > 0$).

Step 1: Cooling the Amounts of Reactants That Reacted from 200°C to 25°C

From Equations 21.17 through 21.19,

$$\Delta H_1 = \Delta H_{cooling} = [(0.143)(18.58)+(0.143)(8.33)] \, (298-473)(0.4536) = -373.2 \text{ (kcal/h)}$$

Step 2: Reactions a and b at 25°C

For ΔH_2, we use the calculated heats of reactions from Example 21.7, noting that the amounts in the table of material balances are in pound-moles.

$$\Delta H_2 = \Delta H_{reaction} = [(0.119)(-35.39) + (0.024)(-54.66)](453.6) = -2500.4 \text{ (kcal/h)}$$

TABLE 21.12
Data Used in the Energy Balance of Example 21.9

	$c_P^*/R = a + bT + cT^2 + dT^3 + eT^4, T(K)$					(c_P^*) (cal/mol)		
Compound	a	$b \cdot 10^3$	$c \cdot 10^5$	$d \cdot 10^8$	$e \cdot 10^{11}$	20°C	25°C	200°C
C_3H_6	3.834	3.893	4.688	−6.0130	2.283	15.217	15.398	21.752
Cl_2	3.056	5.3708	−0.8098	0.5693	−0.15256	8.081	8.100	8.567
$C_3H_5Cl^*$	4.365	9.895	5.366	−7.708	3.12	20.201	20.446	28.732
HCl	3.827	−2.936	0.879	−1.031	0.439	6.944	6.944	7.022
$C_3H_6Cl_2$	1.697	40.582	−2.247	−0.038	0.377	23.212	23.486	31.828

Note: The asterisk indicates approximated as 1-chloropropane.

Step 3: Heating the Products from 25°C to T_p

$$\Delta H_3 = [(0.119)(30.0) + (0.119)(7.1) + (0.024)(32.74)]\,(0.4536)(T_p - 298)$$
$$= 2.36\,(T_p - 298)\,(kcal/h)$$

Step 4: Heating Unreacted Reactants from 200°C to T_p

$$\Delta H_4 = [(0.537)(26.3)+(0.027)(8.7)]\,(0.4536)(T_p - 473) = 6.50\,(T_p - 473)\,(kcal/h)$$

For the final energy balance, since $Q = 0$,

$$0 = 1141.8 - 373.2 - 2034.8 + 2.36\,(T_p - 298) + 6.50\,(T_p - 473)$$

$$8.86\,T_p = 5044$$

$$T_p = 570\,K$$

Using $T_p = 570$ K, the values of c_p that depend on T_p can be revaluated. They are presented in Table 21.13.
Thus,

$$\Delta H_3 = [(0.119)(26.6) + (0.119)(7.0) + (0.024)(29.5)]\,(0.4536)(T_p - 298)$$
$$= 2.135\,(T_p - 298)\,(kcal/h)$$

$$\Delta H_4 = [(0.537)(23.4) + (0.027)(8.6)]\,(0.4536)(T_p - 473) = 5.805\,(T_p - 473)\,(kcal/h)$$

$$0 = 1141.8 - 373.2 - 2034.8 + 2.135\,(T_p - 298) + 5.805\,(T_p - 473)$$

$$7.94\,T_p = 4648.2$$

$$T_p = 585.4\,K$$

This process can be repeated. The convergence is slow, as the direct optimization method gives the value of $T_p = 661$ K.

TABLE 21.13

Average Heat Capacity of Compounds from Example 21.9 as a Function of Temperature T_p and the Approximation Used for Allyl Chloride Data

	$(\bar{c_P^*})$ (cal/mol)					
	$T_p = 570$ K		$T_p = 661$ K		$T_p = 678$ K	
Compound	Step 3	Step 4	Step 3	Step 4	Step 3	Step 4
C_3H_6	—	23.35		24.67		24.89
Cl_2	—	8.63		8.68		8.68
C_3H_5Cl	26.58[a]	—	28.16[a]		24.29[b]	
HCl	7.01	—	7.04		7.04	
$C_3H_6Cl_2$	29.48	—	30.94		31.18	

[a] Approximated as 1-chloropropane.
[b] Benson's group method approximation.

When Benson's approximation for the (c_P^*) value of allyl chloride is used, the result is $T_p = 678$ K. Table 21.13 compares different final values of (c_P^*).

Releasing the first assumption of products being in the gaseous state changes the final result to 647 K. This difference can be understood when we realize that the heats of vaporization of products that have to be subtracted from their respective heats of formation to take into account their liquid state at 25°C in step 2 need to be added again in step 3 when the products are heated above their boiling points. Thus, the only difference between the approaches is heating of liquid products from 25°C to their boiling temperatures. All information required for these calculations can be found in [8,9].

REFERENCES

1. Green, D. W., and Perry, R. H. 2007. *Perry's Chemical Engineers' Handbook*, 8th ed. Blacklick, OH: McGraw-Hill Professional Publishing.
2. Douglas, T. B., and Dever, J. L. 1954. Thermal conductivity of molten materials. II. The heat capacity of anhydrous sodium hydroxide from 0°C to 700°C. *J. Res. Natl. Bur. Stand.* 53(2): Research Paper 2519.
3. Wiberg, K. B., and Fenoglio, R. A. 1968. Heats of formation of C_4H_6 hydrocarbons. *J. Am. Chem. Soc.* 90: 3395–3397.
4. Domalski, E. S., and Hearing, E. D. 1988. Estimation of the thermodynamic properties of hydrocarbons at 298.15 K. *J. Phys. Chem. Ref. Data* (4): 1637–1678.
5. Poling, B. E., Prausnitz J. M., and O'Connell, J. P. 2001. *The Properties of Gases and Liquids*, 5th ed. New York: McGraw-Hill.
6. Varushchenko, R. M., Druzhinina, A. I., Kuramshina, G. M., and Dorofeeva, O. V. 2007. Thermodynamics of vaporization of some freons and halogenated ethanes and propanes. *Fluid Phase Equil.* 256(1–2): 112–122.

7. Smith, J. M., 1981. *Chemical Engineering Kinetics*, 3rd ed. New York: McGraw-Hill.
8. National Center for Biotechnology Information. PubChem compound database. CID = 7850. Available at https://pubchem.ncbi.nlm.nih.gov/compound/7850 (accessed November 16, 2015).
9. National Center for Biotechnology Information. PubChem compound database. CID = 6564. Available at https://pubchem.ncbi.nlm.nih.gov/compound/6564 (accessed November 16, 2015).

22 Adsorption of Gases on Solids

TWO-DIMENSIONAL THERMODYNAMIC APPROACH

There are many different ways of studying adsorption [1–5]. Here, we consider a purely thermodynamic approach. As is usual in thermodynamics, the idea is to systematize the information that can be obtained from the study of the phenomena for pure compounds with the purpose of using it for interpolation and extrapolation, as well as for extension to the treatment of mixtures.

The Gibbs treatment is a simple way of studying the adsorption of a gas phase on a solid surface. The solid surface could be that of a porous solid or that of a granular adsorbent material. The treatment is also applicable, *mutatis mutandis*, to the adsorption of a solute from a liquid mixture or from an inert gas phase.

At atmospheric pressure, or at pressures not higher than 5 atm, gases can be safely considered to behave as an ideal gas. For an ideal gas at pressure, P, and temperature, T, the molar density, d_g, is given by

$$d_g = \frac{n_g}{V_g} = \frac{P}{RT}$$

Consider a vessel in which n_g moles of gas are charged together with a solid adsorbent material. The gas occupies a volume V_g, and the solid occupies a volume V_s. In the case of adsorption in a porous solid, V_g is usually considered to be the void volume of the solid. At equilibrium, the gas phase is at a pressure P at the temperature T of the system.

Due to the effect of adsorption, there is a gradual increase in the density of the gas in the proximity of the solid surface. Thus, it is not a simple matter to decide what is the region of adsorbed gas and what is the region of bulk nonadsorbed gas. The Gibbs treatment of adsorption solves this problem elegantly by defining an ideal nonadsorbing system, indicated by a superscript i. In the ideal nonadsorbing system, the same gas volume V_g, at the pressure P and temperature T of the actual system, has the same molar density d_g as that of the bulk gas, that is, the gas that is far from the adsorbing surface, in the actual system. Thus, the number of moles of gas, n_g^{id}, that would be present in the ideal nonadsorbing system is

$$n_g^{id} = V_g d_g \tag{22.1}$$

As the actual adsorbed system has a higher gas density in the proximity of the solid surface, the number of moles of gas in the actual adsorbed system will be larger than the number of moles of gas in the ideal nonadsorbing system.

$$n_g > n_g^{id}$$

The moles of gas adsorbed, n_g^{ads}, are then given by

$$n_g^{ads} = n_g - n_g^{id} \tag{22.2}$$

For its use in further derivations, this equation is better written as

$$n_g^{id} = n_g - n_g^{ads} \tag{22.2a}$$

From this form of the equation, it is clear that the number of moles of gas in the bulk gas phase, far from the effects of the solid surface, is in fact equal to the number of moles of gas in the ideal nonadsorbing system.

$$n_g^{bulk} = n_g^{id} \tag{22.3}$$

In the adsorbed system there are three phases present: (1) the solid phase, (2) the adsorbed gas phase, and (3) the bulk gas phase. For each phase at internal equilibrium, the Gibbs–Duhem equation, Equation 6.9, applies:

$$\sum n_i d\mu_i = VdP - SdT \tag{22.4}$$

Adding these equations for the three phases present at equilibrium, we obtain

$$n_s d\mu_s + n_g^{ads} d\mu_g^{ads} + n_g^{bulk} d\mu_g^{bulk} = \left(V_s + V_g\right)dP - \left(n_s s_s + n_g^{ads} s_g^{ads} + n_g^{bulk} s_g^{bulk}\right)dT$$

The left-hand side of this equation can be simplified considering that at equilibrium the chemical potential of the gas adsorbed is equal to the chemical potential of the bulk gas. Thus,

$$d\mu_g^{ads} = d\mu_g^{bulk} = d\mu_g$$

In addition, the total number of moles of gas is equal to the sum of the moles of gas adsorbed and gas in the bulk. On the right-hand side, we identify the number of moles of gas in the bulk phase with the number of moles in the ideal nonadsorbing system and recall that the entropy of the gas in the bulk phase is equal to the entropy of the gas at pressure P and temperature T of the system. Therefore, we rewrite this equation as

$$n_s d\mu_s + n_g d\mu_g = \left(V_s + V_g\right)dP - \left(n_s s_s + n_g^{ads} s_g^{ads} + n_g^{id} s_g\right)dT$$

For the ideal nonadsorbing system, the Gibbs treatment considers that although the entropy of the solid does not change by the superficial load of adsorbate, the enthalpy of the solid surface with adsorbate is different from the enthalpy of the solid surface without adsorbate. This difference is evident from the thermal effects of adsorption. The chemical potential of the solid is the difference between its molar enthalpy and the product of the absolute temperature times its molar entropy. Hence, to differentiate the chemical potential of the nonadsorbing solid surface from the chemical potential of the solid with adsorbate existing in the real adsorption system, the former is marked with a superscript 0. Repeating the exercise of adding the Gibbs–Duhem equations for the solid and the gas phase present in the ideal nonadsorbing system, we write

$$n_s d\mu_s^0 + n_g^{id} d\mu_g = \left(V_s + V_g\right)dP - \left(n_s s_s + n_g^{id} s_g\right)dT$$

Subtracting this equation for the ideal nonadsorbing system from the equation describing the actual adsorbing system gives

$$n_s\left(d\mu_s - d\mu_s^0\right) + \left(n_g - n_g^{id}\right)d\mu_g = -n_g^{ads} s_g^{ads} dT$$

Or, using Equation 22.2 and dividing by the number of moles of gas adsorbed, we get

$$\frac{n_s d(\mu_s - \mu_s^0)}{n_g^{ads}} + d\mu_g = -s_g^{ads} dT \qquad (22.5)$$

The Gibbs treatment introduces here three useful definitions. Considering that the total active area of the adsorbing solid is α, the number of moles of gas adsorbed per unit of surface area of adsorbent solid is defined as the *surface excess concentration* by

$$\Gamma \equiv \frac{n_g^{ads}}{\alpha}\left[\frac{\text{moles of adsorbate}}{\text{solid surface area}}\right] \qquad (22.6)$$

The reciprocal value of the surface excess concentration, that is, the area of solid surface per mole of adsorbate, is designated by

$$\bar{\alpha} \equiv \frac{\alpha}{n_g^{ads}} = \frac{1}{\Gamma} \qquad (22.7)$$

Finally, and most importantly, the "spreading pressure," or the energy per unit of surface area of solid that tends to spread the adsorbate over the surface, is defined as

$$\pi \equiv -\frac{n_s(\mu_s - \mu_s^0)}{\alpha} \qquad (22.8)$$

In differential form, this definition is written as

$$-\alpha\, d\pi = n_s\, d(\mu_s - \mu_s^0)$$

Dividing this expression by the number of moles of gas adsorbed, using Equation 22.7 and replacing in Equation 22.5, we obtain

$$-\bar{\alpha}\, d\pi + d\mu_g = -s_g^{ads} dT \qquad (22.5a)$$

So, remembering that at equilibrium the chemical potential of the bulk gas is equal to the chemical potential of the adsorbed gas, we rewrite Equation 22.5a as

$$d\mu_g^{ads} = \bar{\alpha}\, d\pi - s_g^{ads} dT \qquad (22.5b)$$

All the terms in Equation 22.5b correspond to terms of the adsorbed phase. From this equation, we obtain

$$\left(\frac{\partial \mu_g^{ads}}{\partial T}\right)_\pi = -s_g^{ads} \qquad (22.9)$$

and

$$\left(\frac{\partial \mu_g^{ads}}{\partial \pi}\right)_T = \bar{\alpha}\left[\frac{\text{solid surface area}}{\text{moles of adsorbate}}\right] \qquad (22.10)$$

It is illustrative to compare Equations 22.5b, 22.9, and 22.10 with the corresponding equations for a pure compound in "volumetric" thermodynamics. In volumetric thermodynamics, from Equation 7.8, for a pure compound we have

$$d\mu = vdP - sdT$$

$$\left(\frac{\partial \mu}{\partial T}\right)_P = -s$$

and

$$\left(\frac{\partial \mu}{\partial P}\right)_T = v\left[\frac{\text{volume of the container}}{\text{moles of gas}}\right]$$

These analogies suggest the identification of π with a spreading pressure for the case of the adsorbed gas in the two-dimensional system.

GIBBS TREATMENT FOR THE ISOTHERMAL ADSORPTION OF A PURE COMPOUND GAS AT A SOLID SURFACE

For an isothermal process of adsorption, Equation 22.5a can be written as

$$d\pi = \frac{1}{\alpha}\left(d\mu_g\right)_T = \Gamma\left(d\mu_g\right)_T$$

At constant temperature, for a change in the chemical potential of the gas in the bulk phase in terms of its fugacity, f_g, we write

$$(d\mu_g)_T = RT\, d\ln f_{g,T}$$

so,

$$d\pi = RT\, \Gamma d\ln f_{g,T}$$

Integrating between $\pi = 0$ at $f_g = 0, \mu_s = \mu_s^0$, and the value π at $f_g = \phi_g P, \mu_s \neq \mu_s^0$, we obtain

$$\pi = RT \int_0^{f_g} \Gamma d \ln f_{g,T} \qquad (22.11)$$

At this point, it is necessary to introduce some variables of practical use in adsorption studies.

$$\text{Specific surface area of adsorbing solid} = a \left[\frac{\text{solid surface area}}{\text{mass, mole or volume of solid}} \right]$$

$$\text{Loading of adsorbate onto the solid} = c \left[\frac{\text{mass, mole or volume of adsorbate}}{\text{mass, mole or volume of solid}} \right]$$

Both variables, a and c, must be referred to the same units for the amount of solid. For example,

$$c \left[\frac{\text{std cm}^3 \text{ of adsorbate}}{g \text{ of solid}} \right], \quad a \left[\frac{\text{solid surface area}}{g \text{ of solid}} \right]$$

In this case,

$$\Gamma = \frac{c}{a v_0} \left[\frac{\text{moles of adsorbate}}{\text{solid surface area}} \right] \qquad (22.12)$$

with the standard molar volume of gas at a pressure of 1 atm and at a temperature of 0°C,

$$v_0 = 22,400 \left[\frac{\text{std. cm}^3}{g \text{mol of gas}} \right]$$

or alternatively, use

$$c' \left[\frac{g \text{ of adsorbate}}{g \text{ of solid}} \right], \quad a' \left[\frac{\text{solid surface area}}{g \text{ of solid}} \right]$$

In this case,

$$\Gamma = \frac{c'}{a' M_w} \left[\frac{\text{moles of adsorbate}}{\text{solid surface area}} \right] \qquad (22.13)$$

where M_w is a molecular weight of the gas.

Using Equation 22.12, we write Equation 22.11 as

$$\pi = \frac{RT}{av_0} \int_0^{f_g} cd \ln f_{g,T} \tag{22.11a}$$

For convenience in the calculations, it is useful to separate the constant terms from the variables in the adsorption process and define

$$\Psi \equiv T \int_0^{f_g} cd \ln f_{g,T} \left[\frac{\text{std. cm}^3 \text{ of adsorbate} \cdot \text{Kelvins}}{g \text{ of solid}} \right] \tag{22.14}$$

The units of Ψ depend on the choice of units for c. For the case considered here, the relation between this new function and the spreading pressure is

$$\pi = \frac{R}{av_0} \Psi \tag{22.11b}$$

For the treatment of the adsorption of gas mixtures with the approach presented here, it is necessary to have plots of Ψ as a function of pressure for the different compounds forming the gas mixture. The first step for this purpose is to have constant temperature data, as a function of pressure, of the loading of the compounds adsorbing in the solid surface. Data of this type are either collected experimentally for different temperatures close to the temperature of interest [1–4] or obtained from computer simulations [5,6]. Figure 22.1 depicts a schematic plot of two of these

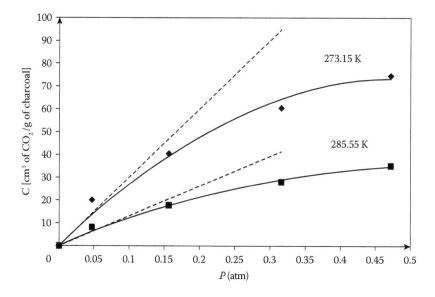

FIGURE 22.1 Example of adsorption isotherms of CO_2 in charcoal.

curves, called *adsorption isotherms*, for adsorption of CO_2 in charcoal. At constant temperature, the loading increases with an increase in pressure. As the temperature increases, the load of adsorbate decreases at constant pressure. From these constant temperature data, smoothed using one of many model equations [1–4], it is possible to evaluate Ψ as a function of pressure. Thus, the graphical representation presented here can be replaced by a numerical approach.

The integration of the right-hand side Equation 22.14 requires some discussion. The logarithm of the fugacity of the gas goes to minus infinity as the fugacity goes to zero at zero pressure. Thus, a transformation is necessary and we write Equation 22.14 as

$$\Psi \equiv T \int_{0}^{f_g} \frac{c}{f_g}\, df_{g,T} \tag{22.14a}$$

As the pressure goes to zero, both c and f_g go to zero, so the limiting value of the ratio can be obtained by the L'Hôpital rule:

$$\lim_{P \to 0} \frac{c}{f_g} = \left(\frac{dc}{dP} \right)_{P \to 0}$$

As shown in Figure 22.1, the value of the lower limit of the integrand is the slope at the origin of the adsorption isotherm. A graphical representation of the integrand at two temperatures is presented in Figure 22.2. The result of this type of integration, at two temperatures, is represented schematically in Figure 22.3. As the carbon dioxide gas used in these measurements behaves as an ideal gas at the conditions of the experiments, it was assumed that $\phi_g = 1$, that is, $f_g = P$

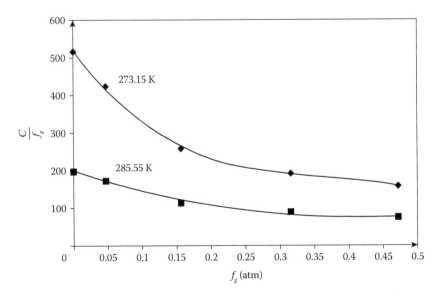

FIGURE 22.2 Plot of the ratio of loading of adsorbate gas over its fugacity as a function of fugacity and temperature in CO_2 in the charcoal system.

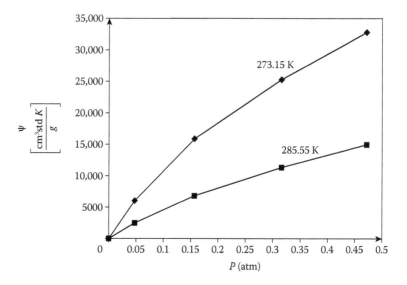

FIGURE 22.3 Pressure and temperature dependence of function ψ, Equation 22.14.

ISOTHERMAL ADSORPTION OF A GAS MIXTURE

For a total active area of the adsorbing solid α, the number of moles of gas i adsorbed per unit of surface area of adsorbent solid is

$$\Gamma_i \equiv \frac{n_i^{ads}}{\alpha} = \frac{n_i - n_i^{id}}{\alpha} \left[\frac{\text{moles of } i \text{ adsorbed}}{\text{solid surface area}} \right] \tag{22.15}$$

The superscript id is used here for the ideal nonadsorbing case, and the subscript i refers to compound i. In the case of a mixture of gases with a mole fraction y_i of compound i in the bulk gas phase, we have

$$n_i^{id} = V_g d_i = V_g y_i d_g$$

Here, we repeat exactly the treatment leading to Equation 22.5. The only difference is that in the Gibbs–Duhem equation, instead of the differential of the chemical potential of a single gas, we have now the sum of the chemical potential changes over all compounds. There is no difference for the term referring to the solid surface area. As in the discussion of the pure compound adsorption, we consider the isothermal case here and write

$$n_s(d\mu_s - d\mu_s^0) + \sum_k (n_k - n_k^{id})d\mu_k = 0\big|_T$$

or

$$n_s(d\mu_s - d\mu_s^0) + \sum_k n_k^{ads} d\mu_k = 0\big|_T$$

Dividing by the total number of moles of gas adsorbed,

$$n_g^{ads} = \sum_j n_j^{ads}$$

we obtain

$$-\bar{\alpha}\, d\pi + \sum_k x_k\, d\mu_k = 0\big|_T \tag{22.16}$$

In Equation 22.16, we have used the same definition of terms given by Equations 22.6 through 22.8. In addition, we have defined the mole fraction of compound k in the adsorbed phase as

$$x_k \equiv \frac{n_k^{ads}}{n_g^{ads}}$$

Equation 22.16 can be rewritten as

$$\sum_k x_k \left(\frac{\partial \mu_k}{\partial \pi}\right)_{T.n} = \bar{\alpha}\big|_T \tag{22.16a}$$

This form suggests the definition

$$\bar{\alpha}_k \equiv \left(\frac{\partial \mu_k}{\partial \pi}\right)_{T.n} \tag{22.17}$$

Thus, Equation 22.16a can be written as

$$\sum_k x_k \bar{\alpha}_k = \bar{\alpha}\big|_T \tag{22.16b}$$

Comparison of Equations 22.16b and 22.17 with the following equations used in volumetric thermodynamics,

$$\bar{v}_k \equiv \left(\frac{\partial \mu_k}{\partial P}\right)_{T.n}$$

and

$$\sum_k x_k \bar{v}_k = v\big|_T$$

suggests the interpretation of $\bar{\alpha}_k$ as a partial molar surface area occupied by the adsorbed compound k at temperature T and spreading pressure π.

Finally, we observe that from the definitions given by Equations 22.6 and 22.15, for the mole fraction of compound i in the adsorbed phase, we get

$$x_i = \frac{\Gamma_i}{\Gamma} \tag{22.18}$$

EXAMPLE OF THE GIBBS TREATMENT FOR THE ADSORPTION OF A GAS MIXTURE

We consider here the case of an ideal adsorption, that is, an adsorption that occurs without heat effects and having the property that the surface covered by the adsorbed mixture corresponds to the sum of the surface areas covered by the pure compounds at the same temperature and spreading pressure. This corresponds exactly with the analogy of forming an ideal mixture in volumetric thermodynamics, which happens without heat effects and without change in volume. In the case of forming an ideal mixture in volumetric thermodynamics, the partial molar volume of compound i in the mixture is equal to the pure compound molar volume at the same temperature and pressure. Thus, in surface thermodynamics, for ideal mixing in an adsorbed phase we write

$$\overline{\alpha}_{i(T,\pi)}^{ads,im} = \overline{\alpha}_{i(T,\pi)}^{*} \tag{22.19}$$

where the superscript * indicates the surface area of solid covered by the pure compound i at the same temperature and spreading pressure of the mixture. The ideal mixture considered here is indicated by a superscript im, and it should not be confused with Gibbs's ideal case of a nonadsorbing system. Following the treatment of an ideal mixture in volumetric thermodynamics, for an ideal adsorbed phase we have

$$f_i^{ads,im} = x_i f_{i(T,\pi)}^{\theta} \tag{22.20}$$

where, using the normal convention, the superscript θ indicates the standard state of pure adsorbed compound i at temperature T and spreading pressure π given by Equation 22.8. For the treatment of a nonideal adsorbed phase, for which experimental information may be available, the introduction of activity coefficients may be necessary. Assuming ideal mixing in the adsorbed phase, from Equation 22.16b, we write

$$\overline{\alpha}_{(T,\pi)}^{im} = \sum_k x_k \overline{\alpha}_{k(T,\pi)} \tag{22.21}$$

STATEMENT OF THE PROBLEM

Assuming ideal adsorption of a gas mixture onto a solid surface, obtain the loading of each adsorbate at a temperature T and a pressure P when the composition of the gas mixture is given in terms of the mole fraction y_i of the components. Adsorption isotherms for the pure compounds are available.

Solution

At equilibrium, assuming an ideal adsorbate phase,

$$f_i^{gas\,mixture} = f_i^{ads,im} = x_i f_{i(T,\pi)}^{\theta}$$

or

$$y_i \phi_i P = x_i f_{i(T,\pi)}^{\theta}$$

At low pressure, the fugacity coefficient ϕ_i can be set to unity. The standard state for the compound i in the adsorbed phase is conveniently set as equal to the pressure (or fugacity, if necessary) $P^*_{i(T,\pi)}$ required to give a spreading pressure π, that is, the corresponding value of ψ, when adsorbed as a pure compound at temperature T. At equilibrium, the spreading pressure is defined by Equation 22.8, and thus it is a property of the solid surface. This means it has the same value for all compounds adsorbed.

$$y_i P = x_i P^*_{i(T,\pi)}$$

or

$$x_i = \frac{y_i P}{P^*_{i(T,\pi)}} \tag{22.22}$$

Thus,

$$\sum \frac{y_i P}{P^*_{i(T,\pi)}} = 1 \tag{22.23}$$

Strategy for the Solution of the Problem

The pressure, P, the temperature, T, and the mole fractions of all compounds in the bulk gas phase, y_i, are known. Assume a value for ψ and from the isotherm corresponding to the temperature of the system (in a figure like Figure 22.3), evaluate the pressure P_i^* for each compound. Test the result obtained from Equation 22.23, and continue by trial and error until this equation is satisfied. If experimental data are smoothed by an analytical expression, the value of ψ can be obtained numerically by solving Equation 22.14a. Subsequently, using Equation 22.22, calculate the values of x_i for each compound in the adsorbed phase. At this point, the mole fractions of all adsorbates are known, but not their loadings c_i. From Equations 22.12 and 22.18, we write

$$c_i = x_i c$$

The value of the total loading c is not known. From Equations 22.7 and 22.12, we write

$$\bar{\alpha}_{(T,\pi)} = \frac{a v_0}{c_{(T,\pi)}} \tag{22.24a}$$

and for a pure compound,

$$\bar{\alpha}_{i(T,\pi)} = \frac{a v_0}{c_{i(T,\pi)}} \tag{22.24b}$$

For an ideal adsorption, according to Equation 22.16b,

$$\alpha^{im}_{(T,\pi)} = \sum_k x_k \bar{\alpha}^*_{k(T,\pi)}$$

Replacing Equations 22.24a and 22.24b in this expression, we obtain

$$\frac{1}{c_{(T,\pi)}} = \sum \frac{x_k}{c^*_{k(T,\pi)}}$$

The values of $c^*_{i(T,\pi)}$ for all pure compounds are obtained from Figure 22.2. With these values using the above equation, the values of the loadings c_i are obtained:

$$c_i = x_i c$$

THERMAL EFFECTS ON ADSORPTION

For a pure compound gas adsorbed onto a solid surface, at constant temperature and pressure,

$$\Delta h^{ads} = \left(h_g^{ads} - h_g \right) + \left(h_s^{ads} - h_s^0 \right) \tag{22.25}$$

In Equation 22.25, all terms are expressed per mole of gas adsorbed. For each phase, we know

$$\mu = g = h - Ts$$

or

$$h = \mu + Ts$$

For the adsorbed gas phase, per mole of gas adsorbed,

$$h_g^{ads} = \mu_g^{ads} + Ts_g^{ads}$$

In an ideal nonadsorbed system,

$$h_g = \mu_g + Ts_g$$

The contribution of the adsorbed phase per mole of adsorbate is

$$\left(h_g^{ads} - h_g \right) = T\left(s_g^{ads} - s_g \right) \tag{22.26}$$

For n_s moles of solid, remembering that adsorption does not change the entropy of the solid phase, we get

$$H_s^{ads} = n_s \mu_s + TS_s$$

$$H_s^0 = n_s \mu_s^0 + TS_s$$

So, subtracting and dividing by n_g^{ads}, we obtain

$$h_s^{ads} - h_s^0 = \frac{n_s\left(\mu_s - \mu_s^0\right)}{n_g^{ads}} = -\bar{\alpha}\pi \qquad (22.27)$$

where

$$\bar{\alpha} = \frac{\alpha}{n_g^{ads}} = \frac{1}{\Gamma}$$

and

$$\pi = \frac{-n_s\left(\mu_s - \mu_s^0\right)}{\alpha}$$

Substituting Equations 22.26 and 22.27 into Equation 22.25, we obtain

$$\Delta h^{ads} = T\left(s_g^{ads} - s_g\right) - \bar{\alpha}\pi$$

This equation is normally written as

$$-\Delta h^{ads} = T\left(s_g - s_g^{ads}\right) + \bar{\alpha}\pi \qquad (22.28)$$

Equation 22.28 can be transformed to a more practical form by eliminating the entropy term.

For the gas phase, we have

$$d\mu_g = -s_g\,dT + v_g\,dP$$

Taking the temperature derivative of both sides at constant spreading pressure π (or, equivalently, at constant Ψ) gives

$$\left(\frac{\partial\mu_g}{\partial T}\right)_\pi = -s_g + v_g\left(\frac{\partial P}{\partial T}\right)_\pi$$

But we know that

$$\left(\frac{\partial\mu_g^{ads}}{\partial T}\right)_\pi = -s_g^{ads} \text{ and } \mu_g = \mu_g^{ads}$$

So,

$$\left(\frac{\partial\mu_g^{ads}}{\partial T}\right)_\pi = -s_g^{ads} = -s_g + v_g\left(\frac{\partial P}{\partial T}\right)_\pi$$

Considering

$$v_g = \frac{zRT}{P}$$

we write

$$s_g - s_g^{ads} = \frac{zRT}{P}\left(\frac{\partial P}{\partial T}\right)_\pi = zRT\left(\frac{\partial \ln P}{\partial T}\right)_\pi$$

So, Equation 22.28 takes the form

$$-\Delta h^{ads} = zRT^2\left(\frac{\partial \ln P}{\partial T}\right)_\pi + \bar{\alpha}\pi \qquad (22.29)$$

This equation is sometimes further transformed using Equations 22.11b and 22.24a to obtain

$$-\Delta h^{ads} = zRT^2\left(\frac{\partial \ln P}{\partial T}\right)_\pi + \frac{R\Psi}{c} \qquad (22.30)$$

This important equation contains only measurable variables.

From Figure 22.3, we get the isosteric temperature–pressure dependence, as shown in Figure 22.4a.

This allows for determination of the derivative of Equation 22.30. If only two isotherms are available, we evaluate $\left(\dfrac{\partial \ln P}{\partial T}\right)_\pi$ as shown in Figure 22.4b. From Figure 22.1, we get c. Thus, $-\Delta h^{ads}$ is evaluated.

If many isotherms are available, Equation 22.30 should be transformed to give

$$-\Delta h^{ads} = zR\left(\frac{\partial \ln P}{\partial\left(\dfrac{1}{T}\right)}\right)_\pi + \frac{R\Psi}{c} \qquad (22.30a)$$

and the slope of the plot of $\ln P$ versus T^{-1} should be used.

HEAT EFFECTS IN IDEAL ADSORPTION OF A MIXTURE OF GASES

We defined ideal adsorption as the case in which there is no change in surface area of solid covered by the effect of mixing. As in volumetric thermodynamics, the definition of ideal adsorption is completed by stating that there are no heat effects

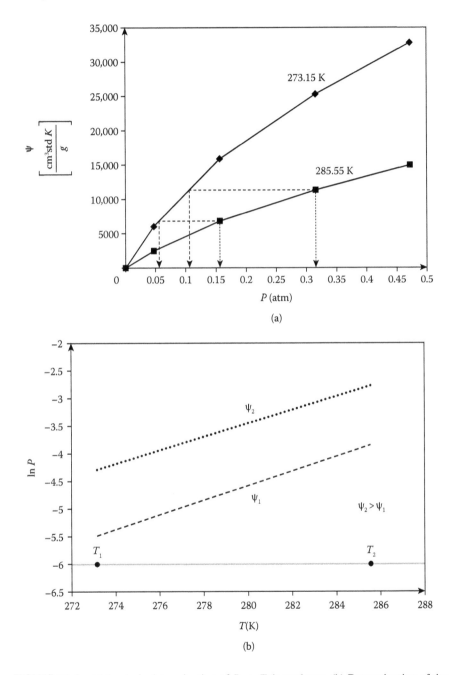

FIGURE 22.4 (a) Isosteric determination of P vs. T dependence. (b) Determination of the enthalpy of adsorption.

by mixing the adsorbed pure compounds at the same temperature T and spreading pressure π. Thus, for the ideal adsorption of a mixture of gases,

$$\left(-\Delta h^{ads}\right) = \sum x_i \left[h_{i(g)}^* - \left(h_i^E\right)_{(T,\pi)}^* \right] + \sum x_i \left[\bar{h}_{i(g)} - h_{i(g)}^* \right]$$

The first term on the right-hand side of the equation is the sum of the individual heats of adsorption of the pure compounds at T and π, which are evaluated as shown previously.

The second term is a minor correction due to variation of pressure from P for $\bar{h}_{i(g)}$ in the bulk gas phase to P_i^* for $h_{i(g)}^*$ in the adsorbed gas phase, both values at the temperature T of the system. For ideal gases, this second term is null, as the enthalpy of an ideal gas is only a function of temperature. For real gases, it can be obtained from equation of state or generalized plots. In both terms, we observe that the mole fractions used refer to the adsorbed phase x_i and not to the gas phase mole fraction y_i.

CONCLUSIONS

Although the thermodynamic approach presented in this chapter is general, it is necessary to remember that it refers to isothermal equilibrium adsorption processes that are reversible; that is, their adsorption and desorption isotherms coincide. This is true for adsorbing solids with pores smaller than 2 nm diameters, but a hysteresis of adsorption can occur if the diameters of the pores are larger [1]. It is recommended to consult the literature listed below for a more in-depth knowledge on adiabatic adsorption. Classical thermodynamics does not focus on the molecular scale of the phenomena studied. An important body of work that exists on this topic is covered extensively in references quoted below.

REFERENCES

1. Myers, A. L. 2004. Thermodynamics of adsorption. In *The Chemical Thermodynamics for Industry*, ed. T. M. Letcher. Letchworth, UK: Royal Society for Chemistry, pp. 243–253.
2. Keller, J. U., and Staudt, R. 2005. *Gas Adsorption Equilibria: Experimental Methods and Adsorptive Isotherms*. New York: Springer.
3. Tien, C. 1994. *Adsorption Calculations and Modeling*. Boston: Butterworth-Heinemann.
4. Ruthven, D. M. 1984. *Principles of Adsorption and Adsorption Processes*. New York: John Wiley & Sons.
5. Pikunic, J., Lastoskie, C. M., and Gubbins, K. E. 2002. Molecular modeling of adsorption from the gas phase. In *Handbook of Porous Solids*, ed. F. Schüth, K. S. W. Sing, and J. Weitkamp. Weinheim: Wiley-VCH, pp. 182–236.
6. Palmer, J. C., Moore, J. D., Roussel, T. J., Brennan, J. K., and Gubbins, K. E. 2011. Adsorptive behavior of CO_2, CH_4 and their mixtures in carbon nanospace. A molecular simulation study. *Phys. Chem. Chem. Phys.* 13(9): 3985–3996.

Section IV

Special Topics

If you do not make mistakes, you're not working on hard enough problems. And that's a big mistake.[†]

[†] The unsolicited advice that Frank Wilczek, recipient of the Nobel Prize in Physics (2004), gives to his graduate students. From Wilczek, F., and Devine, B., *Longing for the Harmonies: Themes and Variations from Modern Physics*, W. W. Norton & Company, New York, 1987.

23 Thermodynamics of Flow of Compressible Fluids

For completeness, in this chapter we present a brief introduction to the thermodynamics of flow of compressible fluids. When a gas flows at steady state in a horizontal adiabatic duct of constant cross section A, the initial velocity at the entrance of the duct determines the kind of flow that the gas will develop. The initial velocity can be subsonic, sonic, or supersonic, depending on whether it is lower than, equal to, or higher than the velocity of propagation of sound in the gas at the prevailing temperature. From basic physics, the velocity of sound is given by

$$\dot{\upsilon}_s = \sqrt{\frac{\gamma R T}{M}} \tag{23.1}$$

where γ is the ratio of the constant pressure heat capacity to the constant volume heat capacity of the gas, R is the universal gas constant, T is the absolute temperature, and M is the molecular weight of the gas. The Mach number, Ma, is defined as the ratio of the linear velocity of the gas in the duct to the velocity of sound:

$$Ma = \frac{\dot{\upsilon}}{\dot{\upsilon}_s}$$

At steady state, the mass flow rate $\dot{m}$ is constant, and it is related to the linear velocity $\dot{\upsilon}$ and the specific volume v of the gas by

$$\dot{m} = \frac{\dot{\upsilon} A}{v} \tag{23.2}$$

SUBSONIC, SONIC, AND SUPERSONIC FLOW

We observe here that Equation 23.2 uses the specific volume and not the molar volume. This is due to the need to combine the thermodynamic equations with fluid mechanics. Experiments show that if the Mach number at the inlet of the duct is smaller than unity, there is a pressure drop down the duct, the gas expands, and its linear velocity increases. If the duct is long enough, eventually the gas reaches sonic velocity and the velocity at the inlet cannot be maintained. If, on the contrary, the Mach number at the inlet is larger than unity, the pressure increases down the duct, the gas is compressed, and the linear velocity decreases as it flows. If the duct is long enough, eventually the sonic velocity is reached and a shock wave is produced. The longer the duct extends beyond the critical length, the further back the shock wave moves until it would eventually reach the source feeding the supersonic gas. Thermodynamics helps us to understand this phenomenon, but to determine the value of the critical length, it is necessary to resort to fluid mechanics, which is beyond the scope of this chapter.

THERMODYNAMIC UNDERSTANDING OF THE LIMITING SPEED

If the horizontal duct is well insulated, there is no heat transfer with the surroundings, but as there is friction in the duct, the irreversibility of the process increases the entropy of the gas as it flows. The fact that in both cases, of subsonic or supersonic entrance velocities, there is a critical length of the duct beyond which it is not possible to keep a stable flow suggests that at that length, the entropy has reached a maximum value. Any further increase in the length of the flow path would require a decrease in entropy that, thermodynamics tells us, is not allowed to happen. Hence, it is of interest to calculate the linear velocity that the gas would have at a maximum value of its entropy.

Assuming ideal gas behavior, according to Equations 5.26 and 5.28b, for a maximum value of the entropy, we need to have

$$ds^* = \frac{c_P^* \, dT}{T} - \left(\frac{\partial v^*}{\partial T}\right)_P dP = \frac{dh^*}{T} - \left(\frac{\partial v^*}{\partial T}\right)_P dP = 0$$

or

$$dh_{s\,max}^* = T\left(\frac{\partial v^*}{\partial T}\right)_P dP$$

Using the ideal gas equation of state for the specific volume,

$$v^* = \frac{RT}{MP} \tag{23.3}$$

we write

$$dh_{s\,max}^* = T\left(\frac{\partial v^*}{\partial T}\right)_P dP = v\,dP \tag{23.4}$$

For simplicity, from here we drop the asterisk with the understanding that we are working with an ideal gas.

An incremental enthalpy change is related to the velocity of the gas by the differential form of the energy balance, which, for the horizontal adiabatic duct from Equations 3.15 and 3.16b, takes the form

$$dh + \frac{1}{2}d\dot{v}^2 = 0 \tag{23.5}$$

Hence, from Equation 23.2,

$$d\dot{v}^2 = 2\left(\frac{\dot{m}}{A}\right)^2 v\,dv$$

Thus, replacing this expression in Equation 23.5 and rearranging, we get

$$dh = -\left(\frac{\dot{m}}{A}\right)^2 v\,dv \tag{23.6}$$

Replacing this expression in Equation 23.4 and rearranging with the help of Equation 23.2, we write

$$\left(\frac{dP}{dv}\right)_{s\,max} = -\left(\frac{\dot{m}}{A}\right)^2 = -\left(\frac{\dot{v}}{v}\right)^2$$

or

$$\dot{v}_{s\,max} = \sqrt{-v^2\left(\frac{dP}{dv}\right)_{s\,max}} \tag{23.7}$$

Although, admittedly, this is a crude approximation for an irreversible process, for the adiabatic expansion of the gas, from Equation 5.34,

$$P v^\gamma = C$$

Differentiating and rearranging, assuming γ to be constant,

$$-v^2\left(\frac{dP}{dv}\right) = \gamma\frac{C}{v^\gamma}v = \gamma Pv = \frac{\gamma RT}{M}$$

or

$$\dot{v}_{s\,max} = \sqrt{\frac{\gamma RT}{M}} \tag{23.8}$$

Comparison of this result with Equation 23.1 confirms that the flow becomes unstable once the speed of sound is reached, and that this speed corresponds to the maximum of the gas entropy. Any further change would cause a decrease in the entropy violating the second law.

Considering that when the gas enters the duct at subsonic velocity there is a pressure drop and the gas expands, its specific enthalpy decreases as the gas flows while its entropy increases.

When the gas enters at supersonic velocity, there is a compression of the gas and its specific enthalpy increases as the gas flows. A plot of enthalpy versus entropy must show that the curves for supersonic and subsonic flow for the same mass flow rate meet at the maximum of the entropy, that is, when the velocity of the gas is equal to the velocity of sound.

ENTHALPY AND ENTROPY CHANGES IN COMPRESSIBLE FLOW

In publications dealing with the flow of compressible fluids, the differential equations for the enthalpy and for the entropy are sometimes written in a form difficult to recognize. For the enthalpy change for an ideal gas, the simple form of Equation 5.28b,

$$dh = c_P \, dT$$

is transformed using the ideal gas relation for specific heats, Equation 5.29,

$$c_P - c_v = \frac{R}{M}$$

and the definition of γ, Equation 5.32,

$$\gamma \equiv \frac{c_P}{c_v}$$

to write

$$c_P = \frac{(c_P - c_v)\, c_P}{(c_P - c_v)} = \frac{(c_P - c_v)\dfrac{c_P}{c_v}}{\left(\dfrac{c_P}{c_v} - 1\right)} = \frac{R\gamma}{M(\gamma - 1)}$$

and

$$dh = \frac{R\gamma}{M(\gamma - 1)} dT \tag{23.9}$$

or using the ideal gas equation of state, Equation 5.27, we write

$$dh = \frac{\gamma}{(\gamma - 1)} d\,(Pv) \tag{23.10}$$

Similarly, for the entropy differential,

$$ds = \frac{R\gamma}{M(\gamma - 1)} \frac{dT}{T} - \frac{R}{M} \frac{dP}{P} \tag{23.11}$$

With the help of the equation of state, it is possible to integrate Equations 23.10 and 23.11 between the conditions at the entrance of the duct and subsequent states, so the curves for h versus s can be calculated.

RELATION OF FLUID PROPERTIES TO MACH NUMBER AT ENTRANCE CONDITION

It is also useful to obtain a relation between the fluid properties down the duct and the properties at the entrance of the duct. Combining Equations 23.6 and 23.10 and rearranging, we get

$$ d\,(Pv) + \left(\frac{\gamma-1}{\gamma}\right)\left(\frac{\dot{m}}{A}\right)^2 v\,dv = 0 $$

Integrating this expression,

$$ Pv + \frac{1}{2}\left(\frac{\gamma-1}{\gamma}\right)\left(\frac{\dot{m}}{A}\right)^2 v^2 = constant $$

Applying this relation for any two states 1 and 2, and rearranging,

$$ \frac{P_2 v_2}{P_1 v_1} = \frac{T_2}{T_1} = 1 + \left(\frac{\gamma-1}{\gamma}\right)\frac{\left(\frac{\dot{m}}{A}\right)^2 v_1^2}{2 P_1 v_1}\left(1 - \frac{v_2^2}{v_1^2}\right) $$

This expression can be conveniently put in terms of the Mach number at the entrance of the duct:

$$ Ma^2 = \frac{v_1^2}{\gamma(R/M)T} = \frac{\left(\frac{\dot{m}}{A}\right)^2 v_1^2}{\gamma P_1 v_1} $$

Thus,

$$ \frac{P_2 v_2}{P_1 v_1} = \frac{T_2}{T_1} = 1 + \left(\frac{\gamma-1}{2}\right)Ma^2\left[1 - \left(\frac{v_2}{v_1}\right)^2\right] \tag{23.12} $$

DETERMINATION OF CRITICAL LENGTH

The determination of the critical length of the duct where the sonic velocity is attained requires using the mechanical energy balance, which in its differential form for compressible fluids flowing at steady state in a horizontal adiabatic duct can be written as

$$v\, dP + \left(\frac{\dot{m}}{A} \right) v\, dv + \delta l_w = 0 \qquad (23.13)$$

    The determination of the lost work by friction belongs to the field of fluid mechanics, and it is beyond the scope of this chapter. More complete treatments of the flow of compressible fluids in horizontal insulated ducts presented in [1–4] or other texts can now be followed after this introduction linking them to the equations discussed in the text.

REFERENCES

1. Cengel, Y. A., and Cimbala, J. M. 2014. *Fluid Mechanics: Fundamentals and Applications*, 3rd ed. New York: McGraw-Hill.
2. Oosthuizen, P. H., and Carscallen, W. E. 2013. *Introduction to Compressible Fluid Flow*, 2nd ed. Boca Raton, FL: CRC Press/Taylor and Francis.
3. Anderson, J. D., Jr. 2003 [1982]. *Modern Compressible Flow*, 3rd ed. New York: McGraw-Hill.
4. Hougen, O. A., Watson, K. M., and Ragatz, R. A. 1959. *Chemical Process Principles. Part II. Thermodynamics*, 2nd ed. New York: John Wiley & Sons.

24 Elements of Statistical Thermodynamics

Statistical thermodynamics is a branch of knowledge that has its own postulates and techniques. We do not attempt to give here even an introduction to the field. In this chapter, we elaborate on the ideas presented by T. H. Hill [1] with the direct purpose of giving a molecular justification to equations of state and excess Gibbs energy models used in practical work.

CANONICAL ENSEMBLE AND PROBABILITY

For this introduction, we consider a closed system having rigid and diathermal walls, immersed in a thermostatic bath at a temperature T. The system has a fixed number of molecules inside and a fixed volume, and it can only exchange energy in the form of heat with its surroundings. The molecules in the system are moving around, hitting each other and also hitting the rigid walls. A detailed description of the internal energy of this system would require the specification of the position and momentum coordinates, that is, the external degrees of freedom, of each molecule at any given instant. In addition, it would be necessary to specify the rotational, vibrational, and electronic coordinates, that is, the internal degrees of freedom, of each molecule at any given instant. Because per each mole inside the system there are 10^{23} molecules, this detailed description becomes an impossible task. On the other hand, macroscopically the system is characterized by three parameters: the temperature T, the total volume V, and the number of moles n or, more properly, the number of molecules N that it contains. Each set of values of the internal and external coordinates of the molecules compatible with the values of the macroscopic parameters T, V, and N represents a possible quantum state of the system. There are very many of such quantum states, each with a value E_i for the internal energy of the macroscopic system. The internal energy U of the system is then a macroscopic average of the internal energies of these very many possible quantum states. The probability $\mathcal{P}_i$ of having a particular quantum state i with an internal energy E_i will be equal to the ratio between the number of quantum states $\mathcal{N}(E_i)$ having that particular value E_i of the internal energy and the total number of quantum states compatible with the macroscopic parameters T, V, and N:

$$\mathcal{P}_i = \frac{\mathcal{N}(E_i)}{\sum_j \mathcal{N}(E_j)} \qquad (24.1)$$

with

$$\sum_j \mathcal{P}_j = 1 \qquad (24.2)$$

INTERNAL ENERGY AND THE PARTITION FUNCTION

Hence, according to Equation 24.1, as the internal energy $U(T, V, N)$ of the system is a weighted average of the internal energy E_i of these quantum states, it takes the form

$$U = \sum_j \mathcal{P}_j E_j \tag{24.3}$$

Considering the large number of possible quantum states, the probability of finding one particular state with internal energy E_i is well represented by a normalized Boltzmann distribution function, namely,

$$\mathcal{P}_i = \frac{e^{\frac{-E_i}{kT}}}{\sum_j e^{\frac{-E_j}{kT}}} \tag{24.4}$$

where k is the Boltzmann constant, given by the ratio of the gas constant over Avogadro's number. In this equation, it is conventional to define the partition function Q as

$$Q \equiv \sum_j e^{\frac{-E_j}{kT}} \tag{24.5}$$

With the purpose of connecting these expressions with classical thermodynamics, we differentiate Equation 24.3 and write

$$dU = \sum_j \mathcal{P}_j \, dE_j + \sum_j E_j \, d\mathcal{P}_j \tag{24.6}$$

The next step is to express the right hand side of this equation in terms of the macroscopic thermodynamic variables. In Classical Thermodynamics, we have

$$\left(\frac{\partial U}{\partial V} \right)_{S,n} = -P$$

So, by analogy, we define

$$\left(\frac{\partial E_j}{\partial V} \right)_{T,n} \equiv -p_j$$

or, at constant T and n,

$$dE_j = -p_j \, dV$$

Multiplying this equation by $\mathcal{P}_j$ and adding over all j, we have

$$\sum_j \mathcal{P}_j \, dE_j = -\left(\sum_j \mathcal{P}_j \, p_j \right) dV = -P \, dV \tag{24.7}$$

In this step, we have identified the pressure of the system with the weighted average of values of p_j.

For the second term of the right-hand side of Equation 24.6, from Equation 24.4 and the definition of the partition function, Equation 24.5, we write for a quantum state j

$$E_j = -kT \ln \mathcal{P}_j - kT \ln Q \tag{24.8}$$

Multiplying this expression by $d\mathcal{P}_j$ and adding over all j,

$$\sum_j E_j d\mathcal{P}_j = -kT \sum_j \ln \mathcal{P}_j d\mathcal{P}_j - kT \ln Q \sum_j d\mathcal{P}_j$$

But according to Equation 24.2,

$$\sum_j d\mathcal{P}_j = d \sum_j \mathcal{P}_j = 0$$

so the second term on the left-hand side vanishes, and we write

$$\sum_j E_j d\mathcal{P}_j = -kT \sum_j \ln \mathcal{P}_j d\mathcal{P}_j$$

This expression can be transformed further considering that

$$d \sum_j \mathcal{P}_j \ln \mathcal{P}_j = \sum_j \ln \mathcal{P}_j \, d\mathcal{P}_j + \sum_j \mathcal{P}_j \, d \ln \mathcal{P}_j$$

but

$$\sum_j \mathcal{P}_j \, d \ln \mathcal{P}_j = \sum_j \mathcal{P}_j \frac{1}{\mathcal{P}_j} d\mathcal{P}_j = \sum_j d\mathcal{P}_j = 0$$

so

$$d \sum_j \mathcal{P}_j \ln \mathcal{P}_j = \sum_j \ln \mathcal{P}_j \, d\mathcal{P}_j$$

Thus,

$$\sum_j E_j d\mathcal{P}_j = -kT \, d \sum_j \mathcal{P}_j \ln \mathcal{P}_j = -T \, d \left(k \sum_j \mathcal{P}_j \ln \mathcal{P}_j \right) \tag{24.9}$$

and replacing Equations 24.7 and 24.9 in Equation 24.6, we obtain

$$dU = -P \, dV + T \, d \left(-k \sum_j \mathcal{P}_j \ln \mathcal{P}_j \right) \tag{24.10}$$

ENTROPY, PROBABILITY AND THE THIRD LAW OF THERMODYNAMICS

From the combination of the first and second laws of thermodynamics, we know that

$$dU = TdS - P\,dV$$

So, by comparison with Equation 24.10, we conclude that the entropy is related to the probability by the Boltzmann equation,

$$S = -k\sum_j \mathcal{P}_j \ln \mathcal{P}_j \qquad (24.11)$$

Equation 24.11 is of importance in information theory.

A perfect crystal at 0 K is in its ground state. Thus, for this single quantum state the value of $\mathcal{P}_j$ is unity and $S = 0$. This is a statement of the Third Law of Thermodynamics.

EQUATION OF STATE, CHEMICAL POTENTIAL, AND PARTITION FUNCTION

After obtaining this first bridge between statistical and classical thermodynamics, we multiply Equation 24.8 by $\mathcal{P}_j$, and add over all quantum states to obtain

$$\sum_j \mathcal{P}_j E_j = T\left(-k\sum_j \mathcal{P}_j \ln \mathcal{P}_j\right) - kT \ln Q \sum_j \mathcal{P}_j$$

Thus, replacing in this expression Equations 24.2, 24.3, and 24.11,

$$U = TS - kT \ln Q$$

or

$$U - TS = -kT \ln Q$$

But from thermodynamics we know that the Helmholtz function is defined by Equation 4.1 as

$$A = U - TS$$

Thus, we conclude that

$$A = -kT \ln Q \qquad (24.12)$$

Equation 24.12 is one of the most important links between statistical thermo-dynamics and classical thermodynamics, as it permits us to obtain expressions for the equation of state and for the chemical potential based on the exact relations.

From Equation 4.15,

$$P = -\left(\frac{\partial A}{\partial V}\right)_{T,n} = kT\left(\frac{\partial \ln Q}{\partial V}\right)_{T,n} \tag{24.13}$$

And from Equation 4.16,

$$\mu_i = \left(\frac{\partial A}{\partial n_i}\right)_{T,V,n_{j\neq i}} = -kT\left(\frac{\partial \ln Q}{\partial n_i}\right)_{T,V,n_{j\neq i}} \tag{24.14}$$

The next step, then, is to say something about the partition function for gases and liquids. This is discussed in the chapters that follow.

REFERENCE

1. Hill, T. L. 1960. *An Introduction to Statistical Thermodynamics*. Reading, MA: Addison-Wesley.

25 Statistical Thermodynamics Basis of Equations of State

The starting point for the statistical thermodynamics justification of equations of state for pure compounds is Equation 24.13, namely,

$$P = -\left(\frac{\partial A}{\partial V}\right)_{T,n} = kT\left(\frac{\partial \ln Q}{\partial V}\right)_{T,n} \tag{24.13}$$

with

$$Q \equiv \sum_j e^{\frac{-E_j}{kT}} \tag{24.5}$$

The next step is to say something about the partition function Q in terms of the temperature T, the total volume V, and the number of identical molecules N of a pure compound in the gas phase. Considering that the number of molecules present is of the order of 10^{23}, it is legitimate to replace quantum statistics by classic statistics and transform the sum over the quantum states with energy E_j into an integral over the range of parameters characterizing the different modes of motion. Thus, Equation 24.5 takes the form

$$Q = \frac{1}{N!} \int_I \int_M \int_r e^{-[E_I + E_M + E_r]/kT} \, dI_N \, dM_N \, dr_N$$

where the integrations are taken over the I internal degrees of motion (rotational, vibrational, and electronic), M momentum coordinates describing the external state of motion (momentum) of the molecules, and r position coordinates describing the translational modes of motion for the N molecules in the three-dimensional space. The inverse factorial term in front of the integrals is included to avoid double counting of the identical molecules. As an approximation, which is acceptable for the gas phase, the different modes of motion are considered to be independent of each other. Thus, the integrals are separable and we write

$$Q = \frac{1}{N!} \int_I e^{-E_I/kT} \, dI_N \int_M e^{-E_M/kT} \, dM_N \int_r e^{-E_r/kT} \, dr_{1 \to 3N}$$

or, in a shorthand nomenclature, designating each integral by the symbol Q with the corresponding subindex,

$$Q = \frac{1}{N!} Q_I Q_M Q_r$$

$$\ln Q = \ln Q_I + \ln Q_M + \ln Q_r - \ln (N!)$$

As indicated by Equation 24.13, for the equation of state we are only interested in the dependence of the partition function on volume. Thus, considering that the internal and external modes of motion of the molecules are only temperature dependent, Equation 24.13 takes the form

$$P = -\left(\frac{\partial A}{\partial V}\right)_{T,n} = kT \left(\frac{\partial \ln Q_r}{\partial V}\right)_{T,n} \tag{25.1}$$

Although it is only a change of formalism sometimes, due to its importance, Q_r is given the special symbol Z_N and called the classical configurational integral,

$$Z_N \equiv Q_r \equiv \int_r e^{-\frac{E_r}{kT}} dr_1 ... dr_{3N} \tag{25.2}$$

so we write,

$$P = -\left(\frac{\partial A}{\partial V}\right)_{T,n} = kT \left(\frac{\partial \ln Z_N}{\partial V}\right)_{T,n} \tag{25.1a}$$

In practical work, we define the compressibility factor z by the general relation

$$z \equiv \frac{Pv}{RT} = \frac{P}{\rho RT} \tag{25.3}$$

where $\rho = 1/v$ is the density of the fluid. The compressibility factor z should not be confused with the classical configurational integral Z_N.

EQUATION OF STATE FOR THE IDEAL GAS

As a first example of application of this treatment, we consider that there are no intermolecular interactions, that is, $E_r = 0$, and that the molecules are just points in space and do not occupy any volume. In this very special case, the integral for each of the N molecules is just the total volume of the system,

$$Z_N^* = \int_r dr_1 ... dr_{3N} = V^N \tag{25.4}$$

and from Equation 25.1a,

$$P^* = \frac{NkT}{V} = \frac{\frac{N}{N_A}(N_A k)T}{V} = \frac{nRT}{V} = \frac{RT}{v} \tag{25.5}$$

where N_A is Avogadro's number and $N_A k = R$ is the gas constant. Thus, according to Equation 25.5, the compressibility factor for the ideal gas is equal to unity

$$z^* \equiv \frac{P^* v}{RT} = 1 \tag{25.5a}$$

VIRIAL EQUATIONS OF STATE

As an improvement toward more realistic systems, we may expand the compressibility factor z, as a Taylor's series around the real gas limit at pressure tending to zero, that is, $P \to 0$; $v \to \infty$; $\rho \to 0$.

$$z = z_{\rho=0} + \left(\frac{\partial z}{\partial \rho}\right)_{\rho=0} (\rho - 0) + \frac{1}{2!}\left(\frac{\partial^2 z}{\partial \rho^2}\right)_{\rho=0} (\rho - 0)^2 + \ldots$$

The first term of the series is given by Equation 25.5, and for convenience, a symbol is given to the value of the coefficients of the density powers

$$B = \left(\frac{\partial z}{\partial \rho}\right)_{\rho=0} \; ; \; C = \frac{1}{2!}\left(\frac{\partial^2 z}{\partial \rho^2}\right)_{\rho=0} \ldots$$

These coefficients are called the second, third, and so forth, virial coefficients. They are evaluated at the limit of the pressure tending to zero for each compound and are a function of temperature only. Thus, we write

$$z = 1 + B\rho + C\rho^2 + \ldots \tag{25.6}$$

Comparison of Equations 25.5a and 25.6 shows that the virial terms are perturbations to the ideal gas case. The term *virial* means force, and it reflects the fact that these perturbations are due to molecular interactions.

It is also possible to expand the compressibility factor as a Taylor's pressure series around the real gas limit,

$$z = z_{P=0} + \left(\frac{\partial z}{\partial P}\right)_{P=0} (P - 0) + \frac{1}{2!}\left(\frac{\partial^2 z}{\partial P^2}\right)_{P=0} (P - 0)^2 + \ldots$$

In this case, we write

$$z = 1 + B'P + C'P^2 + \ldots \tag{25.7}$$

The relationship between the virial coefficients of the two series is obtained considering that combining Equations 25.3 and 25.6, we obtain

$$P = RT \left[\rho + B\rho^2 + C\rho^3 + \ldots \right]$$

Replacing this expression in Equation 25.7, rearranging and comparing the coefficients up to the second power of density with those of Equation 25.6, we obtain

$$B' = \frac{B}{RT}; \quad C' = \frac{C - B^2}{(RT)^2}, \text{ etc.}$$

The second and third virial coefficients of expansion in terms of density, B and C, are known with accuracy for many substances. Higher coefficients are seldom used in practice. Hence, the truncated pressure series of practical use takes the form

$$z = 1 + \frac{B}{RT} P + \frac{C - B^2}{(RT)^2} P^2 + \ldots \tag{25.7a}$$

Equation 25.7a, truncated after the second term, is particularly useful for calculating fugacity coefficients of pure gases at low pressure using Equation 13.7a. The virial equation of state is further discussed in Appendix B.

CUBIC EQUATIONS OF STATE

In these equations, the pressure is a function of the third power of the molar volume. This dependence makes the volume roots of these equations easily solved for particular values of the pressure and temperature. The practical details for working these equations are presented in Appendix C. The need to use more complex expressions for the equation of state appears because as the pressure increases, at fixed volume and temperature, the interactions between molecules become stronger and the intermolecular potential cannot be ignored, nor can the volume occupied by the molecules. In this case, it is convenient to define a mean intermolecular potential φ by the expression

$$e^{-\frac{\varphi}{kT}} \equiv \frac{\int_r e^{-\frac{E_r}{kT}} dr_1 \ldots dr_{3N}}{\int_r dr_1 \ldots dr_{3N}} \tag{25.8}$$

For the case of the ideal gas, the integral appearing in the denominator of this expression is identified with the total volume of the system to the N-th power because for each molecule, the integral runs in the three directions of space to cover the total volume of the system. If the molecules themselves occupy part of the volume, their translational movements can only happen in the free volume, V_f, left unoccupied by the molecules:

$$\int_r dr_1 \ldots dr_{3N} = V_f^N$$

Introducing this expression in Equation 25.8 and combining with Equation 25.2, we write

$$Z_N = \int_r e^{-\frac{E_r}{kT}} dr_1 \ldots dr_{3N} = V_f^N e^{-\frac{\varphi}{kT}} \tag{25.9}$$

Thus, from Equation 25.1a,

$$P = NkT \left(\frac{\partial \ln V_f}{\partial V} \right)_{T,n} - \left(\frac{\partial \varphi}{\partial V} \right)_{T,n} = nRT \left(\frac{\partial \ln V_f}{\partial V} \right)_{T,n} - \left(\frac{\partial \varphi}{\partial V} \right)_{T,n} \tag{25.10}$$

where $N_A k = R$. To continue, it is necessary to introduce some assumptions for the free volume and the intermolecular potential. Designating by b the volume occupied by the molecules per mole of gas, as a first crude approximation we write

$$V_f = V - nb$$

The intermolecular forces are attractive, and by convention, the intermolecular potential is then negative. The total potential will be proportional to the number of moles present, and as it tends to vanish at infinite volume, it will be inversely proportional to the molar volume. As a first approximation, then,

$$\varphi = -\frac{na_c}{v} = -\frac{n^2 a_c}{V}$$

where a_c is the proportionality constant. Introducing these two assumptions in Equation 25.10, we obtain

$$P = \frac{nRT}{V - nb} - \frac{n^2 a_c}{V^2} = \frac{RT}{v - b} - \frac{a_c}{v^2} \tag{25.11}$$

This is the van der Waals equation of state [1].

Next, we can assume that the intermolecular potential is a more complex function of volume and temperature, say

$$\varphi = -n^2 a_c F(V) G(T)$$

Thus, the equation of state would take the form

$$P = \frac{RT}{v - b} + n^2 a_c G(T) \left(\frac{\partial F(V)}{\partial V} \right)_{T,n}$$

For this form, there are the two most successful cubic equations of state:

Redlich–Kwong equation [2]

$$P = \frac{RT}{v - b} - \frac{a_c G_{RK}(T)}{v(v + b)} 1$$

Peng–Robinson equation [3]

$$P = \frac{RT}{v-b} - \frac{a_c G_{PR}(T)}{v^2 + 2bv - b^2} \, |$$

Each of these has a different function of temperature $G(T)$, which at this point we leave unspecified. The use of a cubic equation of state is further discussed in Appendix C. For each of these equations of state, it is possible to back-calculate the form of the function $F(V)$. Although, at our present degree of knowledge, there is not much understanding to gain with this further step for completeness, we include these functions below

$$F_{RK}(V) = \frac{1}{nb} \ln\left(\frac{V+nb}{V}\right)$$

$$F_{PR}(V) = \frac{1}{2nb\sqrt{2}} \ln\left(\frac{V+nb-nb\sqrt{2}}{V+nb+nb\sqrt{2}}\right)$$

RELATION OF INTERMOLECULAR POTENTIALS TO THE SECOND VIRIAL COEFFICIENT

In order to make some progress, we return to the statistical thermodynamics treatment and relate the virial coefficients to intermolecular potentials. For this, we add and subtract one dimensionless unit to the integrand of the configurational integral and write Equation 25.2 as

$$Z_N = \int_r e^{-\frac{E_r}{kT}} dr_1 .....dr_{3N} = V^N\left[1 + \frac{1}{V^N}\int_r\left[e^{-\frac{E_r}{kT}} - 1\right]d\bar{r}_1 .....d\bar{r}_N\right] \quad (25.12)$$

where, in the low-pressure region of applicability of the virial equation, we have used the approximation given by Equation 25.4, and we have introduced the notation $d\bar{r}_i$ to designate the three coordinates of position of molecule i. To continue, let us concentrate attention on the integral appearing in Equation 25.12, namely,

$$I \equiv \frac{1}{V^N}\int_r\left[e^{-\frac{E_r}{kT}} - 1\right]d\bar{r}_1 .....d\bar{r}_N$$

Equation 25.12 can be written as

$$Z_N = V^N[1 + I] \quad (25.12a)$$

or

$$\ln Z_N = N \ln V + \ln[1 + I]$$

Considering that the definition of I includes a denominator V^N, its value is very small, $I \lll 1$, so

$$\ln [I + 1] \approx I$$

and

$$\ln Z_N = N \ln V + I \tag{25.13}$$

So, from Equation 25.1a,

$$P = \frac{NkT}{V} + kT \left(\frac{\partial I}{\partial V} \right)_{T,N} \tag{25.14}$$

At low pressure, interactions of three or more molecules are extremely rare. Thus, considering only pair interactions, the number of possible arrays is

$$\frac{N(N-1)}{2} \approx \frac{N^2}{2}$$

For each array, only the coordinates of the two interacting molecules $\bar{r}_1$ and $\bar{r}_2$ are important for the intermolecular potential $E(\bar{r}_1, \bar{r}_2)$, while the rest of the noninteracting molecules will produce a term of the form

$$\int_r d\bar{r}_3 ..... d\bar{r}_N = V^{N-2}$$

Thus,

$$I = \frac{N^2 V^{N-2}}{2V^N} \int_r \left[e^{-\frac{E(\bar{r}_1, \bar{r}_2)}{kT}} - 1 \right] d\bar{r}_1 d\bar{r}_2 = \frac{N^2}{2V^2} \int_r \left[e^{-\frac{E(\bar{r}_1, \bar{r}_2)}{kT}} - 1 \right] d\bar{r}_1 d\bar{r}_2$$

In addition, as the interaction potential between two molecules is a function of the distance r between the two molecules, considering a central molecule 1, we may consider the element of volume in which there is interaction with molecule 2 as $4\pi r^2 dr$ and integrate over $d\bar{r}_1$ to cover the whole volume V.

$$I = \frac{N^2}{2V^2} \left\{ \int \left[\int_r \left[e^{-\frac{E(r)}{kT}} - 1 \right] 4\pi r^2 dr \right] V \right\} = \frac{2\pi N^2}{V} \left\{ \int_r \left[e^{-\frac{E(r)}{kT}} - 1 \right] r^2 dr \right\}$$

In order to simplify the nomenclature, let us define

$$B \equiv -2\pi N_A \int_r \left[e^{-\frac{E(r)}{kT}} - 1 \right] r^2 dr \tag{25.15}$$

or

$$I = -\frac{\left(N^2/N_A\right)}{V}B$$

and

$$\left(\frac{\partial I}{\partial V}\right)_{T,N} = \frac{\left(N^2/N_A\right)}{V^2}B$$

Hence, Equation 25.14 takes the form

$$P = \frac{NkT}{V} + kT\frac{(N^2/N_A)B}{V^2} = \frac{(N/N_A)(N_A k)T}{V} + (N_A k)T\frac{(N/N_A)^2 B}{V^2} \qquad (25.14a)$$

$$P = \frac{RT}{\mathrm{v}} + \frac{RT}{\mathrm{v}^2}B$$

So,

$$z = 1 + \frac{B}{\mathrm{v}} = 1 + B\rho$$

The comparison of this result with Equation 25.6 shows that Equation 25.15 gives the relation of the second virial coefficient with the two-body intermolecular potential. In practice, Equation 25.15 is used to study potential functions starting from experimentally measured second virial coefficients.

SECOND VIRIAL COEFFICIENTS OF BINARY MIXTURES

For a binary mixture of two dissimilar compounds i and j, the configurational integral takes the form

$$Z_N = V^N\left[1 + \frac{1}{V^N}\sum_i\sum_j\int_r\left[e^{-\frac{E_{ij(r)}}{kT}} - 1\right]d\bar{r}_i d\bar{r}_j\right] \qquad (25.16)$$

where again we have considered only pair interactions and integrated over r for the $(N-2)$ noninteracting molecules. As above, we may consider again a central molecule i, integrate over $d\bar{r}_i$ to obtain V and express $d\bar{r}_j$ as $4\pi r^2 dr$ and integrate over dr to obtain

$$Z_N = V^N\left[1 + \frac{4\pi}{V}\sum_i\sum_j\int_r\left[e^{-\frac{E_{ij(r)}}{kT}} - 1\right]r^2 dr\right] \qquad (25.17)$$

In the sums, we distinguish three types of interactions: i–i interactions, repeated $N_i^2/2$ times; i–j interactions, repeated $N_i N_j$ times; and j–j interactions, repeated $N_j^2/2$ times. Thus, we write

$$Z_N = V^N \left[1 + \frac{2\pi N_i^2}{V} \int_r \left[e^{-\frac{E_{ii(r)}}{kT}} - 1 \right] r^2 dr + \frac{4\pi N_i N_j}{V} \int_r \left[e^{-\frac{E_{ij(r)}}{kT}} - 1 \right] r^2 dr \right.$$

$$\left. + \frac{2\pi N_j^2}{V} \int_r \left[e^{-\frac{E_{ii(r)}}{kT}} - 1 \right] r^2 dr \right]$$

In order to identify the second virial coefficient for the binary mixture, there is no need to derive the equation of state. In short notation, we write

$$Z_N = V^N \left[1 + \frac{N^2 B_m}{N_A V} \right]$$

According to Equation 25.15,

$$B_{ii} \equiv -2\pi N_A \int_r \left[e^{-\frac{E_{ii}(r)}{kT}} - 1 \right] r^2 dr$$

$$B_{ij} \equiv -2\pi N_A \int_r \left[e^{-\frac{E_{ij}(r)}{kT}} - 1 \right] r^2 dr$$

$$B_{jj} \equiv -2\pi N_A \int_r \left[e^{-\frac{E_{jj}(r)}{kT}} - 1 \right] r^2 dr$$

Thus,

$$B_m = x_i^2 B_{ii} + 2 x_i x_j B_{ij} + x_j^2 B_{jj} \tag{25.18}$$

with $x_i = N_i/N$.

The term B_{ij} is called the cross second virial coefficient of compounds i and j.

MIXING RULE FOR CUBIC EQUATIONS OF STATE

Expanding Equation 25.11, or any of the cubic equations of state of the van der Waals family, into a series in terms of density gives

$$z = 1 + \frac{B}{v} + \dots = 1 + B\rho + \dots$$

with

$$B = b - \frac{a}{RT}$$

Comparison of this result with Equation 25.18 suggests the use of the following mixing rules for the extension of cubic equations to mixtures:

$$b_m = x_i^2 b_{ii} + 2x_i x_j b_{ij} + x_j^2 b_{jj} \tag{25.19}$$

and

$$a_m = x_i^2 a_{ii} + 2x_i x_j a_{ij} + x_j^2 a_{jj} \tag{25.20}$$

For the parameter b, which represents the bulkiness of the molecules, it seems logical to eliminate the double subindex and take the cross-term as an arithmetic average of the values for the pure compounds. Thus,

$$b_m = x_i^2 b_i + 2x_i x_j \frac{b_i + b_j}{2} + x_j^2 b_j = x_i(x_i + x_j)b_i + x_j(x_i + x_j)b_j$$

or

$$b_m = x_i b_i + x_j b_j \tag{25.21}$$

For the parameter a, which is related to the intermolecular potentials, the form of Equation 25.20 is kept and the cross-term a_{ij} is approximated by

$$a_{ij} = (a_{ii} a_{jj})^{0.5} (1 - \delta_{ij}) \tag{25.22}$$

where δ_{ij} is an empirical adjustable parameter.

NONCUBIC EQUATIONS OF STATE OF THEORETICAL INTEREST

There is a very large number of cubic and noncubic equations of state proposed in the literature. Normally, the variables fixed for calculation of fugacities in phase equilibrium problems are the pressure and the temperature, so due to the additional iterations necessary to find the roots of volume, noncubic equations of state are not of common use in practice. In this chapter, we only give a general discussion of the subject based on our original work.

The *free volume* term in van der Waals's generalized model, presented above, represents the repulsive forces arising from the interactions of the hard core of the molecules, while the intermolecular potential represents the attractive forces between molecules. Van der Waals's model considers that the pressure is determined by the difference of these two contributions. Another equally valid assumption is that of Dieterici, in which the potential perturbs directly the hard-core contribution. The Dieterici equation [4] of state has the form

$$P = \frac{RT}{v-b} \exp\left(-\frac{a}{RTv}\right) \qquad (25.23)$$

As discussed elsewhere by Polishuk and Vera [5,6], both potentials are not mutually exclusive, and they can be combined in a single expression. Van der Waals potential represents the cohesive energy keeping the molecules together, while the Dieterici potential considers that only those molecules with enough kinetic energy to escape the cohesive field contribute to the pressure exerted by the fluid over the walls of the system. Perhaps the most serious theoretical shortcoming of both the van der Waals and Dieterici equations is that their repulsive term fails to reproduce well-known results for hard-sphere fluids. Stated shortly, in a close-packed system of hard spheres of equal diameter, it is impossible to fill all the volume of the system. In round numbers, the hard spheres fill 0.74 of the total volume and the rest is empty space between the close-packed hard spheres. Thus, the compressibility factor goes to infinity when the hard spheres are close packed but not all the volume of the system is filled. Wang et al. [7] proposed a solution to this problem using the scaled particle theory with a lattice formed by Kelvin's tetrakaidecahedron cells [8,9]. The following two basic forms were studied:

$$z = \frac{\sum\limits_{i=1} C_i \xi^{i-1}}{(1-\xi)^n} \qquad (25.24)$$

and

$$z = \frac{1}{\xi} \sum\limits_{i=1} C_i \left(\frac{\xi}{1-\xi}\right)^i \qquad (25.25)$$

with

$$\xi = \frac{V_0}{V} \qquad (25.26)$$

where V_0 is the close-packed volume of the hard spheres.

The incorporation of an intermolecular potential for a system of hard spheres of equal diameter (pure compounds) was studied by Khoshkbarchi and Vera [10]. Equations of state for mixtures of hard spheres of different diameters were studied by Khoshkbarchi and Vera [11], with corrections that were noted by Taghikhani et al. [12]. Further studies were presented by Ghotbi and Vera [13–15]. Although we consider the details of this material too specialized to be included here, we quote the above references as a guide for those interested in pursuing further the study of this subject.

REFERENCES

1. van der Waals, J. D, 1873. *Over de Continuiteit van den Gas- en Vloeistoftoestand.* Leiden: A. W. Sijthoff.
2. Redlich, O., and Kwong, J. N. S. 1949. On the thermodynamics of solutions. V. An equation of state. Fugacities of gaseous solutions. *Chem. Rev.* 44(1): 233–244.
3. Peng, D. Y., and Robinson, D. B. 1976. A new two-constant equation of state. *Ind. Eng. Chem. Fundam.* 15: 50–64.
4. Dieterici, C. 1899. Über den kritischen Zustand. *Ann. Phys. Chem. Wiedemanns Ann.* 69: 685–705.
5. Polishuk, I., and Vera, J. H. 2005. A novel EOS that combines the van der Waals and the dieterici potentials. *AIChE J.* 51: 2077–2088.
6. Polishuk, I., and Vera, J. H. 2005. A novel equation of state for the prediction of thermodynamic properties of fluids. *J. Phys. Chem. B* 109: 5977–5984.
7. Wang W., Khoshkbarchi, M. K., and Vera, J. H. 1996. A new volume dependence for the equations of state of hard spheres. *Fluid Phase Equilib.* 115: 25–38.
8. Lord Kelvin. 1894. On homogeneous division of space. *Proc. R. Soc. (Lond.)* 55: 1–16.
9. Smith, V. J. 1982. *Geometrical and Structural Crystallography.* New York: Wiley & Sons.
10. Khoshkbarchi, M. K., and Vera, J. H. 1997. A simplified hard-sphere equation of state meeting the high and low density limits. *Fluid Phase Equilib.* 130: 189–194.
11. Khoshkbarchi, M. K., and Vera, J. H. 1998. A generalized mixing rule for hard-sphere equations of state of Percus–Yevick type. *Fluid Phase Equilib.* 142: 131–147.
12. Taghikhani, V., Khoshkbarchi, M. K., and Vera, J. H. 1999. On the expression for the chemical potential in mixtures of hard spheres. *Fluid Phase Equilib.* 165: 141–146, 279.
13. Ghotbi, C., and Vera, J. H. 2001. Performance of three mixing rules using different equations of state for hard-spheres. *Fluid Phase Equilib.* 187–188: 321–336.
14. Ghotbi, C., and Vera, J. H. 2001. Extension to mixtures of two robust hard-sphere equations of state satisfying the ordered close-packed limit. *Can. J. Chem. Eng.* 79: 678–686.
15. Ghotbi, C., and Vera, J. H. 2002. A general expression for the ordered-packed volume fraction of hard spheres of different diameters. *Ind. Eng. Chem. Res.* 41(5): 1122–1128.

26 Statistical Thermodynamics Justification of Some Commonly Used Expressions for the Excess Gibbs Energy

Models of mixtures are analytical expressions that, based on molecular concepts, aim to represent the thermodynamic behavior of solutions. As discussed in Chapter 17, the thermodynamic behavior of organic mixtures depends strongly on the nature of the functional groups present and on the length of the hydrocarbon chains of the molecules. The thermodynamic behavior of aqueous mixtures depends on the nature of the solute. A preliminary classification of the nature of organic mixtures helps in deciding what kind of modeling would be appropriate to describe their behavior. Considering Equation 16.12a,

$$g_\theta^E = h_\theta^E - Ts_\theta^E \qquad (16.12a)$$

we may distinguish two extreme cases.

In the first one, there are no strong interactions between the functional groups forming the molecules and the predominant effect is due to difference in the hydrocarbon length of the compounds. In this case, the mixing will be almost athermal; that is, it will not be necessary to have heat exchanged to keep the temperature constant while mixing compounds at the same temperature. As the change in enthalpy is related to heat exchange, we associate this athermal process with an entropic effect. Thus, for athermal mixing, as an approximation, we write

$$g_\theta^E \approx -Ts_\theta^E$$

$$h_\theta^E = \sum x_i \bar{h}_i^E \approx 0$$

So, $\bar{h}_i^E \approx 0$, and from Equation 16.17,

$$\bar{h}_i^E = -RT^2\left(\frac{\partial \ln \gamma_{i,\theta}}{\partial T}\right) \approx 0 \tag{26.1}$$

Hence, $\gamma_{i,\theta}$ is independent of temperature.

In the second case, the heat effect of mixing is present and the process is called nonathermal, which means that it is thermal. This situation, when the enthalpic contribution dominates, is the most commonly encountered in practice,

$$g_\theta^E \approx h_\theta^E$$

$$s_\theta^E = \sum x_i \bar{s}_i^E \approx 0$$

So, $\bar{s}_i^E \approx 0$, and from Equation 16.18,

$$\bar{s}_i^E = -RT\left(\frac{\partial \ln \gamma_{i,\theta}}{\partial T}\right) - R\ln\gamma_{i,\theta} = -\left(\frac{\partial RT\ln\gamma_{i,\theta}}{\partial T}\right) \approx 0 \tag{26.2}$$

Thus, in this case, $(RT\ln\gamma_{i,\theta})$ is independent of temperature.

In actual mixtures, the two extreme situations considered above are only approximations of reality. In any case, the relation between the excess enthalpy and the excess Gibbs energy is given by the Gibbs–Helmholtz equation, Equation 16.13a:

$$h_\theta^E = -RT^2\left(\frac{\partial\left(g_\theta^E/RT\right)}{\partial T}\right)_{P,n} \tag{16.13a}$$

Therefore, any model that would represent well the excess Gibbs energy as a function of temperature could be used to obtain values for the excess enthalpy, that is, the heats of mixing. Reciprocally, information on heats of mixing can be used with advantage to fit parameters for the excess Gibbs energy as a function of temperature.

Van Laar Model

One of the first successful models of solutions was proposed by van Laar [1,2], who, being a student of van der Waals, based his work on the van der Waals equation of state. As this equation can be linked to basic statistical thermodynamics, it is only fair to start this chapter with a presentation of the van Laar model that is useful for nonathermal, that is, enthalpic, mixtures.

In the van Laar model, it is assumed that the mixing in the liquid phase occurs without change in volume; thus,

$$g_0^E \approx h_0^E = u_0^E + Pv_0^E \approx u_0^E = u - u^{id,0} = \Delta u$$

Subsequently, the two pure compounds are decompressed isothermally from the liquid phase to the ideal gas state, the two ideal gases are mixed at constant temperature, and the mixture is isothermally compressed from the ideal gas state to the liquid phase. As discussed in Chapter 11, forming an ideal gas mixture happens without change in internal energy. Therefore, only the first and third steps contribute to the change in internal energy per mole of mixture,

$$g_0^E = \sum x_i \Delta u_i^{vaporization} - \Delta u_{mixture}^{vaporization}$$

From Equation 5.21, for an isothermal change assuming the validity of the van der Waals equation (Equation 25.11),

$$\left(\frac{\partial u}{\partial v} \right)_T = \frac{a}{v^2}$$

Thus, integrating between the liquid volume and an infinite volume for the ideal gas state, we get

$$\Delta u^{vaporization} = \frac{a}{v^L} \tag{26.3}$$

and

$$g_0^E = \left(x_1 \frac{a_1}{v_1^L} + x_2 \frac{a_2}{v_2^L} \right) - \frac{a_m}{v_m^L} \tag{26.4}$$

As van Laar assumed no volume change, for a binary mixture, we can write

$$v_m^L = x_1 v_1^L + x_2 v_2^L$$

For a_m, he used the mixing rule given by Equation 25.20:

$$a_m = x_1^2 a_1 + 2 x_1 x_2 \sqrt{a_1 a_2} + x_2^2 a_2$$

Replacing and rearranging gives

$$g_0^E = \frac{\left(x_1 v_1^L \right)\left(x_2 v_2^L \right)}{x_1 v_1^L + x_2 v_2^L} \left(\frac{\sqrt{a_1}}{v_1^L} - \frac{\sqrt{a_2}}{v_2^L} \right)^2 = \left(x_1 v_1^L + x_2 v_2^L \right) \Phi_1 \Phi_2 \left(\frac{\sqrt{a_1}}{v_1^L} - \frac{\sqrt{a_2}}{v_2^L} \right)^2 \tag{26.4a}$$

with

$$\Phi_i = \frac{x_i v_i^L}{x_1 v_1^L + x_2 v_2^L} \tag{26.5}$$

The activity coefficients obtained from this expression for the excess Gibbs energy using Equation 16.20 are

$$\ln \gamma_1 = \frac{A}{\left(1 + \dfrac{A}{B} \dfrac{x_1}{x_2} \right)^2} \tag{26.6}$$

$$\ln \gamma_2 = \frac{B}{\left(1 + \dfrac{B}{A}\dfrac{x_2}{x_1}\right)^2} = \frac{B}{\left(\dfrac{B}{A}\dfrac{x_2}{x_1}\right)^2 \left(1 + \dfrac{A}{B}\dfrac{x_1}{x_2}\right)^2} \tag{26.7}$$

with

$$A = \frac{v_1^L}{RT}\left(\frac{\sqrt{a_1}}{v_1^L} - \frac{\sqrt{a_2}}{v_2^L}\right)^2 \tag{26.8}$$

and

$$B = \frac{v_2^L}{RT}\left(\frac{\sqrt{a_1}}{v_1^L} - \frac{\sqrt{a_2}}{v_2^L}\right)^2 \tag{26.9}$$

van Laar replaced the liquid volume by the van der Waals parameter b. For all practical purposes, this is immaterial. Independently of the values of the parameters, according to Equation 26.4a this model always gives positive values of the excess Gibbs energy; so, it can only predict positive deviations from ideality. In addition, being a thermal model, the activity coefficients satisfy Equation 26.2. In practice, the van Laar equation is used with adjustable parameters A and B. Thus, Equations 26.8 and 26.9 are irrelevant and the van Laar equation can represent positive or negative deviations from ideality. The van Laar equation is particularly useful for back-of-the-envelope calculations as the two parameters A and B can be directly evaluated from the knowledge of the coordinates of a single vapor–liquid equilibrium point using Equation 16.2 or Equation 16.4. For example, if the binary system has an azeotropic point, it is enough to know the coordinates of this point to calculate both parameters. For this direct evaluation of the parameters, it is useful to observe that from Equations 26.6 and 26.7,

$$\frac{\ln \gamma_1}{\ln \gamma_2} = \frac{B}{A}\left(\frac{x_2}{x_1}\right)^2 \tag{26.10}$$

REGULAR SOLUTION THEORY

The regular solution theory, independently developed by Hildebrand [3] and Scatchard [4], follows closely the van Laar equation. This theory defines the solubility parameter δ_i as

$$\delta_i \equiv \sqrt{\frac{\Delta u_i^{vap}}{v_i^L}} \tag{26.11}$$

Then from Equation 26.3,

$$\delta_i = \frac{\sqrt{a_i}}{v_i^L}$$

Direct replacement into Equation 26.4 gives

$$g_0^E = \left(x_1 v_1^L + x_2 v_2^L\right)\Phi_1\Phi_2\left(\delta_1 - \delta_2\right)^2 \tag{26.12}$$

and then

$$\ln\gamma_1 = \frac{v_1^L \Phi_2^2}{RT}\left(\delta_1 - \delta_2\right)^2 \tag{26.13}$$

and

$$\ln\gamma_2 = \frac{v_2^L \Phi_1^2}{RT}\left(\delta_1 - \delta_2\right)^2 \tag{26.14}$$

The form of Equations 26.12 through 26.14 explains why the parameter δ was called the solubility parameter. If δ has the same value for both compounds, the solution is ideal and the two compounds are totally soluble one in the other. If a solvent has a value of the solubility parameter close to that of a solute, it is a good solvent for that solute. Solubility parameters for different compounds are available in the literature [5,6].

The extension to mixtures of the regular solution theory follows directly from a generalization of Equation 26.4 with van der Waals mixing rules:

$$g_0^E = \left(\sum_i x_i \frac{a_i}{v_i^L}\right) - \frac{a_m}{\sum_k x_k v_k^L} \tag{26.15}$$

with

$$a_m = \sum_j \sum_k x_j x_k a_{jk}$$

The activity coefficient for a component i in a regular solution mixture is given by

$$\ln\gamma_i = \frac{v_i^L}{RT}\left(\delta_i - \overline{\delta}\right)^2 \tag{26.16}$$

with

$$\overline{\delta} = \sum_j \Phi_j \delta_j \tag{26.17}$$

There are two facts worth observing from these results. There is a major simplification in the equations caused by the introduction of the definition of the solubility parameter. Although both the van Laar and the regular solution treatments appear to be the same, the latter provides a more clear connection of the parameters with intermolecular forces through Equation 26.11. The regular solution theory interprets the ratio of the molar internal energy change for the total vaporization over the liquid molar volume as a cohesive energy density, that is, the energy that holds together the molecules in the liquid phase. In fact, as discussed in Chapter 25, the relation of the parameter a with intermolecular forces was already made in the derivation of the van der Waals equation of state.

TWO-SUFFIX MARGULES EQUATION AND THE REDLICH–KISTER POLYNOMIAL

As a corollary to this comparison between the van Laar and regular solution treatments, it is interesting to see their connection with empirical forms, which were used in the first half of the twentieth century. Considering the molar volumes of compounds 1 and 2 to be the same, say v^L, and defining a new parameter A_0 by

$$A_0 \equiv v^L \left(\delta_1 - \delta_2\right)^2 = v^L \left(\frac{\sqrt{a_1}}{v^L} - \frac{\sqrt{a_2}}{v^L}\right)^2 \tag{26.18}$$

Equation 26.4a can be written as

$$g_0^E = A_0 x_1 x_2 \tag{26.19}$$

and Equations 26.6a and 26.7a become

$$\ln \gamma_1 = \frac{A_0}{RT} x_2^2 \tag{26.20}$$

and

$$\ln \gamma_2 = \frac{A_0}{RT} x_1^2 \tag{26.21}$$

Considering the definition of A_0 by Equation 26.18, this is again a thermal model. Nevertheless, Equation 26.19 was an empirical proposal independent of the van Laar model or the regular solution theory. Thus, parameter A_0 was adjusted freely, and it could be positive or negative depending on the deviation of the solution from ideality. One problem with Equation 26.19, known as Margules two-suffix form [7], is that it gives a symmetric excess Gibbs energy. In order to account for asymmetry, extra terms were introduced in what it was called a Redlich–Kister [8] expansion of the form,

$$g_0^E = x_1 x_2 \left[A_0 + B_0 \left(x_1 - x_2\right) + C_0 \left(x_1 - x_2\right)^2\right] \tag{26.22}$$

These empirical forms were used until the early fifties.

FLORY HUGGINS AND GUGGENHEIM–STAVERMAN MODELS FOR ATHERMAL MIXTURES

In many cases of solutions of polymers in organic solvents, functional group interactions are either not present or weak. In these cases, we expect the entropic contribution to the excess Gibbs energy to be predominant. Huggins [9] and Flory [10] recognized this fact and, almost simultaneously, proposed a combinatorial expression based on the array of segments of the molecules in a two-dimensional lattice. Without entering into the details of the derivations, we give here the results for the excess entropy that, in this case, directly relates to the excess Gibbs energy. For the Flory–Huggins model,

$$\left(\frac{g_0^E}{RT}\right)_{FH} = -\frac{s_0^E}{R} = \sum_j x_j \ln \frac{\Phi_j}{x_j} \tag{26.23}$$

with

$$\Phi_j = \frac{x_j r_j}{\sum_k x_k r_k} \tag{26.24}$$

We observe that Equation 26.24 is just a generalization of Equation 26.5, considering $r_j = v_j^L/v_0$, where v_0 is the volume of a standard segment and r_j is then the number of segments of the molecule j. The activity coefficients obtained from the Flory–Huggins model are of the form

$$\left(\ln \gamma_i\right)_{FH} = 1 - \frac{\Phi_i}{x_i} + \ln \frac{\Phi_i}{x_i} \tag{26.25}^\dagger$$

For the Guggenheim–Staverman model,

$$\left(\frac{g_0^E}{RT}\right)_{GS} = -\frac{s_0^E}{R} = \sum_j x_j \ln \frac{\Phi_j}{x_j} + \sum_j x_j q_j \ln \frac{\theta_j}{\Phi_j} \tag{26.26}$$

with

$$\theta_j = \frac{x_j q_j}{\sum_k x_k q_k} \tag{26.27}$$

---

† In the study of polymer solutions, Equation 26.25 is written in an almost unrecognizable way. It is easy to show that for a solvent s having in solution a polymer of r_p segments, $\dfrac{\Phi_s}{x_s} = \Phi_s + \dfrac{1}{r_p}\Phi_p = 1 - \left(1 - \dfrac{1}{r_p}\right)\Phi_p$. Thus, Equation 26.25 can be written as

$$\ln \gamma_s = \left(1 - \frac{1}{r_p}\right)\Phi_p + \ln\left[1 - \left(1 - \frac{1}{r_p}\right)\Phi_p\right] \text{ and } \ln a_s \equiv \ln(x_s \gamma_s) = \left(1 - \frac{1}{r_p}\right)\Phi_p + \ln\left(1 - \Phi_p\right).$$

In Equation 26.27, q_i is a surface area parameter of molecule i. As detailed in a review by Sayegh and Vera [11], Guggenheim [12] considered that the surface area of the segments was important for the combinatorial array in a lattice, and Staverman [13] demonstrated that Guggenheim's combinatorial expression was applicable not only to open-chain r-mers but also to bulky molecules.

When a new theory is presented, it is not easy to get rid of older nomenclature and symbols that are unnecessary in the new approach. In his combinatorial analysis, Guggenheim retained the nomenclature coming from lattice models. Thus, in Equations 26.26 and 26.27, we have taken the liberty of modifying Guggenheim's nomenclature and used the symbol q_j for what in Guggenheim's treatment appears as $zq_j/2$, with z being the coordination number, that is, the number of closest neighbors to a site in a lattice. This change is only formal, as it does not change the concepts or the algebra involved. The activity coefficients obtained from the Guggenheim–Staverman model are of the form

$$\left(\ln\gamma_i\right)_{GS} = \left(1 - \frac{\Phi_i}{x_i} + \ln\frac{\Phi_i}{x_i}\right) - q_i\left(1 - \frac{\Phi_i}{\theta_i} + \ln\frac{\Phi_i}{\theta_i}\right) \tag{26.28}$$

If q_i is directly proportional to r_i, Equation 26.27 reduces to Equation 26.24 and Equation 26.28 reduces to Equation 26.25.

COMPLETE REGULAR SOLUTIONS AND FLORY–HUGGINS EXPRESSIONS

It did not take long for researchers to realize that both the regular solution theory and the Flory–Huggins theory were using only half of Equation 16.12a.

$$g_\theta^E = h_\theta^E - Ts_\theta^E \tag{16.12a}$$

For a binary polymer–solvent mixture, the Flory–Huggins complete form is

$$\frac{g_0^E}{RT} = \left(x_s\ln\frac{\Phi_s}{x_s} + x_p\ln\frac{\Phi_p}{x_p}\right) + v_s^L\left(x_s + x_pr_p\right)\Phi_s\Phi_p\chi_F \tag{26.29}$$

and for a binary mixture, the regular solution theory complete form is

$$\frac{g_0^E}{RT} = \left(x_1\ln\frac{\Phi_1}{x_1} + x_2\ln\frac{\Phi_2}{x_2}\right) + \left(x_1v_1^L + x_2v_2^L\right)\Phi_1\Phi_2\frac{\left(\delta_1-\delta_2\right)^2}{RT} \tag{26.30}$$

or, defining $r_2 = v_2^L/v_1^L$,

$$\frac{g_0^E}{RT} = \left(x_1\ln\frac{\Phi_1}{x_1} + x_2\ln\frac{\Phi_2}{x_2}\right) + v_1^L\left(x_1 + x_2r_2\right)\Phi_1\Phi_2\frac{\left(\delta_1-\delta_2\right)^2}{RT} \tag{26.30a}$$

Comparison of Equation 26.29 and Equation 26.30a shows that the Flory parameter is directly related to the solubility parameters. However, in practice the Flory parameter is used as an adjustable parameter and sometimes is considered to be a function of temperature. The activity coefficient of a compound i obtained from the complete form of both theories is just the sum of the two contributions. For the complete regular solution theory, for example, from Equations 26.16 and 26.25,

$$\ln \gamma_i = \left(1 - \frac{\Phi_i}{x_i} + \ln \frac{\Phi_i}{x_i} \right) + \frac{v_i^L}{RT} \left(\delta_i - \bar{\delta} \right)^2 \tag{26.31}$$

MODERN TIMES

The second half of the twentieth century was a period of intense search for models of the excess Gibbs energy or, equivalently, of activity coefficients for the calculation of phase equilibria involving nonideal liquid mixtures. One stream followed the quasi-chemical theory of Guggenheim [14–18]. Another stream developed models based either in the two-fluid approach [19] or in the local composition concept proposed by Wilson [20] and elaborated further by Prausnitz et al. [21,22], among others. These later models were and are extremely successful, and with the advent of computers, they were easy to use. Both kinds of treatments eventually evolved into approaches based on the functional groups forming the molecules. The quasi-chemical theory evolved into the form proposed by Kehiaian et al. [15] while the local composition models into the analytical solution of groups (ASOG) [23–25] and universal functional activity coefficient (UNIFAC) [26]. We present here a brief description of the quasi-chemical theory and the two-fluid and local composition models.

QUASI-CHEMICAL THEORY

In this presentation, we distinguish between the Guggenheim combinatorial partition function and the Guggenheim quasi-chemical partition function, and we also get rid of the coordination number that was included in this treatment as a leftover from old lattice approaches. Following the work of Kehiaian et al. [15], we adopt from the beginning the group approach, which reduces to a molecular approach when one considers the molecules of different compounds as individual groups. The Guggenheim partition function is based on the idea that in a liquid mixture, what is important is the way in which different molecules or the segments of the molecules contact each other. The segments are characteristic functional groups, so the whole liquid mixture can be considered a mixture of groups. The importance of the functional groups forming the molecules has been discussed in Chapter 17. The interaction between functional groups is determinant in the thermodynamic behavior of the mixture. Thus, Guggenheim considered that the mixture behavior is characterized by the way in which the contacts between groups will arrange in the liquid phase. Based on the combinatorial array for the number of contacts of each type, Guggenheim's combinatorial partition function is written as

$$Q_{CG} = \frac{\left(\sum\limits_{k=1}^{G} \eta_{kk} + \sum\limits_{k=1}^{G}\sum\limits_{l \neq k}^{G} \frac{1}{2}\eta_{kl}\right)!}{\prod\limits_{k=1}^{G}(\eta_{kk})! \prod\limits_{k \neq l}^{G}\left(\frac{1}{2}\eta_{kl}\right)!} \qquad (26.32)$$

In this expression, η_{kl} represents the "number of contacts" between groups k and l, considered to be proportional to the surface area of group k available for contacts with group l. This availability for contacts will certainly depend on the energetic compatibility of the groups. Interactions between groups are always attractive, but the interactions between groups of different kinds can be stronger or weaker than the interaction between groups of the same kind. Stronger interactions will tend to produce negative deviations from ideality, and weaker interactions will tend to produce positive deviations from ideality. If the interactions are similar, the array will be random and the mixture will approach the behavior of an ideal mixture. The summation of the number of contacts will be the same for the random and nonrandom arrays, as the surface areas of the groups do not change. Only the way in which the contacts are arranged changes. Hence,

$$2\eta_{kk} + \sum\limits_{l \neq k}^{G} \eta_{kl} = \alpha_k \qquad (26.33a)$$

and

$$\left(\sum\limits_{k=1}^{G} \eta_{kk} + \sum\limits_{k=1}^{G}\sum\limits_{l \neq k}^{G} \frac{1}{2}\eta_{kl}\right) = \left(\sum\limits_{k=1}^{G} \eta_{kk}^0 + \sum\limits_{k=1}^{G}\sum\limits_{l \neq k}^{G} \frac{1}{2}\eta_{kl}^0\right) = \sum\limits_{k=1}^{G} \frac{\alpha_k}{2} \qquad (26.33b)$$

where η_{kl}^0 represents the number of contacts of type k-l in a random array and α_k is the total surface area of groups k available for external contacts. For compatibility with the counting of the number of contacts, α_k is better interpreted as the total number of contact points of group k, each point representing a unit of surface area. This total number is proportional to the surface area available for external contacts in group k. The factor 1/2, which is immaterial, is introduced to avoid double counting of contacts. Hence, Equation 26.32 for a random array can be written as

$$Q_{CG}^0 = \frac{\left(\sum\limits_{k=1}^{G} \frac{\alpha_k}{2}\right)!}{\prod\limits_{k=1}^{G}(n_{kk}^0)! \prod\limits_{k \neq l}^{G}(n_{kl}^0)!} \qquad (26.32a)$$

In his quasi-chemical approach, Guggenheim normalized the combinatorial partition function to modify the total random array of contacts and multiplied the combinatorial partition function by a factor g_R of the form

$$g_R = \frac{\prod_{k=1}^{G}(n_{kk}^0)! \prod_{k \neq l}^{G}\left(\frac{1}{2}n_{kl}^0\right)!}{\prod_{k=1}^{G}(n_{kk})! \prod_{k \neq l}^{G}\left(\frac{1}{2}n_{kl}\right)!}$$
(26.34)

So, for the complete quasi-chemical partition function he obtained

$$Q_{QC} = Q_{CG}^0 g_R \exp\left[\frac{-E}{RT}\right]$$
(26.35)

with

$$-E = \sum_{k=1}^{G} \eta_{kk}\varepsilon_{kk} + \sum_{k=1}^{G-1}\sum_{l>k}^{G} \eta_{kl}\varepsilon_{kl}$$
(26.36)

where ε_{kl} is the interaction energy between groups k and l. By minimization of the Helmholtz function for the equilibrium condition, Guggenheim obtained the following relation:

$$\frac{4\eta_{kk}\eta_{ll}}{\eta_{kl}^2} = \exp\left[\frac{\Delta\varepsilon_{kl}}{RT}\right]$$
(26.37)

Due to the similitude of this expression with the form of a chemical equilibrium constant, Guggenheim called this treatment the quasi-chemical theory. A complete derivation of the thermodynamic functions generated from this theory was presented by Wilczek-Vera and Vera [18]. Panayiotou and Vera [27] identified the rigorous use of the concept of local composition in the quasi-chemical theory.

The quasi-chemical theory of Guggenheim, attractive as it seems from a theoretical point of view, has some serious drawbacks. On the theoretical side, Prigogine et al. [28] and also Kemeny et al. [29] have argued that Guggenheim's normalization of the partition function satisfying the condition for total randomness of the contacts between groups make this theory valid only for cases close to total randomness and not for cases of very specific interactions between groups. On the practical side, the quasi-chemical theory has two major drawbacks. First, by Equation 26.37, it ties the interactions between different groups kl to the interactions of the individual groups kk and ll. Thus, it limits the treatment to a single binary parameter when it is well known that the thermodynamic treatment of binary mixtures needs two parameters due to the different behaviors of a mixture in the two regions of high dilution. The second drawback is that for more than two groups, the quasi-chemical relation, Equation 26.37, is far from simple to solve. A general solution method was developed by Abusleme and Vera [30], but it is certainly not easy to apply. Notably, both problems presented by the quasi-chemical treatment arise from the form of the normalization function g_R used by Guggenheim, Equation 26.34, which is not part of the combinatorial expression, Equation 26.32a.

WILSON EQUATION FOR THE EXCESS GIBBS ENERGY

The structure of the equation for the excess Gibbs energy proposed by Wilson [20] is best seen in the context of Flory treatment, Equations 26.23 through 26.25. Wilson argued that due to specific interactions in a solution of components i and j, the local mole fraction of compound j around a central molecule i, x_{ij}, would be different from the overall mole fraction of compound j, x_j, in the mixture. For this local composition, he proposed

$$\frac{x_{ij}}{x_{ii}} = \frac{x_j}{x_i}\exp\left(\frac{-\left(\lambda_{ij}-\lambda_{ii}\right)}{RT}\right) = \frac{x_j}{x_i}\tau_{ij} \tag{26.38}$$

So, instead of Equation 26.24, Wilson wrote a "local" volume, or segment, fraction as

$$\Phi_j^w = \frac{x_{jj}r_j}{\displaystyle\sum_k x_{jk}r_k} = \frac{x_j r_j}{\displaystyle\sum_k x_k r_k \tau_{jk}} = \frac{\Phi_j}{\displaystyle\sum_k \Phi_k \tau_{jk}} \tag{26.39}$$

with Φ_j given by Equation 26.5. Replacing this expression in Equation 26.23,

$$\left(\frac{g_0^E}{RT}\right)_w = \sum_j x_j \ln\frac{\Phi_j^w}{x_j} = \sum_j x_j \ln\frac{\Phi_j}{x_j \displaystyle\sum_k \Phi_k \tau_{jk}} \tag{26.40a}$$

$$= \sum_j x_j \ln\frac{\Phi_j}{x_j} - \sum_j x_j \ln\sum_k \Phi_k \tau_{jk}$$

On the right-hand side of Equation 26.40a, we recognize a combinatorial contribution and the introduction of an enthalpic term. The Wilson equation is usually written in a more compact form as

$$\left(\frac{g_0^E}{RT}\right)_w = -\sum_j x_j \ln\sum_k x_k \Lambda_{jk} \tag{26.40b}$$

with

$$\Lambda_{jk} = \frac{r_k}{r_j}\tau_{jk} \tag{26.41}$$

We observe here that by their definition in Equations 26.38 and 26.41, both τ_{jk} and Λ_{jk} are considered to be temperature dependent.

The expression for the activity coefficient obtained from Equation 26.40a is

$$\ln \gamma_i = \left(1 - \frac{\Phi_i}{x_i} + \ln \frac{\Phi_i}{x_i}\right) + \left(\frac{\Phi_i}{x_i} - \ln \sum_k \Phi_k \tau_{ik} - \frac{\Phi_i}{x_i} \sum_k \frac{x_k \tau_{ki}}{\sum_j \Phi_j \tau_{kj}}\right) \qquad (26.42a)$$

And the activity coefficient obtained from Equation 26.40b is

$$\ln \gamma_i = 1 - \ln \sum_k x_k \Lambda_{ik} - \sum_k \frac{x_k \Lambda_{ki}}{\sum_j x_j \Lambda_{kj}} \qquad (26.42b)$$

Equations 26.42a and 26.42b are just a transformation one of the other. The form shown in Equation 26.42a explicitly shows the combinatorial contribution and the contribution due to interactions between molecules. Realizing that Equation 26.40 (a or b) was unable to describe the case of instability of a liquid mixture, that is, liquid–liquid phase separation, Wilson [20] added a third parameter, C_W, which should be greater than unity for the representation of mixtures splitting into two liquid phases.

$$\left(\frac{g_0^E}{RT}\right)_w = -C_W \sum_j x_j \ln \sum_k x_k \Lambda_{jk} \qquad (26.43)$$

This three-parameter form of the Wilson equation satisfies the condition for instability.

Flemr [31] and also McDermott and Ashton [32] questioned the theoretical basis of Wilson's local composition expression. A redefinition of local composition and local surface area fractions compatible with the quasi-chemical theory was later presented by Panayiotou and Vera [27]. A new approach to this problem is discussed at the end of this chapter.

BIRTH OF THE SOLUTION OF GROUPS METHOD TO CALCULATE ACTIVITY COEFFICIENTS IN LIQUID MIXTURES

Superior minds sometimes can foresee relations that normal intellects will take time to justify. This is the case with the proposal of the Wilson equation, Wilson and Deal's proposal of a group contribution method to evaluate the activity coefficients in organic mixtures, the formulation of ASOG by Derr, and the proposal of universal quasi-chemical (UNIQUAC). We briefly review these groundbreaking advances that followed Wilson's equation. We use our own nomenclature.

Wilson and Deal [23], based on some previous studies on heats of mixing [33,34], and obviously aware of the two contributions to the excess Gibbs energy shown by Equation 16.12a, proposed that the activity coefficient of a compound in an organic mixture be expressed as the sum of a Flory combinatorial term and an "enthalpic term" arising from the interactions between the groups:

$$\ln \gamma_i = (\ln \gamma_i)_{combinatorial} + (\ln \gamma_i)_{interactions}$$

$$\ln\gamma_i = \left(1 - \frac{r_i}{\sum_k x_k r_k} + \ln\frac{r_i}{\sum_k x_k r_k}\right) + \sum_k \nu_{ki}\left[\ln\Gamma_k - \ln\Gamma_k^{(i)}\right] \tag{26.44}$$

The presentation of this expression, so familiar to us now, is so impressive that it is necessary to say something about each of the two terms. For the Flory combinatorial term, Wilson and Deal, combining Equations 26.24 and 26.25, explicitly showed the size parameter r_i of compound i, which they proposed to identify with the number of atoms different from hydrogen in molecule i. This way, they clearly were counting the functional groups present in compound i. In the second term, ν_{ki} is the number of groups of type k in molecule i, Γ_k is the activity coefficient of a group of type k with respect to an arbitrary but well-defined standard state, and $\Gamma_k^{(i)}$ is the activity coefficient of a group of type k in pure compound i, with respect to the same standard state as Γ_k.

Here, we observe that Γ_k is a function of the (group) composition of the mixture, while $\Gamma_k^{(i)}$ is a function of the group composition in pure compound i. The (mole) fraction of group k is given by

$$X_k = \frac{\sum_j x_j \nu_{kj}}{\sum_l \sum_m x_m \nu_{lm}} \tag{26.45a}$$

and in pure compound i,

$$X_k^{(i)} = \frac{\nu_{ki}}{\sum_l \nu_{li}} \tag{26.45b}$$

Thus, according to Equation 26.44 the activity coefficient $\gamma_{i,0}$ is normalized to unity at the condition of a pure compound i. The group activity coefficient Γ_k is normalized to unity in a system formed only by group k. This is the case of the group CH_2 in alkanes, as no distinction is made between the methyl (CH_3-) and methylene ($-CH_2-$) groups. This normalization was assumed to also be valid for group $-OH$ in pure water. Water in mixtures was considered to be 1.5 ($-OH$) groups. To test the model, they studied several binary and ternary systems: alcohol–alkane, alcohol–alcohol, water–alcohol, and so forth. With the assumption that the group activity coefficient is only a function of the group composition of the mixture, and it does not depend on the way that the groups are attached in the molecules, data from one system could be used for a different system at the same group composition. Activity coefficient data of water in alcohol–water systems, from methanol to decyl alcohol, provided the information necessary to calculate Γ_{OH} at different X_{OH} values in CH_2/OH systems. Similarly, from activity coefficient data of the hydrocarbon in hydrocarbon–alcohol systems, it is possible to calculate Γ_{CH_2} at different X_{CH_2} values in CH_2/OH systems. The method was direct and simple. In the evaluation of the group activity coefficients, the authors did not use any model, other than Equation 26.44, which does not contain any adjustable parameter. The combinatorial contribution in Equation 26.44 is calculated and subtracted from the logarithm of the activity coefficient γ. Plots of group activity coefficients Γ_k against the group k mole fraction, coming from very different systems at the same temperature, all fell over a single curve for a given group. In addition, it was observed that the curves of group activity coefficients against group fraction presented a remarkable symmetry, with curves of compound activity coefficients versus mole fraction of the compound. Figure 26.1 depicts the values of the activity coefficient of group OH in the system methanol–hexane at 60°C. Figure 26.2 presents the values of the activity coefficient of methanol in the same system.

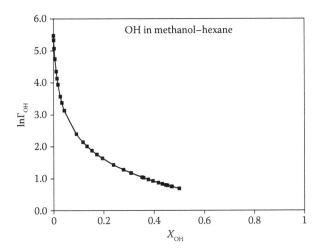

FIGURE 26.1 Activity coefficient of group OH in the system methanol–hexane at 60°C. (From Koennecke, H. G., *J. Prakt. Chem.* 311: 974–982, 1969.)

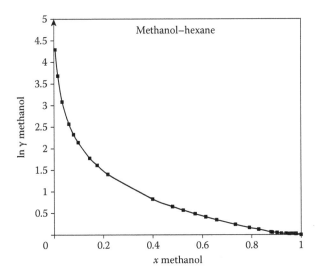

FIGURE 26.2 Activity coefficient of methanol in the system methanol–hexane at 60°C. (From Koennecke, H. G., *J. Prakt. Chem.*, 311, 974–982, 1969.)

Studies of other alkane–alcohol systems gave values of the activity coeffi-cient of the OH group overlapping and extending the curve shown in Figure 26.1. These results are so impressive that they should be shown in any lecture dealing with group methods. Further results were presented by Deal and Derr [24], who announced that Derr had developed ASOG, to be published later. The publication of Derr, based on the use of the Wilson equation for the excess Gibbs energy, came a year later [25].

ANALYTICAL SOLUTION OF GROUPS (ASOG)

After obtaining plots of $\ln \Gamma_k$ versus X_k, without using a single model and observ-ing that the curves looked like curves of $\ln \gamma_i$ versus x_i, Derr and Deal [25] used the Wilson equation for the correlation of the experimental values of $\ln \Gamma_k$. From Equation 26.42b, in terms of group fractions, in its more general form we write

$$\ln \Gamma_k = 1 - \ln \sum_l X_l A_{kl} - \sum_l \frac{X_l A_{lk}}{\sum_m X_m A_{lm}} \tag{26.46a}$$

and

$$\ln \Gamma_k^{(i)} = 1 - \ln \sum_l X_l^{(i)} A_{kl} - \sum_l \frac{X_l^{(i)} A_{lk}}{\sum_m X_m^{(i)} A_{lm}} \tag{26.46b}$$

Values for the group interaction parameters were published much later by Kojima and Tochigi [35a,b], using the form

$$A_{kl} = \exp\left(m_{kl} + \frac{n_{kl}}{T} \right) \tag{26.47}$$

Group parameters m_{kl} and n_{kl} for this version of ASOG, usually designated as ASOG-KT, are available in the literature [35a,b], and they are reproduced with per-mission in Appendix D. Other applications of ASOG-KT for high-pressure or poly-mer solutions have also been developed [36,37]. For the counting of groups (v_{kj}), the ASOG-KT method recommends using 1.6 OH for water, 0.8 CH_2 for the CH group, and 0.5 CH_2 for the C group; otherwise, v_{kj} is the total number of atoms different from hydrogen in all the groups of type k in molecule i.

As the proposal of ASOG started from studies of heats of mixing, for complete-ness, we give below the expressions for the excess enthalpy and the excess entropy obtained from ASOG [38].

$$h^E = \sum_j x_j \sum_k v_{kj} \left[H_k - H_k^{(j)} \right]$$

with

$$\frac{H_k}{RT^2} = \frac{\partial \ln \Gamma_k}{\partial T} = \frac{\sum_l X_l B_{kl}}{\sum_j X_j A_{kj}} + \sum_j \frac{X_j B_{jk}}{\sum_l X_l A_{jl}} - \sum_j \frac{X_j A_{jk}\left(\sum_l X_l B_{jl}\right)}{\left(\sum_l X_l A_{jl}\right)^2}$$

where

$$B_{kl} = \frac{\partial A_{kl}}{\partial T} = -A_{kl}\frac{n_{kl}}{T^2}$$

and

$$s^E = \sum_j x_j \sum_k v_{kj}\left[S_k - S_k^{(j)}\right]$$

with

$$\frac{S_k}{R} = \frac{H_k}{RT} - \ln \Gamma_k$$

NONRANDOM TWO-LIQUID EQUATION
FOR THE EXCESS GIBBS ENERGY

Based on Scott's two-fluid model [39], coupled with Wilson's local composition concept [20], Renon and Prausnitz [40] proposed a three-parameter expression for the excess Gibbs energy of the form

$$\left(\frac{g_0^E}{RT}\right)_{NRTL} = \sum_k x_k \frac{\sum_j x_j G_{jk}\tau_{jk}}{\sum_l x_l G_{lk}} \qquad (26.48)$$

with

$$G_{jk} = \exp(-\alpha_{jk}\tau_{jk}) \qquad (26.49)$$

and

$$\tau_{jk} = \frac{\Delta g_{jk}}{RT} \qquad (26.50)$$

In these expressions, the adjustable parameters in a binary system are Δg_{12}, Δg_{21}, and $\alpha_{12} = \alpha_{21}$. No additional parameters are needed for multicomponent systems. The difference between the nonrandom two-liquid equation (NRTL) and the three-parameter Wilson equation is that in the latter one, once the third parameter is fixed, it has the same value for all the binary systems in the multicomponent system under study. As it was observed by Renon and Prausnitz [19], following a suggestion by R. L. Scott, had they made their two-fluid derivation for the excess enthalpy instead of the excess Gibbs energy, they would have obtained Wilson's three-parameter equation.

The expressions for the activity coefficients obtained from Equation 26.48 take the form

$$\ln \gamma_i = \frac{\sum_j x_j G_{ji} \tau_{ji}}{\sum_l x_l G_{ji}} + \sum_j \frac{x_j G_{ij}}{\sum_l x_l G_{lj}} \left[\tau_{ij} - \frac{\sum_r x_r G_{rj} \tau_{rj}}{\sum_l x_l G_{lj}} \right] \quad (26.51a)$$

For a binary system, the expressions for the activity coefficients take a friendlier form,

$$\ln \gamma_1 = x_2^2 \left[\tau_{21} \left(\frac{G_{21}}{x_1 + x_2 G_{21}} \right)^2 + \frac{\tau_{12} G_{12}}{\left(x_2 + x_1 G_{12}\right)^2} \right] \quad (26.51b)$$

and

$$\ln \gamma_2 = x_1^2 \left[\tau_{12} \left(\frac{G_{12}}{x_2 + x_1 G_{12}} \right)^2 + \frac{\tau_{21} G_{21}}{\left(x_1 + x_2 G_{21}\right)^2} \right] \quad (26.51c)$$

Although the theoretical basis of both the local composition concept, as formulated by Wilson, and the two-fluid approach have been questioned [41], the NRTL equation has been found to be extremely successful for the correlation and prediction of vapor–liquid and liquid–liquid equilibria.

UNIQUAC EQUATION

As stated by Prausnitz et al. [42a], the UNIQUAC equation proposed by Abrams and Prausnitz [21] was derived by Maurer and Prausnitz [22] using phenomenological arguments based on the local composition concept and the two-liquid approach, in a similar way as these concepts were used in the derivation of NRTL. These authors conclude that UNIQUAC, being a very useful two-parameter equation, "cannot claim as much theoretical foundation as was hoped originally" [22, p.96]. Looking in retrospect, the name UNIQUAC was a misnomer. The main advantage of UNIQUAC is that it is unrelated to the quasi-chemical theory, so it does not have any of its limitations. An alternative derivation of UNIQUAC relying only on the local composition concept was proposed by Vera et al. [43]. This derivation does not need the two-liquid approach. In addition, it can be related with a renormalization of Guggenheim's partition function proposed by Wang and Vera [44]. The main point distinguishing Guggenheim's from Flory–Huggins's combinatorial term was that the important factor for the interaction of the molecules (or segments of the molecules) was the surface area available for contact. Thus, the key innovation introduced by UNIQUAC was the incorporation of the surface area fraction, as defined by Equation 26.27. Here, we follow the simpler derivation suggested by Vera et al. [43], and following Wilson, we define a local surface area by the equivalent to Equation 26.42:

$$\xi_j^{LS} = \frac{x_{jj}q_j}{\sum_k x_{jk}q_k} = \frac{x_j q_j}{\sum_k x_k q_k \tau_{jk}} = \frac{\theta_j}{\sum_k \theta_k \tau_{jk}} \tag{26.52}$$

Replacing this expression in Equation 26.26 gives

$$\left(\frac{g_0^E}{RT}\right)^{LS} = \left(\sum_j x_j \ln\frac{\Phi_j}{x_j} + \sum_j x_j q_j \ln\frac{\theta_j}{\Phi_j}\right) - \sum_j x_j q_j \ln\sum_k \theta_k \tau_{jk} \tag{26.53}$$

This equation, called the local surface Guggenheim (LSG) form, only differs from UNIQUAC [42b] by a factor $(z/2)$ that was left in the combinatorial part of the latter from its historical lattice roots. The two first terms of the right-hand side of Equation 26.53 correspond to the combinatorial expression coming from Equation 26.26. Models of solutions keep changing with time, so when using parameters reported in the literature, it is mandatory to use the exact analytical form that was used for the evaluation of the parameters.

The expression for the activity coefficient derived from LSG, Equation 26.53, has the form

$$(\ln \gamma_i) = \left(1 - \frac{\Phi_i}{x_i} + \ln\frac{\Phi_i}{x_i}\right) - q_i\left(1 - \frac{\Phi_i}{\theta_i} + \ln\frac{\Phi_i}{\theta_i}\right)$$
$$+ q_i\left[1 - \ln\sum_k \theta_k \tau_{ik} - \sum_k \frac{\theta_k \tau_{ki}}{\sum_j \theta_j \tau_{kj}}\right] \tag{26.54a}$$

The first two brackets in Equation 26.54a correspond to the combinatorial contribution given by Equation 26.28. Equation 26.54a can be written in a more compact form as

$$(\ln \gamma_i) = \left(1 - \frac{\Phi_i}{x_i} + \ln\frac{\Phi_i}{x_i}\right) + q_i\left[\frac{\Phi_i}{\theta_i} - \ln\left(\frac{\Phi_i}{\theta_i}\sum_k \theta_k \tau_{ik}\right) - \sum_k \frac{\theta_k \tau_{ki}}{\sum_j \theta_j \tau_{kj}}\right] \tag{26.54b}$$

For UNIQUAC, the activity coefficients are given by

$$(\ln \gamma_i) = \left(1 - \frac{\Phi_i}{x_i} + \ln\frac{\Phi_i}{x_i}\right) - \frac{zq_i}{2}\left(1 - \frac{\Phi_i}{\theta_i} + \ln\frac{\Phi_i}{\theta_i}\right)$$

$$+ q_i\left[1 - \ln\sum_k \theta_k\tau_{ik} - \sum_k \frac{\theta_k\tau_{ki}}{\sum_j \theta_j\tau_{kj}}\right]$$
(26.55)

with $z = 10$. Notably, if for all compounds q_i is directly proportional to r_i, then $\theta_i = \Phi_i$ and both Equation 26.54a or Equation 26.54b and Equation 26.55 reduce to

$$\ln\gamma_i = \left(1 - \frac{\Phi_i}{x_i} + \ln\frac{\Phi_i}{x_i}\right) + q_i\left[1 - \ln\sum_k \Phi_k\tau_{ik} - \sum_k \frac{\Phi_k\tau_{ki}}{\sum_j \Phi_j\tau_{kj}}\right]$$

This form is similar but not identical to Wilson's equation, Equation 26.42a. As discussed by Vera et al. [43], Wilson's equation is only recovered when both Equations 26.39 and 26.52 are combined with Guggenheim's combinatorial form and for all compounds q_i is directly proportional to r_i.

UNIFAC

There are many versions of UNIFAC. We give here the original form only. Basically, UNIFAC follows the scheme of ASOG, and it considers the activity coefficient of a compound given by a combinatorial contribution and the contribution of the group interactions. For the combinatorial term, UNIFAC uses Guggenheim's form, and for the group contribution, UNIFAC follows the suggestion of Wilson and Deal [23] and writes

$$\ln\gamma_i = \left(1 - \frac{\Phi_i}{x_i} + \ln\frac{\Phi_i}{x_i}\right) - \frac{zq_i}{2}\left(1 - \frac{\Phi_i}{\theta_i} + \ln\frac{\Phi_i}{\theta_i}\right) + \sum_k v_{ki}\left[\ln\Gamma_k - \ln\Gamma_k^{(i)}\right]$$
(26.56)

with ϕ_i and θ_i given by Equations 26.24 and 26.27, respectively, but using

$$r_i = \sum_k v_{ki}R_k$$
(26.57)

and

$$q_i = \sum_k v_{ki}Q_k$$
(26.58)

where R_k and Q_k are the size and surface area parameters of group k. These values are tabulated for the different groups.

For the activity coefficients of the groups, UNIFAC uses the same form as the term for interactions in UNIQUAC but written in terms of groups, namely,

$$\ln \Gamma_k = Q_k \left[1 - \ln \sum_l \Theta_l a_{lk} - \sum_l \frac{\Theta_l a_{lk}}{\sum_m \Theta_m a_{lm}} \right] \tag{26.59}$$

with

$$\Theta_k = \frac{\sum_j x_j v_{kj} Q_k}{\sum_l \sum_m x_m v_{lm} Q_m} \tag{26.60}$$

and in pure compound i,

$$\Theta_k^{(i)} = \frac{v_{ki} Q_k}{\sum_l v_{li} Q_l} \tag{26.60a}$$

The values of the group interaction parameters are considered to be a function of temperature,

$$a_{kl} = \exp \left[\frac{B_{kl}}{T} \right] \tag{26.61}$$

Comparisons between ASOG and UNIFAC are scarce and depend on the systems chosen for the comparison. See, for example, a publication by Gmehling et al. [45].

WANG'S RENORMALIZATION OF GUGGENHEIM'S PARTITION FUNCTION

As stated in the presentation of the quasi-chemical theory, its major drawbacks come from the normalization factor introduced by Guggenheim. Thus, Wang and Vera [44] proposed to normalize Guggenheim's combinatorial term with a different function meeting the limits of total randomness, indicated with a superscript 0, and total order, indicated with a superscript *, and wrote

$$Q = Q_{CG} F_\alpha \tag{26.62}$$

For total disorder, Guggenheim's combinatorial term gives the right limiting behavior, that is, $Q = Q_{CG}^0$, so

$$F_\alpha^0 = 1 \tag{26.63}$$

For total order, indicated with an asterisk, $Q^* = 1$, because when all contacts are of type k-k, and all $\eta_{kl}^* = 0$ for $l \neq k$, there is one single array for the system. Thus, for complete order,

$$F_\alpha^* = \frac{1}{Q_{CG}^*} = \frac{\displaystyle\prod_{k=1}^{G} (\eta_{kk}^*)!}{\left(\displaystyle\sum_{k=1}^{G} \frac{\alpha_k}{2}\right)!}$$

From Equation 26.33, when all $\eta_{kl}^* = 0$ for $l \neq k$, $2\eta_{kk}^* = \alpha_k$, and we write

$$F_\alpha^* = \frac{\displaystyle\prod_{k=1}^{G} \left(\frac{\alpha_k}{2}\right)!}{\left(\displaystyle\sum_{k=1}^{G} \frac{\alpha_k}{2}\right)!}$$

Using Stirling's approximation for the factorials, we obtain

$$F_\alpha^* = \prod_{k=1}^{G} \left[\frac{\dfrac{\alpha_k}{2}}{\displaystyle\sum_{k=1}^{G} \frac{\alpha_k}{2}} \right]^{\frac{\alpha_k}{2}} \tag{26.64}$$

The function satisfying both limits proposed by Wang is

$$F_\alpha = \prod_{k=1}^{G} \left[\frac{\displaystyle\sum_{l=1}^{G} \frac{\alpha_k}{2}\tau_{lk}}{\displaystyle\sum_{k=1}^{G} \frac{\alpha_k}{2}} \right]^{\frac{\alpha_k}{2}} \tag{26.65}$$

with $\tau_{kk} = \tau_{ll} = 1$ and, in general, $\tau_{kl} \neq \tau_{lk}$. For totally disordered systems, $\tau_{kl}^0 = \tau_{lk}^0 = 1$, and for totally ordered systems, $\tau_{kl}^* = \tau_{lk}^* = 0$. The values of the parameters τ_{kl} and τ_{lk} characterize the degree of order in the mixture and are not related to any local composition.

We observe that Wang transferred the temperature dependence of the partition function from the exponential term used in Guggenheim's formulation, Equation 26.35, to the temperature dependence of the interaction parameters. Guggenheim's formulation needed the exponential term; otherwise, Equation 26.37 would not have had any adjustable parameter to represent energetic interactions. In Wang's formulation, these energy interactions are represented by the parameters characterizing the interactions of groups. According to Equation 24.14, the chemical potential is obtained by differentiation of the logarithmic form of Equation 26.62 with respect to n_i at a constant volume and constant number of moles of all other compounds different from i, so we write the logarithm of Equation 26.65 as

$$\ln F_\alpha = \sum_{k=1}^{G} \frac{\alpha_k}{2} \ln \sum_{l=1}^{G} \frac{\alpha_l}{2} \tau_{lk} - \sum_{k=1}^{G} \frac{\alpha_k}{2} \ln \sum_{l=1}^{G} \frac{\alpha_l}{2} \qquad (26.66)$$

Because the only reason to have the factor 1/2 correcting the surface area of interaction was to avoid the double counting of contacts, we eliminate it here and write

$$\frac{\alpha_k}{2} = \sum_{j=1}^{N} n_j v_{kj} Q_k \qquad (26.67)$$

where n_j is the number of moles of compound j in the mixture, v_{kj} is the number of groups of type k in molecule j, and Q_k is the surface parameter of group k, already introduced in Equation 26.58. Following the standard thermodynamic path to obtain activity coefficients,

$$\frac{\mu_i}{RT} = -\left(\frac{\partial \ln Q_{CG}}{\partial n_i} \right)_{T,V,n_{j\neq1}} - \left(\frac{\partial \ln F_\alpha}{\partial n_i} \right)_{T,V,n_{j\neq1}}$$

$$\ln(x_i \gamma_i) = \frac{\mu_i}{RT} - \frac{\mu_i^{(i)}}{RT}$$

and using Equation 26.28 for the combinatorial term, for the activity coefficient of compound i in terms of groups, we write

$$\ln \gamma_i = \left(1 - \frac{\Phi_i}{x_i} + \ln \frac{\Phi_i}{x_i} \right) - q_i \left(1 - \frac{\Phi_i}{\theta_i} + \ln \frac{\Phi_i}{\theta_i} \right) + \sum_k v_{ki} \left[\ln \Gamma_k - \ln \Gamma_k^{(i)} \right] \qquad (26.68)$$

$$\ln \Gamma_k = Q_k \left[1 - \ln \sum_l \Theta_l a_{kl} - \sum_l \frac{\Theta_l a_{lk}}{\sum_m \Theta_m a_{lm}} \right] \qquad (26.69)$$

Working in terms of molecules, the result is

$$\ln \gamma_i = \left(1 - \frac{\Phi_i}{x_i} + \ln \frac{\Phi_i}{x_i}\right) - q_i\left(1 - \frac{\Phi_i}{\theta_i} + \ln \frac{\Phi_i}{\theta_i}\right)$$

$$+ q_i\left[1 - \ln \sum_k \theta_k \tau_{ik} - \sum_k \frac{\theta_k \tau_{ki}}{\sum_j \theta_j \tau_{kj}}\right] \qquad (26.70)$$

For the equations for activity coefficients in terms of molecules, it is interesting to compare Equation 26.70 with Equation 26.54a for LSG and with Equation 26.55 for UNIQUAC. If for all k, $r_k = q_k = 1$, equation (26.70) reduces to Wilson equation (26.42b). For the equations in terms of groups, it is worth comparing Equations 26.68 and 26.69 with Equations 26.56 and 26.59 for UNIFAC. The important point to notice is that Equations 26.68 through 26.70 come from a proper normalization of Guggenheim's combinatorial partition function and are unrelated to the concept of local composition or the concept of local surface area fraction.

SAFT

The statistical associated fluid theory (SAFT) [46] is perhaps one of the most active international efforts in our days aiming at the understanding of the behavior of fluids. As Prausnitz et al. [42c] state, "The literature on SAFT is complex and confusing." Basically, SAFT is a successful computer modeling of fluid behavior. It considers that molecules are formed by groups, i.e. by repulsive hard core segments that are affected by attractive potentials. These potentials arise from chain attachments and also from interactions with other segments. These interactions can be only physical or also include association. By adding parameters to account for the different effects, SAFT has been very successful in reproducing and predicting the behavior of pure compounds and even complex mixtures. The whole simulation of the behavior of pure compounds and mixtures is computer based, and so it is out of the scope of this book. A good overall view of SAFT has been presented by Prausnitz et al. [42d], as well as by Müller and Gubbins [47].

CONCLUSIONS

Models come and go. There are several versions of UNIFAC and ASOG in the literature—too many to mention. In our view, what is important is to have a perspective of the historical path of the development of models of solutions so when new ideas are generated, they can be seen in context and the experience of past developments is not lost.

Summary of Some Useful Equations for the Correlation of Activity Coefficients

VAN LAAR MODEL

$$\ln \gamma_1 = \frac{A}{\left(1 + \dfrac{A\,x_1}{B\,x_2}\right)^2} \tag{26.6}$$

$$\ln \gamma_2 = \frac{B}{\left(1 + \dfrac{B\,x_2}{A\,x_1}\right)^2} = \frac{B}{\left(\dfrac{B\,x_2}{A\,x_1}\right)^2\left(1 + \dfrac{A\,x_1}{B\,x_2}\right)^2} \tag{26.7}$$

where A and B are adjustable parameters.

REGULAR SOLUTION THEORY

$$\ln \gamma_1 = \frac{v_1^L \Phi_2^2}{RT}(\delta_1 - \delta_2)^2 \tag{26.13}$$

and

$$\ln \gamma_2 = \frac{v_2^L \Phi_1^2}{RT}(\delta_1 - \delta_2)^2 \tag{26.14}$$

$$\ln \gamma_i = \frac{v_i^L}{RT}(\delta_i - \bar{\delta})^2 \tag{26.16}$$

$$\bar{\delta} = \sum_j \Phi_j \delta_j \tag{26.17}$$

with

$$\Phi_i = \frac{x_i r_i}{\sum_k x_k r_k} \tag{26.24}$$

The regular solution theory uses $r_i = v_i^L$. Solubility parameters δ_j are tabulated for many compounds [5].

WILSON EQUATION

$$\ln \gamma_i = 1 - \ln \sum_k x_k \Lambda_{ik} - \sum_k \frac{x_k \Lambda_{ki}}{\sum_j x_j \Lambda_{kj}} \tag{26.42b}$$

where Λ_{ik} are adjustable parameters obtained from binary systems data (Λ_{12} and Λ_{21}).

NRTL EQUATION

$$\ln \gamma_i = \frac{\sum_j x_j G_{ji} \tau_{ji}}{\sum_l x_l G_{ji}} + \sum_j \frac{x_j G_{ij}}{\sum_l x_l G_{lj}} \left[\tau_{ij} - \frac{\sum_r x_r G_{rj} \tau_{rj}}{\sum_l x_l G_{lj}} \right] \qquad (26.51a)$$

For a binary system,

$$\ln \gamma_1 = x_2^2 \left[\tau_{21} \left(\frac{G_{21}}{x_1 + x_2 G_{21}} \right)^2 + \frac{\tau_{12} G_{12}}{(x_2 + x_1 G_{12})^2} \right] \qquad (26.51b)$$

and

$$\ln \gamma_2 = x_1^2 \left[\tau_{12} \left(\frac{G_{12}}{x_2 + x_1 G_{12}} \right)^2 + \frac{\tau_{21} G_{21}}{(x_1 + x_2 G_{21})^2} \right] \qquad (26.51c)$$

with

$$G_{jk} = \exp(-\alpha_{jk} \tau_{jk}) \qquad (26.49)$$

and

$$\tau_{jk} = \frac{\Delta g_{jk}}{RT} \qquad (26.50)$$

The adjustable parameters in a binary system are Δg_{12}, Δg_{21} and $\alpha_{12} = \alpha_{21}$.

UNIQUAC EQUATION

$$\ln \gamma_i = \left(1 - \frac{\Phi_i}{x_i} + \ln \frac{\Phi_i}{x_i} \right) - \frac{zq_i}{2} \left(1 - \frac{\Phi_i}{\theta_i} + \ln \frac{\Phi_i}{\theta_i} \right)$$

$$+ q_i \left[1 - \ln \sum_k \theta_k \tau_{ik} - \sum_k \frac{\theta_k \tau_{ki}}{\sum_j \theta_j \tau_{kj}} \right] \qquad (26.55)$$

with

$$\theta_i = \frac{x_i q_i}{\sum_k x_k q_k} \qquad (26.27)$$

and Φ_i given by Equation 26.5 above. The adjustable binary parameters are τ_{12} and τ_{21}. The values of r_i and q_i are tabulated, $z = 10$.

REFERENCES

1. van Laar, J. J. 1910. The Vapor Pressure of Binary Mixtures. *Z. Phys. Chem.* 72: 723.
2. van Laar, J. J. 1929. Über die Verschiebung des heterogenen Gleichgewichts von zwei geschmolzenen Metallen mit ihren Salzen durch einen dritten, indifferenten Stoff. *Z. anorgan. allg. Chem.* 185: 35–48.
3. Hildebrand, J. H. 1947. The entropy of solution of molecules of different sizes. *J. Chem. Phys.* 15: 225–228.
4. Scatchard, G. 1949. Equilibrium in nonelectrolyte mixtures. *Chem. Rev.* 44: 7–35.
5. Barton, A. F. M. 1991. *Handbook of Solubility Parameters and Other Cohesion Parameters*, 2nd ed. Boca Raton, FL: CRC Press.
6. Hansen, Ch. M. 2007. *Hansen Solubility Parameters: A User's Handbook*, 2nd ed. Boca Raton, FL: CRC Press, Taylor & Francis Group.
7. Margules, M. 1895. Über die Zusammensetzung der gesättigten Dämpfe von Misschungen. *Sitzungsberichte der Kaiserliche Akademie der Wissenschaften Wien Mathematisch-Naturwissenschaftliche Klasse II* 104: 1243–1278. Available at https://archive.org/details/sitzungsbericht10wiengoog.
8. Redlich, O., and Kister A. T. 1948. Algebraic representation of thermodynamic properties and the classification of solutions. *Ind. Eng. Chem.* 40(2): 345–348.
9. Huggins, M. L. 1942. Thermodynamic properties of solutions of long-chain compounds. *Ann. N.Y. Acad. Sci.* 43: 1–32.
10. Flory, P. J. 1942. Thermodynamics of high polymer solutions. *J. Chem. Phys.* 10: 51–61.
11. Sayegh, S. G., and Vera, J. H. 1980. Lattice-model expressions for the combinatorial entropy of liquid mixtures: A critical discussion. *Chem. Eng. J.* 19: 1–10.
12. Guggenheim, E. A. 1944. Statistical thermodynamics of mixtures with zero energies of mixing. *Proc. R. Soc. Lond.* A 183: 203–212.
13. Staverman, A. J. 1950. The entropy of polymer solutions. *Rec. Trav. Chim. Pays. Bas.* 69: 163–174.
14. Guggenheim, E. A. 1944. Statistical thermodynamics of mixtures with nonzero energies of mixing. *Proc. R. Soc. Lond.* A 183: 213–227.
15. Kehiaian, H. V., Grolier, J.-P. E., and Benson, G. C. 1978. Thermodynamics of organic mixtures—Generalized quasi-chemical theory in terms of group surface interactions. *J. Chim. Phys. Phys. Chim. Biol.* 75(11–1): 1031–1048.
16. Panayiotou, C., and Vera, J. H. 1980. The quasi-chemical approach for non-randomness in liquid mixtures. Expressions for local surfaces and local compositions with an application to polymer solutions. *Fluid Phase Equilib.* 5(1–2): 55–80.
17. Abusleme, J. A., and Vera, J. H. 1985. The quasi-chemical group solution theory for organic mixtures. *Fluid Phase Equilib.* 22(2): 123–138.
18. Wilczek-Vera, G., and Vera, J. H. 1990. A general derivation of thermodynamic functions from the quasichemical theory and their use in terms of non-random factors. *Fluid Phase Equilib.* 59(1): 15–30.
19. Renon, H., and Prausnitz, J. M. 1969. Derivation of the three-parameter Wilson equation for the excess Gibbs energy of liquid mixtures. *AIChE J.* 15(5): 785–787.
20. Wilson, G. M. 1964. Vapor-liquid equilibrium. XI. A new expression for the excess free energy of mixing. *J. Am. Chem. Soc.* 86(2): 127–130.
21. Abrams, D. S., and Prausnitz, J. M. 1975. Statistical thermodynamics of liquid mixtures. New expression for the excess Gibbs energy of partly or completely miscible systems. *AIChE J.* 21(1): 116–128.
22. Maurer, G., and Prausnitz, J. M. 1978. On the derivation and extension of the UNIQUAC equation. *Fluid Phase Equilib.* 2(2): 91–99.
23. Wilson, G. M., and Deal, C. H. 1962. Activity coefficients and molecular structure. *Ind. Eng. Chem. Fundam.* 1: 20–23.

24. Deal, C. H., and Derr E. L. 1968. Group contributions in mixtures. *Ind. Eng. Chem.* 60(4): 28–38.

25. Derr, E. L., and Deal, C. H., Jr. 1969. Analytical solutions of groups. Correlation of activity coefficients through structural group parameters. *Proc. Int. Symp. Distill.* 3: 40–51.

26. Fredenslund, Aa., Jones, R. L., and Prausnitz, J. M. 1975. Group contribution estimation of activity coefficients in nonideal solutions. *AIChE J.* 21: 1086–1099.

27. Panayiotou, C., and Vera, J. H. 1981. Local compositions and local surface area fractions: A theoretical discussion. *Can. J. Chem. Eng.* 59(4): 501–505.

28. Prigogine, I., Mathot-Sarolea, L., and van Hove L. 1952. Combinatory factor in regular assemblies. *Trans. Faraday Soc.* 48: 485–492.

29. Kemeny, S., Balog, Gy., Radnai, Gy., Sawinsky, J., and Rezessy, G. 1990. An improved quasilattice expression for liquid phase order-disorder. *Fluid Phase Equilib.* 54: 247–275.

30. Abusleme, J. A., and Vera, J. H. 1985. A generalized solution method for the quasichemical local composition equations. *Can. J. Chem. Eng.* 63: 845–849.

31. Flemr, V. 1976. A note on excess Gibbs energy equations based on local composition concept. *Coll. Czech. Chem. Commun.* 41: 3347–3349.

32. McDermott, C., and Ashton, N. 1977. Note on the definition of local composition. *Fluid Phase Equilib.* 1: 33–35.

33. Redlich, O., Derr, E. L., and Pierotti, G. J. 1959. Group interaction. I. A model for interaction in solutions. *J. Am. Chem. Soc.* 81(10): 2283–2285.

34. Papadopoulos, M. N., and Derr, E. L. 1959. Group interaction. II. A test of the group model on binary solutions of hydrocarbons. *J. Am. Chem. Soc.* 81(10): 2285–2289.

35. (a) Kojima, K., Tochigi, K. 1979. Prediction of vapor-liquid equilibria by ASOG method. Tokyo: Kodasha-Elsevier. (b) Tochigi, K., Tiegs, D., Gmeling, J., and Kojima, K. 1990. Determination of new ASOG parameters. *J. Chem. Eng. J. Jpn.* 23: 453–463.

36. Tochigi, K. 1995. Prediction of high-pressure vapor-liquid equilibria using ASOG. *Fluid Phase Equilib.* 104: 253–260.

37. Tochigi, K. 1998. Prediction of vapor–liquid equilibria in non-polymer and polymer solutions using an ASOG-based equation of state (PRASOG). *Fluid Phase Equilib.* 144: 59–68.

38. Ashraf, F. A., and Vera J. H. 1980. A simplified group method analysis. *Fluid Phase Equilib.* 4: 211–228.

39. Scott, R. L. 1956. Corresponding states treatment of nonelectrolyte solutions. *J. Chem. Phys.* 25: 193–205.

40. Renon, H., and Prausnitz J. M. 1968. Local composition in thermodynamic excess functions for liquid mixtures. *AIChE J.* 14: 135–144.

41. Vera, J. H. 1982. On the two-fluid local composition expressions. *Fluid Phase Equilib.* 8(3): 315–318.

42. Prausnitz, J. M., Lichtenthaler, R. N., and Gomes de Azevedo, E. 1999. *Molecular Thermodynamics of Fluid Phase Equilibria*, 3rd ed. Upper Saddle River, NJ: Prentice Hall, (a) p. 346, (b) p. 290, (c) p. 368, (d) pp. 390–400.

43. Vera, J. H., Sayegh, S. G., and Ratcliff, G. A. 1977. A quasi lattice-local composition model for the excess Gibbs free energy of liquid mixtures. *Fluid Phase Equilib.* 1: 113–135.

44. Wang, W., and Vera, J. H. 1993. Statistical thermodynamics of disordered and ordered systems. A properly normalized local order theory. *Fluid Phase Equilib.* 85: 1–18.

45. Gmehling, J., Tiegs, D., and Knipp, U. 1990. A comparison of the predictive capability of different group contribution methods. *Fluid Phase Equilib.* 54: 147–165.

46. Chapman, W. G., Jackson, G., and Gubbins, K. E. 1988. Phase equilibria of associating fluids: Chain molecules with multiple bonding sites. *Molec. Phys.* 65: 1057–1079.

47. Müller, E. A., and Gubbins, K. E. 2001. Molecular-based equations of state for associating fluids: A review of SAFT and related approaches. *IEC Res.* 40: 2193–2211.

27 The Activity of Individual Ions
Measuring and Modeling

Although the material included here should have been part of Chapter 18, keeping with the structure of this book, the permanent structure of classical thermodynamics has been kept separate from material that may change with time. The measurability of the activity of individual ions may not be considered to be definitely settled, as there are people who still maintain that these measurements are impossible to achieve in concentrated solutions above 0.01 m. A historical perspective of roots of that point of view is given elsewhere [1] and discussed briefly in the closing comments of this chapter. In addition, two discussions available in the literature illustrate the extent of the disagreement on this matter [2–10]. In both cases, the editor of the respective journal closed the debate once the informed reader could judge how much each of the two sides knew about the subject. Being a matter of experimental measurements, this topic can only be resolved with independent laboratory tests. The arguments in favor of the measurability of the activity of individual ions have been presented elsewhere [1,2,6,11–30]. This chapter presents the main points supporting these views, compares experimental results obtained by different researchers in concentrated electrolyte solutions, and briefly mentions recent theoretical advances in the field.

CRUX OF THE PROBLEM

Chapter 18 discussed the direct electrochemical measurement of the mean ionic activity coefficient of an electrolyte in aqueous solution by measuring the potential difference between the responses of an ion-selective electrode (ISE) sensitive to the cation and an ISE sensitive to the anion. For clarity, some of the information given in Chapter 18 is repeated here. It is an accepted fact that the voltage response $E_{i,k}$ of an ISE sensitive to ion i at a molality $m_{i,k}$ in aqueous solution is given by the Nernst equation:

$$E_{i,k} = E_{i,0} + \frac{RT}{Z_i F} \ln(a_{i,k}) \qquad (18.23)$$

Symbols used in this equation are detailed in Chapter 18. When an ISE for the cation and an ISE for the anion of an electrolyte are both immersed in the same solution at the same time and their responses are measured directly one against the other in a voltmeter, the voltage difference is

$$\Delta E_k = \Delta E_0 + \frac{RT}{F} \ln \left[a_{+,k}^{\frac{1}{Z_+}} \, a_{-,k}^{\frac{1}{|Z_-|}} \right] \tag{18.24a}$$

Using Equations 18.1b, 18.1c, and 18.7, Equation 18.4a, at a dimensionless value $\tilde{m}_k$ of the solution molality, can be written as

$$\Delta E_k = \Delta E_0 + \frac{RT}{(\nu_+ Z_+)F} \ln \left[\tilde{m}_k^\nu \gamma_{\pm,k}^\nu \left(\nu_+^{\nu_+} \nu_-^{\nu_-} \right) \right] \tag{18.24b}$$

The term ΔE_0 is a constant if measurements are done in continuous runs over a short period of time by adding electrolyte into the sample solution. Thus, ΔE_0 can be evaluated by calibration measuring several points in the dilute region where the mean ionic activity coefficient is given by the Debye–Hückel equation:

$$\ln \gamma_\pm^{dil} = -\frac{A_D |Z_+ Z_-| I^{1/2}}{1 + I^{1/2}}$$

It has been repeatedly proved that the mean ionic activity coefficient values for different electrolytes obtained by these direct electrochemical measurements agree with the values obtained by independent measurements of osmotic coefficients, also presented in Chapter 18.

An interesting result is obtained if instead of connecting directly both ISEs to the voltmeter and reading one response against the other, each of the two voltage responses is separately measured against a half-cell single-junction Ag/AgCl reference electrode, as shown in Figure 18.4. The half-cell reference electrode is immersed in an internal standard solution that leaks into the sample solution. A 4 M KCl solution is commonly used as internal standard. A liquid junction potential $E_{J,k}$ is created at the tip of the reference electrode, where the reference solution is in contact with the sample solution. This contact of the two solutions is necessary for closing the electric circuit.

In this case, the voltage response of an ISE takes the form

$$E_{i,k}^R = E_{i,0}^R + \frac{RT}{Z_i F} \ln(a_{i,k}) + E_{J,k} \tag{27.1}$$

So for each of the electrodes,

$$E_{+,k}^R = E_{+,0}^R + \frac{RT}{Z_+ F} \ln(a_{+,k}) + E_{J,k}$$

$$E_{-,k}^R = E_{-,0}^R - \frac{RT}{|Z_-| F} \ln(a_{-,k}) + E_{J,k}$$

The superscript R indicates that in this case, the voltage response is measured against a reference electrode. The junction potential has the same value for both responses in simultaneous measurements. Taking the difference of both independent readings, we obtain

$$\Delta E_k^R = \Delta E_0^R + \frac{RT}{F}\ln\left[a_{+,k}^{\frac{1}{z_+}}\, a_{-,k}^{\frac{1}{|z_-|}} \right] \tag{27.2}$$

Equation 27.2 is formally identical to Equation 18.24a. By independent calibration of the values of $E_{+,0}^R$ and $E_{-,0}^R$, the value of ΔE_0^R is obtained, so values for the mean ionic activity coefficients of the electrolyte can be calculated from the independent measurements of the two ISEs. These values of the mean ionic activity coefficient agree with the values obtained by the direct electrochemical procedure and also with the values obtained from measurements of the osmotic coefficient. This result is not surprising at all, since it is clear that the new element $E_{J,k}$ introduced by the presence of a junction potential cancels out in this exercise. Although the value of $E_{J,k}$ changes from one experimental point to the next because the sample solution changes composition, in simultaneous measurements its value is the same for both ISEs. Even accepting this fact, one cannot avoid the feeling that the expression for the individual responses of the cation and the anion ISEs, written as

$$E_{i,k}^R = E_{i,0}^R + \frac{RT}{z_i F}\ln(\tilde{m}_{i,k}\gamma_{i,k}) + E_{J,k} \tag{27.1a}$$

contains the information of the individual activity of the ions $\tilde{m}_{i,k}\,\gamma_{i,k}$, or more specifically, about the individual activity coefficient of the ions $\gamma_{i,k}$. Almost a century ago, Harned (1926) [31] called attention to the main problem, stating at the beginning of his manuscript, "Since the difficulty in estimating liquid junction potentials *is the only cause* of our not being able to estimate exactly the individual activity of ions, an analysis of the problem of liquid junction potentials will be made" (emphasis ours). Admittedly, the agreement of the results of the independent measurement of the two ISEs with the known values of the mean ionic activity coefficients of the electrolyte is a necessary condition for accepting the goodness of the measurements, but it is not a sufficient condition for extracting from that information the value of the activity coefficients of the individual ions. Harned's difficulty is resolved only if it can be demonstrated that the value of the junction potential present in the experimental measurements of the individual ISE, against a half-cell single-junction reference electrode, does not significantly affect the results obtained. This is the important point to keep in mind. Before addressing this point, however, it is necessary to dispel a wrong concept that seems to be behind the objection to the measurability of the activity of the individual ions.

The chemical potential is defined as the partial derivative of the Gibbs energy with respect to the number of moles of species i at constant temperature and pressure, and the number of all other species in the system:

$$\mu_i = \left(\frac{\partial G}{\partial n_i} \right)_{T,P,n_{j \neq i}} \tag{7.1}$$

On the other hand, for an ion i, the chemical potential is related to its activity by

$$\mu_i = \mu_i^\theta + RT \ln(a_{i,\theta}) \tag{12.12a}$$

In an aqueous solution of an electrolyte, due to the condition of electroneutrality, it is not possible to change the number of moles of one ion and keep constant the moles of all other ions. Thus, one may argue that it is not possible to measure the activity of an individual ion. This is a seemingly convincing argument, indeed.

The inconsistency of this argument is better understood considering the case of the measurement of the activity coefficient of species i in a mixture of nonelectrolytes. As discussed in Chapter 16, this is done by measuring the equilibrium pressure and the compositions of the vapor and liquid phases of a binary system at constant temperature. Equation 16.2 can be written as

$$\ln(x_i \gamma_{i,0}) = \ln\left(\frac{y_i P}{P_i^s} \right) - \left[\ln\left(\frac{\varphi_i^{0,s}}{\varphi_i^V} \right) + \frac{v_i^{0,L}(P - P_i^s)}{RT} \right] \tag{16.2b}$$

In a more detailed consideration, however, in vapor–liquid equilibrium of a binary mixture, according to the phase rule, there are two degrees of freedom. Thus, only two variables can be fixed or changed independently. At constant temperature, the pressure cannot be kept constant if the number of moles of one of the species is changed and the number of moles of the second species is kept constant. In spite of this fact, nobody would argue that because one cannot change the number of moles of one compound and keep the pressure and temperature constant, as required by the definition of the chemical potential by Equation 7.1, the activity of a compound cannot be obtained from Equation 16.2b. Even more, one observes that in the use of Equation 16.2b there is no change of composition required to obtain the activity of species i. The measurement is carried out without a change in composition. In complete analogy, in the case of the measurement of the activity of species i in an aqueous solution of an electrolyte, Equation 27.1a can be written as

$$\ln(\tilde{m}_{i,k} \gamma_{i,k}) = \frac{Z_i F}{RT} \left(E_{i,k}^R - E_{i,0}^R \right) - \left[\frac{Z_i F}{RT} E_{J,k} \right] \tag{27.1b}$$

Again, in this case there is no change of composition required to obtain the activity of species i. In both Equations 16.2b and 27.1b, the first term of the right-hand side contains the measured variables and gives the main contribution to the activity of species i, while the second term represents a small correction, and it is usually evaluated using appropriate correlations. The truth is that this later statement is accepted without discussion for the case of nonelectrolytes, and it is precisely the matter under discussion for the case of electrolytes. The next section presents arguments demonstrating the negligible effect of the junction potential on the experimental values of the individual ionic activities.

Results presented from here on are those obtained by members of the McGill University research group under the direction of the senior author of this text. These results are compared with literature values from other researchers in the later part of this chapter. Measurements were performed in continuous isothermal runs carried out in short time spans. It is important to realize that ISEs are transient devices. They change with time and so does their response. The same ISE immersed in the same electrolyte aqueous solution at the same temperature will give different responses on different days.

METHODS USED TO REDUCE THE EXPERIMENTAL DATA

Equation 27.1a can be written as

$$\xi_{i,k} \equiv (E_{i,k}^R - E_{J,k}) = E_{i,0}^R + S_i \ln(\tilde{m}_{i,k}\gamma_{i,k}) \tag{27.1c}$$

For a Nernstian ISE, the slope is $S_i = RT/Z_iF$, and for a set of measurements done in a reasonably short time, $E_{i,0}^R$ is a constant value. If an approximate value for $E_{J,k}$ is entered, the value of $\xi_{i,k}$ can be calculated for each experimental point from the known measured values of $E_{i,k}^R$. Equations for the junction potential are discussed in the last section of this chapter. The first studies [11] used an approximation to the junction potential proposed by Henderson [32]. For the purposes of this presentation, at this point, the analytical form of this approximation is immaterial. At first, the following indirect approach was used.

INDIRECT ELECTROCHEMICAL APPROACH

As a way of avoiding the evaluation of $E_{i,0}^R$ and imposing the Nernstian value for S_i, two arbitrary points, of the same experimental run, at molalities $\tilde{m}_{i,p}$ and $\tilde{m}_{i,q}$ were selected and Equation 27.1c was used to eliminate these two terms, thus obtaining

$$\zeta_{i,k} = \frac{\xi_{i,k} - \xi_{i,p}}{\xi_{i,q} - \xi_{i,p}} = \alpha_i + \beta_i \ln(\tilde{m}_{i,k}\gamma_{i,k}) \tag{27.3}$$

with

$$\alpha_i = \frac{-\ln(\tilde{m}_{i,p}\gamma_{i,p})}{\ln(\tilde{m}_{i,p}\gamma_{i,p}/\tilde{m}_{i,q}\gamma_{i,q})}$$

and

$$\beta_i = \frac{1}{\ln(\tilde{m}_{i,p}\gamma_{i,p}/\tilde{m}_{i,q}\gamma_{i,q})}$$

For each experimental run, the values of α_i and β_i, although unknown, are constant. These two values were then obtained by calibration of the response of the electrode in the region below 0.01 m. In this very dilute region, data of ionic activities exist [33], and it has long been recognized that these values are well represented by the Guntelberg equation [34]. For monovalent ions, this equation gives

$$\ln \gamma_{i,k}^{(\tilde{m}<0.01)} = \frac{-1.1762\tilde{m}^{0.5}}{1+\tilde{m}^{0.5}}$$

The important point here is that, as shown later in this chapter, the values obtained for the activity of the individual ions in concentrated solutions are repeatable in different experimental runs and independent of the two points selected in each run to eliminate $E_{i,0}^R$ and S_i.

DIRECT ELECTROCHEMICAL APPROACH

Realizing that the above approach to reduce the data assumes that $E_{i,0}^R$ and S_i are constant values for a particular run, Equation 27.1c was used directly to evaluate $E_{i,0}^R$ and S_i for each run of experiments. Measurements were repeated [14], and instead of eliminating these two constants, their values were obtained by calibration of the response of the electrode for compositions below 0.01 m. Within the accuracy suggested by the first indirect approach, the results were in total agreement. An agreement in the activity coefficient around ±0.02 is normally obtained.

As both methods handle differently the contribution of the junction potential, the close agreement of the results obtained with the two methods for reducing the data was considered an indication of the negligible influence of the approximation used for the junction potential.

EFFECT OF AN ERROR IN THE SIGN OF THE JUNCTION POTENTIAL

However, it may be argued that although both methods handle the data in different ways, both still rely on the same approximation to the junction potential. At this point, one fortuitous event provided an additional element for estimating the interference introduced by the approximation used to calculate the junction potential. Armin C. Schneider [35], working independently at Institut für Energie- und Umweltverfahrenstechnik, Universität Duisburg, Germany, realized that the approximation for the junction potential used in the previous calculations was written in the "potentiometry convention," while the ISE and the reference electrode were connected in the "pH convention" [2]. That it, the correction for the junction potential had been used with the wrong sign. This caused a recalculation of all measured values [2]. The most notable observation coming out of this exercise was that the error of sign in the approximation to the junction potential did not change the results up to concentrations of 2 m for the case of the ions potassium and chloride in solutions of KCl. For the ions sodium and chloride in NaCl solutions, on the other hand, errors were only slightly above the expected accuracy up to concentrations of 0.5 m and increased thereafter. This unplanned event had three positive outcomes. In the first place, it confirmed previous theoretical predictions [20] that an error in the estimate of the junction potential would affect the results more in NaCl solutions than in KCl solutions. It also suggested new ways of carrying out the experimental measurements, as well as promoted the development of better approximations for the junction potential. These three aspects are addressed in what follows.

The measurement of the activity of the individual ions in KCl aqueous solutions using a half-cell single-junction reference electrode filled with a 4 M KCl standard solution is a very special case of a homoionic system. The largest junction potential happens when the concentrated standard KCl solution leaks into the most dilute sample solutions in the calibration range of the ISE. As the activity coefficient is set to its theoretical value in this range of composition, this error gets absorbed in the value of $E_{i,0}^R$ [20]. As the sample solution gets more concentrated in KCl, the value of the junction potential decreases, and it would be null at a sample solution concentration of 4 M.

ADDITIONAL EXPERIMENTAL STUDIES

The previous reasoning suggested the idea of measuring the activity of different ions using not only 4 M KCl but also 2 M NH_4Cl as standard solution of the reference electrode [22]. Additional studies reported results of measurements of the activity of sodium and chloride ions in NaCl solutions using NaCl at different concentrations, 0.5, 1, 1.5, 2, and 3 m, as standard solution of the reference electrode [24]. Finally, in this series of experimental studies, measurements of the activities of potassium and chloride ions in KCl solutions using KCl, NaCl, and CsCl in the reference electrode and of sodium and chloride in NaCl solutions using CsCl in the reference electrode [6] confirmed the fact that the method of calibration of the ISEs absorbs the effect introduced by the approximation made for the junction potential. The negligible effect of the junction potential in the results for potassium and chloride ions in KCl solutions is shown in Figure 27.1, and for the ions sodium and chloride in NaCl solutions is shown in Figure 27.2.

On the side of theory, a study of the activities of potassium and chloride ions in KCl solutions [29] showed that the use of the theoretically sound approximation to the junction potential described below does not change significantly the value of the activity of the individual ions reported previously.

DERIVATION OF EQUATIONS TO CALCULATE THE JUNCTION POTENTIAL

Those only interested in the practical use of the activity of individual ions can go directly to the next section. The treatment that follows is reproduced [2] with

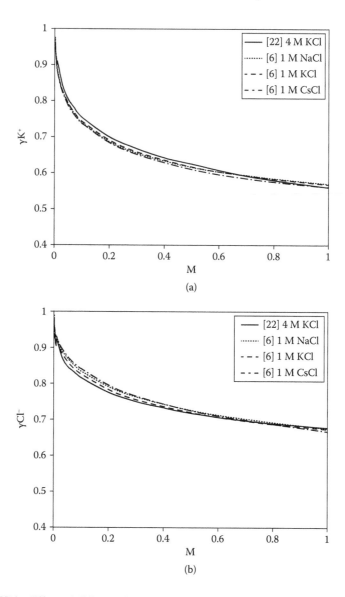

FIGURE 27.1 Effect of different filling solutions on the measurements of the individual activity of ions in aqueous KCl system at 298.15 K. Square brackets indicate the number of the reference from which data were taken. (a) Activity coefficients of K^+ ion. (b) Activity coefficients of Cl^- ion.

permission from John Wiley & Sons. Equations to calculate the junction potential between two solutions of different concentrations are obtained considering that no net electric current is transported by the ions through the junction. As the current would be carried by ions in solution, we consider first the molar flux of ions i, J_i, through a unit surface area A. This flux is proportional to the molar concentration

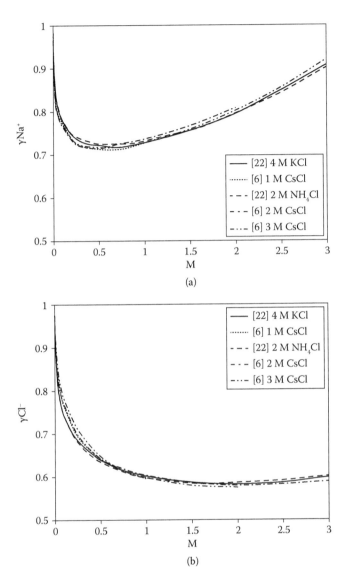

FIGURE 27.2 Effect of different filling solutions on the measurements of the individual activity of ions in aqueous NaCl system at 298.15 K. Square brackets indicate the number of the reference from which data were taken. (a) Activity coefficients of Na^+ ion. (b) Activity coefficients of Cl^- ion.

of ions i, C_i, and to the negative of the gradient of the electrochemical potential of species i, μ_i, over the path l, that is, the thermodynamic force acting over the ions i. Thus, we write

$$J_i = \frac{D_i}{RT} C_i \left(-\frac{\partial \mu_i}{\partial l} \right) \qquad (27.4)$$

In Equation 27.4, D_i is the diffusion coefficient of i, R is the gas constant, and T is the absolute temperature. The electrochemical potential of species i has the form

$$\mu_i = \mu_i^o + RT \ln a_i + Z_i F \Psi \tag{27.5}$$

where a_i and Z_i are the activity and the charge of ion i, respectively; F is the Faraday constant; and Ψ is the electric potential of the liquid phase. Thus, the gradient of the electrochemical potential takes the form

$$\frac{\partial \mu_i}{\partial l} = RT \frac{\partial \ln a_i}{\partial l} + Z_i F \frac{\partial \Psi}{\partial l} \tag{27.6}$$

To continue, it is necessary to express the diffusion coefficient in terms of the molar ionic conductivity of ions moving under the effect of an electric potential only. For this, it is useful to define the drift speed s_i attained by ion i when the drag forces balance the forces acting over the ion. This drift speed is proportional to the gradient of the applied electrical field; thus,

$$s_i = \frac{|J_i|}{C_i} = u_i \left| \frac{\partial \Psi}{\partial l} \right| \tag{27.7}$$

The proportionality constant u_i, known as the electrical mobility of ion i, represents the limiting steady-state velocity attained by ion i under a unit of potential gradient. From Equations 27.4 and 27.6, for an ion moving under the effect of an electric field only, in the absence of a concentration gradient in the solution, the drift speed takes the form

$$s_i = \frac{|J_i|}{C_i} = \frac{D_i}{RT} |Z_i| F \left| \frac{\partial \Psi}{\partial l} \right| \tag{27.8}$$

Hence, combining Equations 27.7 and 27.8 gives

$$u_i = \frac{D_i}{RT} |Z_i| F \tag{27.9}$$

Equation 27.9 is known as the Einstein relation [36]. In electrochemistry, the molar ionic conductivity λ_i is defined as the electric current transported per mole of species i, per unit electrical potential gradient, through a unit cross section. Thus,

$$\lambda_i = |Z_i| F u_i \tag{27.10}$$

Combining Equations 27.9 and 27.10 gives

$$\frac{D_i}{RT} = \frac{\lambda_i}{|Z_i|^2 F^2} \tag{27.11}$$

Equation 27.11 is known as the Nernst–Einstein relation [36]. Combining Equations 27.4, 27.6, and 27.11, after rearrangement, gives an expression for the current density $(Z_i F J_i)$ transported by ions of type i:

$$z_i FJ_i = -\left[\frac{RT}{F} \frac{C_i\lambda_i}{Z_i} \frac{\partial \ln a_i}{\partial l} + C_i\lambda_i \frac{\partial \Psi}{\partial l} \right] \tag{27.12}$$

For the calculation of the junction potential between two solutions of different composition, the sum of the current densities transported by all ions is set to zero. Thus, setting to zero the sum over all ions of the right-hand side of Equation 27.12 gives

$$\frac{\partial \Psi}{\partial l} = -\frac{RT}{F} \sum_i \frac{1}{Z_i} \frac{C_i\lambda_i}{\sum_j C_j\lambda_j} \frac{\partial \ln a_i}{\partial l} \tag{27.13}$$

Finally, integrating Equation 27.13 over the thickness l of the liquid junction gives

$$E_J = \Psi_2 - \Psi_1 = -\frac{RT}{F} \sum_i \int_1^2 \frac{t_i}{Z_i} d \ln a_i \tag{27.14}$$

where, to follow the conventional nomenclature, we have used the symbol E_J to indicate the potential difference $\Delta\Psi = \Psi_2 - \Psi_1$, across the junction, and introduced the symbol t_i for the transference number of ion i, defined as

$$t_i = \frac{C_i\lambda_i}{\sum_j C_j\lambda_j} = \frac{\tilde{C}_i\tilde{\lambda}_i}{\sum_j \tilde{C}_j\tilde{\lambda}_j} = \frac{|Z_i|C_i\tilde{\lambda}_i}{\sum_j |Z_j|C_j\tilde{\lambda}_j} \tag{27.15}$$

It is interesting to note that the transference number t_i appears naturally in Equation 27.13. This term, which has caused some confusion in the literature, can be written in alternative ways. In Equation 27.15, the equivalent conductivity $\tilde{\lambda}_i$ and the concentration of ions i, $\tilde{C}_i$, in equivalent per liter, are related to the corresponding molar quantities, λ_i and C_i, by

$$\lambda_i = |Z_i|\tilde{\lambda}_i \tag{27.16}$$

and

$$C_i = \frac{\tilde{C}_i}{|Z_i|} \tag{27.17}$$

An erroneous interpretation of these terms found in the literature is discussed in the section dealing with the Henderson approximation.

The next step is to obtain an integrated form of the equation for the junction potential. For the purposes of this work, we follow the potentiometric convention. In this convention, the ISE electrode is written at the right and the reference electrode at the left. In agreement with the Stockholm convention, the junction potential in this case corresponds to the potential of the sample solution (k) minus the potential of the solution at the reference electrode (r). Thus, the lower limit of the integral in Equation 27.14 is the state of the ion at the reference solution and the upper limit is the state of the ion in the sample solution. Hence, Equation 27.14 takes the form

$$E_{J,k} = \Psi_k - \Psi_r = -\frac{RT}{F} \sum_i \int_r^k \frac{1}{Z_i} \frac{C_i \lambda_i}{\sum_j C_j \lambda_j} \frac{da_i}{a_i} \qquad (27.18)$$

For simplicity we rewrite Equation 27.18 as

$$E_{J,k} = +\frac{RT}{F} \sum_i \tau_i \qquad (27.19)$$

with τ_i defined as

$$\tau_i = \frac{1}{Z_i} \int_k^r \frac{C_i \lambda_i}{\sum_j C_j \lambda_j} \frac{da_i}{a_i} \qquad (27.20)$$

The key problem to determine the activity of ions is the evaluation of τ_i. This topic is discussed below.

HENDERSON'S APPROXIMATION

Henderson's approximation [32] is the best-known equation for the junction potential. Thus, we discuss here its relation with the equations presented above. In order to avoid the problem of the undefined terms in the case that ion i is not present at one of the two solutions, Henderson introduced the auxiliary variable α and considered a linear variation of the concentration of the ions with α across the junction:

$$C_i = C_{i,k} + \alpha \Delta C_i \qquad (27.21)$$

with

$$\Delta C_i = C_{i,r} - C_{i,k} \qquad (27.22)$$

In addition, Henderson introduced two stringent simplifications. He assumed that the activity of ion i was equal to its molar concentration and that the molar conductivity was independent of concentration and equal to the molar conductivity of the ion at infinite dilution, λ_i^∞. Thus, setting $a_i = C_i$ and also $\lambda_{i,k} = \lambda_{i,r} = \lambda_i^\infty$ in Equation 27.20 gives

$$\tau_i^H = \frac{\lambda_i^\infty \Delta C_i}{Z_i \left[\sum_i C_{i,r} \lambda_i^\infty - \sum_i C_{i,k} \lambda_i^\infty \right]} \ln \frac{\sum_i C_{i,r} \lambda_i^\infty}{\sum_i C_{i,k} \lambda_i^\infty}$$ (27.23)

Hence, for the schematics of the electrochemical cell written in the potentiometric convention, as it is considered in the case of an ISE, we write

$$E_{J,k}^H = \frac{RT}{F} \left[\frac{\sum_i \frac{1}{Z_i} C_{i,r} \lambda_i^\infty - \sum_i \frac{1}{Z_i} C_{i,k} \lambda_i^\infty}{\sum_i C_{i,r} \lambda_i^\infty - \sum_i C_{i,k} \lambda_i^\infty} \right] \ln \frac{\sum_i C_{i,r} \lambda_i^\infty}{\sum_i C_{i,k} \lambda_i^\infty}$$ (27.24)

The usual form for the Henderson equation found in the literature [37a,b] is written for the pH convention, which considers the ISE to be written at the left and the reference electrode at the right. In addition, it is normally written in terms of the equivalent limiting conductivity, $\tilde{\lambda}_i$, of the ion instead of the molar limiting conductivity, λ_i These two limiting conductivities are related by Equation 27.16. Hence, the form of the Henderson equation found in the literature is as follows:

$$\left(E_{J,k}^H \right)^* = -E_{J,k}^H = \frac{RT}{F} \left[\frac{\sum_i \frac{|Z_i|}{Z_i} C_{i,k} \tilde{\lambda}_i^\infty - \sum_i \frac{|Z_i|}{Z_i} C_{i,r} \tilde{\lambda}_i^\infty}{\sum_i |Z_i| C_{i,k} \tilde{\lambda}_i^\infty - \sum_i |Z_i| C_{i,r} \tilde{\lambda}_i^\infty} \right] \ln \frac{\sum_i |Z_i| C_{i,k} \tilde{\lambda}_i^\infty}{\sum_i |Z_i| C_{i,r} \tilde{\lambda}_i^\infty}$$ (27.25)

The ratio of the absolute value of the charge over the value of the charge of ion i, appearing in the numerator inside the brackets of the right-hand side of Equation 27.25, is +1 for cations and −1 for anions. Hence, this term is usually not explicitly written and the sum over all ions is written as the sum over the cations minus the sum over the anions.

$$\left(E_{J,k}^H \right)^* = \frac{RT}{F} \frac{\left[\sum C_+ \tilde{\lambda}_+^\infty - \sum C_- \tilde{\lambda}_-^\infty \right]_k - Y_1}{\left[\sum C_+ \tilde{\lambda}_+^\infty |Z_+| + \sum C_- \tilde{\lambda}_-^\infty |Z_-| \right]_k - Y_2}$$ (27.26)

$$\times \ln \frac{\left[\sum C_+ \tilde{\lambda}_+^\infty |Z_+| + \sum C_- \tilde{\lambda}_-^\infty |Z_-| \right]_k}{Y_2}$$

In Equation 27.26, the symbols Y_1 and Y_2 represent numerical values that depend on the solution r used in the reference electrode. In Henderson's approximation, these constants are obtained as

$$Y_1 = C_{+,r}\tilde{\lambda}^\infty_{+,r} - C_{-,r}\tilde{\lambda}^\infty_{-,r} \tag{27.27}$$

and

$$Y_2 = C_{+,r}\tilde{\lambda}^\infty_{+,r}\left|Z_+\right|_r + C_{-,r}\tilde{\lambda}^\infty_{-,r}\left|Z_-\right|_r \tag{27.28}$$

At 298.15 K, for a 4 M KCl reference electrode solution, $Y_1 = -11.6$ and $Y_2 = 623$. At this temperature, the value of the ratio (RT/F) is 25.693 (mV). However, more important is to note that the terms in each sum contain the product of the molar concentration times the equivalent limiting conductivity. Standard references in the literature are in error here. In some places, it is indicated that in Equation 27.26 both the concentration and the conductivity are in molar units [37a], while in others it is stated that both the concentration and the conductivity are in terms of equivalents [37b]. Both of these statements are wrong. Although these errors do not affect the value of the junction potential for 1:1 electrolytes, they give wrong results for 1:2 and 2:1 electrolytes. Values tabulated in the literature for these latter cases are in error.

New Equation to Calculate Liquid Junction Potentials

Two early attempts were made to improve Henderson's equation for the calculation of the liquid junction potential. In 1926, Harned [31] proposed a modification that corrects for the variation of the ionic conductivity with concentration, and in 1985, Harper [38] included the correction for nonideality of the solutions and for the variation of ionic conductivity. While the method of Harned resulted in an analytical solution that does not correct for activities, the method of Harper produced complex expressions that required numerical solutions. In more recent times, a new analytical equation to calculate the value of the liquid junction potential that includes corrections for nonideality and for the variation of conductivity with concentration was proposed [2]. The same mathematical artifice used by Henderson was employed for the integration of Equation 27.20. The integration path followed a linear variation of the product $(C\lambda)_i$ with α. Thus,

$$C_i\lambda_i = (C\lambda)_{i,k} + \alpha\Delta(C\lambda)_i \tag{27.29}$$

with

$$\Delta(C\lambda)_i = (C\lambda)_{i,r} - (C\lambda)_{i,k} \tag{27.30}$$

As the integral in Equation 27.20 depends only on the two limiting electrolyte solutions, the following form, meeting the conditions of being zero at the electrolyte solution k and unity at the reference solution r, was used for the dummy variable α:

$$\alpha = \frac{a_i - a_{i,k}}{\Delta a_i} \tag{27.31}$$

with

$$\Delta a_i = a_{i,r} - a_{i,k} \tag{27.32}$$

With these definitions, Equation 27.20 can be written as

$$\tau_i = \frac{\Delta a_i}{Z_i} \int_0^1 \frac{[(C\lambda)_{i,k} + \alpha\Delta(C\lambda)_i]d\alpha}{[a_{i,k} + \alpha\Delta a_i][\beta_k + \alpha\Delta\beta]} \tag{27.33}$$

with

$$\beta_k = \sum_j C_{j,k}\lambda_{j,k} \tag{27.34}$$

and

$$\Delta\beta = \beta_r - \beta_k \tag{27.35}$$

where

$$\beta_r = \sum_j C_{j,r}\lambda_{j,r} \tag{27.36}$$

The integration of Equation 27.33 is analytical, and after rearrangement, it produces the following general expression for τ_i for the case when ion i is present in both solutions k and r:

$$\tau_i = \frac{1}{Z_i}\left[\frac{[a_{i,k}(C\lambda)_{i,r} - a_{i,r}(C\lambda)_{i,k}]}{[a_{i,k}\beta_r - a_{i,r}\beta_k]}\ln\frac{a_{i,r}}{a_{i,k}} + \frac{\Delta a_i}{\Delta\beta}\frac{[(C\lambda)_{i,k}\beta_r - (C\lambda)_{i,r}\beta_k]}{[a_{i,k}\beta_r - a_{i,r}\beta_k]}\ln\frac{\beta_r}{\beta_k}\right] \tag{27.37}$$

As expected, the integrated form of the term τ_i is only a function of the compositions of the end solutions, and it is independent of the integration path. If an ion i is not present in one of the two solutions, the first term inside the brackets of the right-hand side of Equation 27.37 vanishes. For the case of ion i present only in the reference electrode solution r, its composition and activity are equal to zero in solution k and Equation 27.37 simplifies to

$$\tau_{i,r} = +\frac{1}{Z_i}\frac{C_{i,r}\lambda_{i,r}}{\Delta\beta}\ln\frac{\beta_r}{\beta_k} \tag{27.38}$$

Similarly, for the case of ion i present only in the sample solution k, Equation 27.37 simplifies to

$$\tau_{i,k} = -\frac{1}{Z_i}\frac{C_{i,k}\lambda_{i,k}}{\Delta\beta}\ln\frac{\beta_r}{\beta_k} \tag{27.39}$$

Notably, these two expressions are independent of the activity of the ions. Equations 27.19 and 27.37 suffice for the calculation of the liquid junction potential to be used in the measurement of the activity of individual ions. All that is required is to have experimentally determined values of the molar conductivity λ_i of the ions and good initial estimates of the activity of the individual ions present in both solutions forming the liquid junction. There is a clear advantage in using homoionic systems for measuring the activity of ions.

The interesting point to observe here is that to measure the activity of individual ions, it is necessary to know these activities to calculate the junction potential. This is not a "vicious circle," as stated by Guggenheim [39], but is just an iteration loop in the calculations. In fact, recent work [29] used exactly this iteration scheme. The first estimate of the activity coefficient of the individual ions was obtained reducing the data assuming Henderson's approximation for the junction potential. Another alternative is to follow Harper's suggestion [38] and obtain a first crude estimate of the activity coefficients of the individual ions from the mean ionic activity coefficient of the electrolyte at the concentration of the solution. By direct application of the Debye–Hückel equation,

$$\ln\gamma_{i,k} = \frac{Z_i^2}{|Z_+Z_-|}\ln\gamma_{\pm,k} \tag{27.40}$$

USE OF THE ACTIVITY OF INDIVIDUAL IONS IN MULTI-ION AQUEOUS SOLUTIONS

It is an established fact that in concentrated electrolyte solution, the activity of an individual ion depends on the nature of the counterion present in the solution. Thus, the key problem when dealing with a multi-ion concentrated aqueous solution is how to combine information gathered studying the behavior of each ion in systems with a single counterion present. Lin and Lee [40], in an impressively insightful derivation, elegantly solved this problem. Although their derivation was coupled with a new equation the authors were proposing, their final result is totally general. Lin and Lee considered the ionic composition of the multi-ion aqueous solution to be known by an independent chemical analysis. In their mathematical treatment, the solvent-free mixture of ions is brought from infinite dilution in the solvent to the ionic strength of the multi-ion solution, as defined by Equation 18.17. In all this process, the composition of the solvent-free mix of ions is constant. Lin and Lee clearly presented the algebra in detail, so it is not necessary to repeat it here. The Lin and Lee equation for the osmotic coefficient of a multi-ion aqueous solution at an ionic strength I_k takes the form

$$\varphi_k = 1 + \sum_i Y_{i,k} \left[\ln \gamma_{i,k} - \frac{1}{I_k} \int_0^{I_k} (\ln \gamma_i) \, dI \right] \tag{27.41}$$

with

$$Y_{i,k} = \frac{m_{i,k}}{\sum_j m_{j,k}} \tag{27.42}$$

The activity of water is then obtained from Equation 18.11 for a multi-ion solution, written as

$$\ln a_{w,k} = -\frac{M_w \sum_j \tilde{m}_{j,k}}{1000} \varphi_k$$

A particularly useful correlation for the activity coefficient of individual ions proposed in the literature [2] has the form

$$\ln \gamma_{i,k} = \frac{-AZ_i^2 I_k^{0.5}}{1 + I_k^{0.5}} + \frac{b_i I_k}{(1 + 1.5 I_k)^2} + d_i I_k \tag{27.43}$$

For aqueous solutions at 298.15 K, $A = 1.1762$ and b_i and d_i are adjustable parameters, the values of which are given in the publication referenced above [2]. The first term of the right-hand side of this equation represents the contribution of long-range interactions, and the second term represents the contribution of the short-range (specific) interactions. Further study using the Lin and Lee equation [26] showed that a particularly simple expression for the osmotic coefficient is obtained with a minor modification of Equation 27.43, namely,

$$\ln \gamma_{i,k} = -AZ_i^2 \left[\frac{I_k^{0.5}}{1 + I_k^{0.5}} + 2 \ln \left(1 + I_k^{0.5} \right) \right] + \frac{b_i I_k}{(1 + 1.5 I_k)^2} + d_i I_k \tag{27.44}$$

The second term for the long-range interactions finds justification in the expression for the mean ionic activity coefficients obtained from the radial distribution function,[†] and it has little effect on the fitting of the activities of the individual ions.

[†] See Equation 16.42 from [40], "pressure equation," for the mean ionic activity coefficient written as,

$$\ln \gamma_{\pm} = -\frac{b}{3a} \left[\frac{Ka}{1 + Ka} + 2 \ln(1 + Ka) \right]$$ where according to Equation 16.17 of [41], $Ka \approx I_k^{0.5}$.

If necessary, the parameters b_i and d_i can be refitted from those reported in the literature [2]. The expression for the osmotic coefficient obtained using Equation 27.43 and the Lin and Lee equation is

$$\varphi_k = 1 + \sum_i Y_{i,k} \left\{ \begin{array}{l} -AZ_i^2 \left[1 - 2I_k^{-0.5} + \dfrac{2}{I_k}\ln\left(1+I_k^{0.5}\right)\right] \\ + \bar{b}_{i,k}\left[\dfrac{3I_k+1}{1.5(1+1.5I_k)^2} - \dfrac{1}{1.5^2 I}\ln(1+1.5I_k)\right] + \dfrac{\bar{d}_{i,k}}{2}I_k \end{array}\right\}$$ (27.45)

while using Equation 27.44, it is

$$\varphi_k = 1 + \sum_i Y_{i,k} \left\{ \begin{array}{l} -AZ_i^2 \left[\dfrac{I_k^{0.5}}{1+I_k^{0.5}}\right] \\ +\bar{b}_{i,k}\left[\dfrac{3I_k+1}{1.5(1+1.5I_k)^2} - \dfrac{1}{1.5^2 I}\ln(1+1.5I_k)\right] + \dfrac{\bar{d}_{i,k}}{2}I_k \end{array}\right\}$$ (27.46)

For the parameters $\bar{b}_{i,k}$ and $\bar{d}_{i,k}$ of the term representing short-range interactions in a multi-ion aqueous solution, the following mixing rules are used:

$$\bar{b}_{i,k} = \sum_j X_{j,k} b_{i/j}^0$$ (27.47)

$$\bar{d}_{i,k} = \sum_j X_{j,k} d_{i/j}^0$$ (27.48)

where for an ion i the sums run over all the counterions present, with the counterion charge mole fractions defined as

$$X_{j,k} = \frac{m_{j,k}|Z_j|}{\sum_l m_{l,k}|Z_l|}$$ (27.49)

In Equations 27.47 and 27.48, $b_{i/j}^0$ and $d_{i/j}^0$ are the values of the parameters in the binary i/j ion–counterion system. These parameters are independent of composition. Figure 27.3 presents the results of the prediction of the equilibrium pressure over aqueous solutions of mixed electrolytes with a common ion in comparison with experimental data available in the literature [26]. It is important to point out that parameters used for the prediction were evaluated at 298.15 K only.

The importance of this new thermodynamic approach for the treatment of electrolyte solutions will be emphasized after demonstrating the consistency of the experimental information.

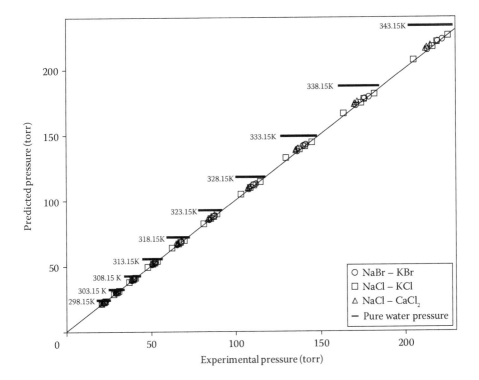

FIGURE 27.3 Comparison of predicted and experimental equilibrium pressure over aqueous solutions of mixed electrolytes with a common ion. The horizontal lines cutting the 45° line indicate the vapor pressure of pure water at each experimental temperature. (Reprinted from Wilczek-Vera, G., and Vera, J. H., *Ind. Eng. Chem. Res.* 48(13): 6436–6440, 2009. With permission. Copyright 2016 American Chemical Society.)

COMPARISON OF EXPERIMENTAL RESULTS FOR THE ACTIVITY OF INDIVIDUAL IONS OBTAINED BY DIFFERENT RESEARCHERS[†]

Experimental values of the activity of individual ions in concentrated solutions had been measured by several researchers [42]. These experimental results were questioned by some, arguing that such measurements are "impossible" to make. The negative attitude against the mere concept of measuring ionic activities was so strong that previous researchers publishing these experimental results were forced to pay lip service to the prevailing opinion and admit, as the paradigm required, that theory did not support the concept of individual activity of ions [43–45].

First, we consider the results for aqueous solutions of sodium salts. As the measurements reported were performed at different concentrations, and any smoothing

---

[†] Reprinted and partially modified from *The Journal of Chemical Thermodynamics*, 99, Wilczek-Vera, G., and Vera J. H., How much do we know about the activity of individual ions?, 65–69, Copyright 2016, with permission from Elsevier.

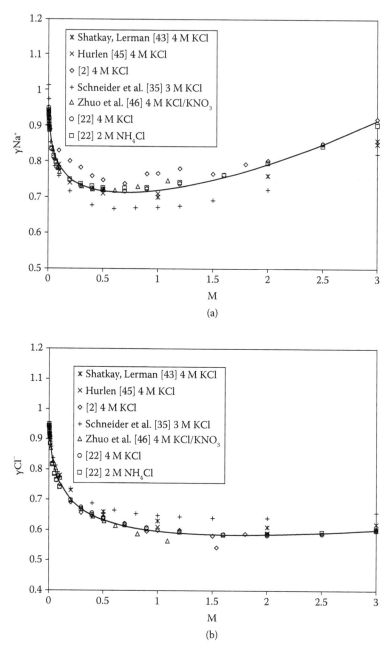

FIGURE 27.4 Comparison of different experimental results obtained for the individual ionic activity coefficients in the aqueous solutions of NaCl. (a) Individual activity coefficients of Na^+. (b) Individual activity coefficients of Cl^-. (Inner solutions of reference electrodes are listed in the corresponding reference.)

equation would introduce its own bias, a graphical comparison is probably the most direct way of proceeding. Figure 27.4 presents the comparison of experimentally measured individual activity coefficients of Na^+ and Cl^- in the aqueous NaCl system determined in five different laboratories with significantly different experimental techniques (see Table 27.1). In addition to our original results from 2004 [2], we also include later measurements done with the improved experimental setups [22] and the results obtained with a Ag/AgCl reference electrode filled with 2 M NH_4Cl instead of a saturated KCl solution, as was done before.

Figure 27.5a depicts comparison of our data with results reported by Hurlen and Breivik [48] for aqueous solutions of NaBr. We used Na ISE and Br ISE. Our measurements were performed at different times and with different inner electrolytes in the single-junction Ag/AgCl reference electrode. Hurlen's measurements of anion data were done with a Ag/AgBr electrode, while the activity coefficients for the Na^+ ions were calculated from the mean ionic activity coefficient data. Figure 27.5b presents data for the system of aqueous Na_2SO_4. In our study [2], we used the Na^+-sensitive glass electrode to measure the activity of the sodium ion. The activities of the sulfate ion were calculated from the mean ionic activity coefficient. Hurlen and Breivik [48], on the other hand, measured only the anion using a Hg/Hg_2SO_4 electrode, and the activity of cation was calculated.

Figures 27.6 and 27.7a depict a comparison of results for 1:1 chlorides LiCl, CsCl, and NH_4Cl, while Figure 27.7b presents the results for the LiBr system. In all these cases, the measurements were done for anions only, while the missing information for cations was calculated from the mean ionic activity coefficient data. Due to the use of Ag/AgCl and Ag/AgBr electrodes, Hurlen [45,48] could extend his measurements up to 4 m, while the use of a Cl ISE and Br ISE limited the range of measurements to 3 m.

Figures 27.8 and 27.9 show comparisons for the aqueous system's alkaline–earth chlorides. As before, only the activity coefficients for anions were determined experimentally. The corresponding information for cations was calculated from the mean ionic activity coefficient data. It is interesting to observe that in the systems of aqueous $MgCl_2$ and $CaCl_2$ at lower concentrations, the activity coefficients of cations are smaller than the activity coefficients of anions, but at higher concentrations this trend is reversed. As the activity coefficients for the $BaCl_2$ were determined below 2 m only, it is difficult to say if the same behavior would be observed in this system as well.

Due to the main assumption that $\gamma_- = \gamma_\pm$ for saturated KCl aqueous solution used by Hurlen [45,47–49] to calibrate the results of his method, we can expect a discrepancy between his and our data. The comparison presented in Figure 27.10 confirms this prediction. On the other hand, the agreement of our data with data of Schneider et al. [35] is quite satisfactory.

In Figure 27.10, we have included our oldest data from 2004 [2], as well as newer data [22] in which two different inner solutions were used in the single-junction Ag/AgCl reference electrode. It is clear that with the increasing experience of the experimenters, the quality of our data improved.

Figure 27.11 presents comparison of data measured by different authors in the years 1935–2011. The lines on the graph correspond to our smoothed data. It is clear

TABLE 27.1
Summary of Different Techniques Used for the Measurements of Ionic Activity Coefficients

ISE Cation	ISE Anion	Reference Electrode	Runs	Calibration
Szabo, 1935 [50]				
Pt/H$_2$, HCl	Hg/HgCl, HCl	Liquid chains of cells of identical electrodes were measured	Discontinuous	At low concentration range, junction potential evaluated through diffusion potentials measured in the same study
Shatkay and Lerman, 1969 [43,44]				
Glass sodium electrode	Ag/AgCl	Standard calomel electrode (SCE), KCl bridge	Discontinuous	By taking $\gamma_\pm = 0.903$ for 0.01 m NaCl; E_J Henderson [32]
Hurlen, 1979–1983 [45,47–49]				
Calculated	Ag/AgCl electrode [45,47] Ag/AgBr electrode [48] Hg/Hg$_2$SO$_4$ electrode [49]	SCE, SJ, inner solution sat. KCl	Discontinuous	$\gamma_- = \gamma_\pm$ for sat. KCl aq.; E_J Henderson [32]
Khoshkbarchi and Vera, 1996 [11]				
Na ISE K ISE	Cl ISE, Br ISE	Ag/AgCl, SJ, inner solution sat. KCl	Continuous	Based on 2 points at intermediate concentrations; E_J Henderson [32]; wrong sign
Schneider et al., 2003–2004 [35,51]				
Na ISE, K ISE Ca ISE	Cl ISE	Ag/AgCl, inner solution: 3 M KCl [35] 3 M KCl [51]	Continuous	Extended Debye–Hückel equation used up to $I_m = 0.01$ m; (1) E^0 and S fitted; (2) S Nernstian, E^0 fitted; E_J Henderson [32]
Vera et al., 1999–2009 [2,6,22,24]				
K ISE, Na ISE	Br ISE, Cl ISE, F ISE, NO$_3$ ISE	Ag/AgCl, SJ, inner solutions: KCl, NaCl, CsCl, or NH$_4$Cl	Continuous	Based directly on least-squares fit of several points in the dilute region
Zhuo et al., 2008 [46]				
Na glass ISE	Br ISE, Cl ISE, F ISE	Ag/AgCl, DJ, inner solution: KCl, outer solution: 10% KNO$_3$	Continuous	Based on one intermediate point and a model equation with 2 parameters; uses all points of the run for the fit; assumes E_J constant
Sakaida and Kakiuchi, 2011 [52]				
Pt/H$_2$,H$^+$	Ag/AgCl	Ionic liquid salt bridge	Discontinuous	Use of DHL at 0.01 M HCl aq.

Note: All measurements were performed at 25°C. SCE, standard calomel electrode; SJ, single junction; DJ, double junction; DHL, Debye–Hückel limiting law; aq., aqueous; sat., saturated.

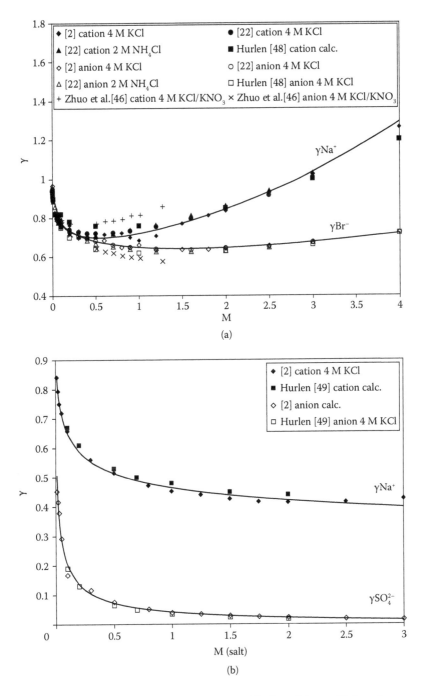

FIGURE 27.5 Comparison of different experimental ionic activity coefficients in aqueous solutions of (a) NaBr and (b) Na_2SO_4. (Inner solutions of reference electrodes are listed in the corresponding reference.)

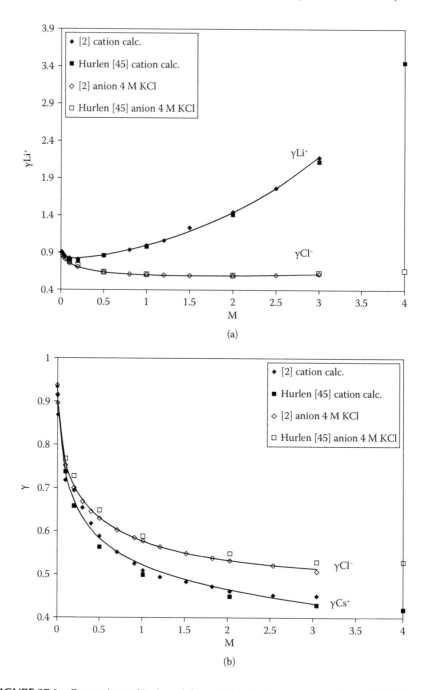

FIGURE 27.6 Comparison of ionic activity coefficients in aqueous solutions of (a) LiCl and (b) CsCl. Note that the activity coefficients of cations were calculated.

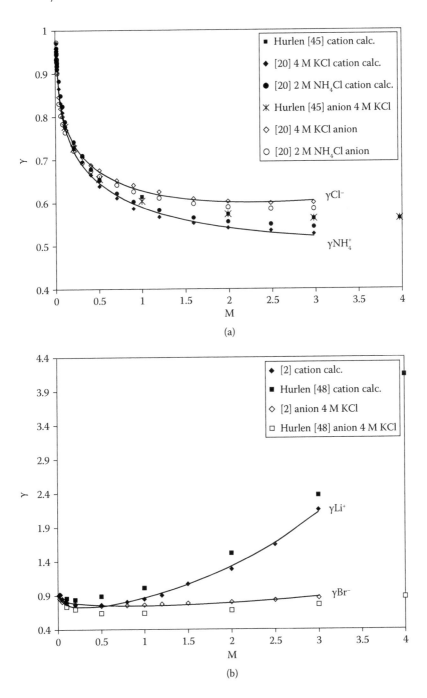

FIGURE 27.7   Comparison of ionic activity coefficients in aqueous solutions of (a) NH_4Cl and (b) LiBr. Note that the activity coefficients of cations were calculated.

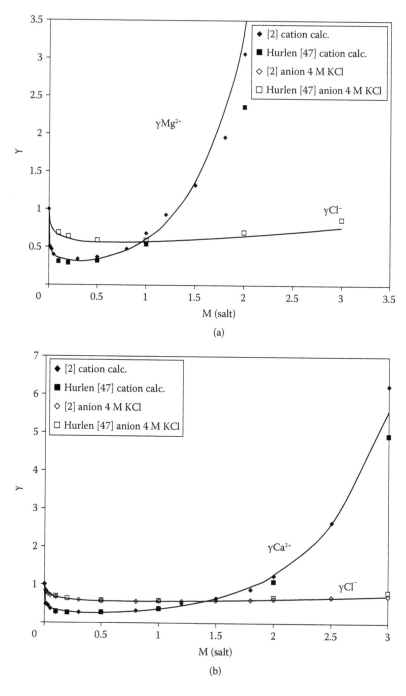

FIGURE 27.8 Comparison of ionic activity coefficients in aqueous solutions of (a) $MgCl_2$ and (b) $CaCl_2$. Note that the activity coefficients of cations were calculated.

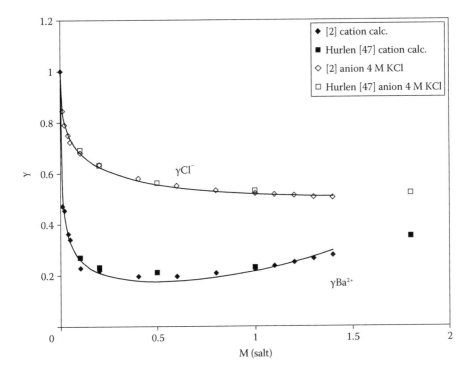

FIGURE 27.9 Comparison of ionic activity coefficients in aqueous solutions of BaCl$_2$. Note that the activity coefficients of cations were calculated.

that there are big discrepancies of different data sets. The data of Sakaida and Kakiuchi [52] plotted together with their error limits and the data of Schneider et al. [51] seemed to be in mutual agreement. The data of Szabo [50], on the other hand, show a reverse relation compared with all other data between the activity coefficients of H$^+$ and those of Cl$^-$.

Summarizing, with the exception of data for the HCl system, there is good agreement between the values of the activity coefficients of individual ions reported by different laboratories using very different methods for collecting the data. In some cases, such as NaCl and KCl, for which several reports coincide, while some measurements are clearly out of agreement, it is obvious which measurements should be considered for future comparisons with theory. The next desirable step in the development of the field of electrolyte solutions would be to find a theoretical expression for the relation between ionic activities, the mean ionic activity coefficient, and the transfer number of the ions [24]. This relation, until now, has only been explored in a correlational form [7,10,28].

EMERGING THEORIES OF ELECTROLYTE SOLUTIONS

It has been the norm in science that a theory becomes validated once it is confirmed by experimental results. Hence, it is remarkable that in the case of the individual

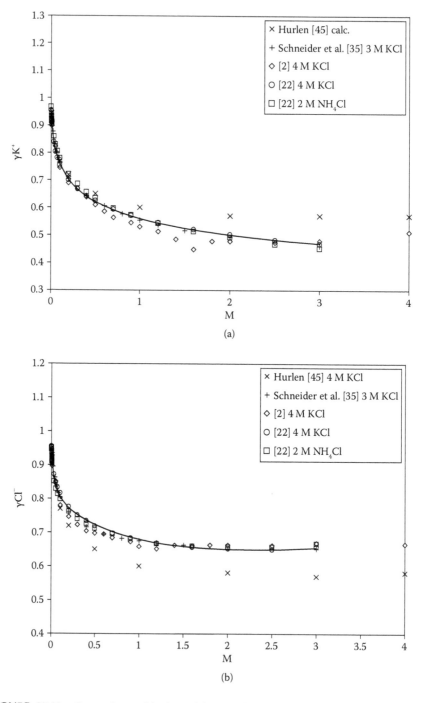

FIGURE 27.10 Comparison of ionic activity coefficients in aqueous solutions of KCl. (a) Individual activity coefficients of K⁺. (b) Individual activity coefficients of Cl⁻. (Inner solutions of reference electrodes are listed in the corresponding reference.)

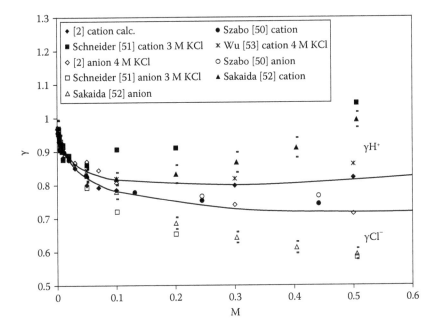

FIGURE 27.11 Comparison of ionic activity coefficients in aqueous solutions of HCl. (Inner solutions of reference electrodes are listed in the corresponding reference.)

activity of ions, the situation has been the inverse. Notably, recent theoretical developments [54,55] using well-defined, physically meaningful parameters have closely reproduced the experimentally measured ionic activities in aqueous solutions. In this case, the theory is confirming the validity of the experimental measurements. As expected, the same voices that had objected to the experimental results are now objecting to the theories. A detailed discussion of these theories is beyond the scope of this text. A brief description of both theories is given below, quoting only what, from the point of view of this chapter, we consider to be the most significant manuscript of each of these two series of publications. Another comparison of theoretical predictions versus experimental data can be found in [56], but the reported discrepancies between the theory and experiment are much greater than those presented in [54,55].

FRAENKEL'S SMALLER-ION SHELL THEORY

Fraenkel's treatment [54] is an extension of the Debye–Hückel theory (DHT). Like the DHT, it uses a linearized Poisson–Boltzmann equation, but unlike the DHT, it considers the ion and the counterion as having two shells of different diameters.

In his paper [54], Fraenkel presented successful comparisons with our data for the following systems: NaCl, LiCl, $MgCl_2$, $CaCl_2$, and K_2SO_4, as well as for the Zhuo et al. data [46] for the aqueous solution of NaBr. Using Fraenkel's Excel files, sent kindly to us by him, we show here the NaCl and $CaCl_2$ systems (Figure 27.12a and b), also treated by Liu and Eisenberg [55]. We are also including the K_2SO_4 system.

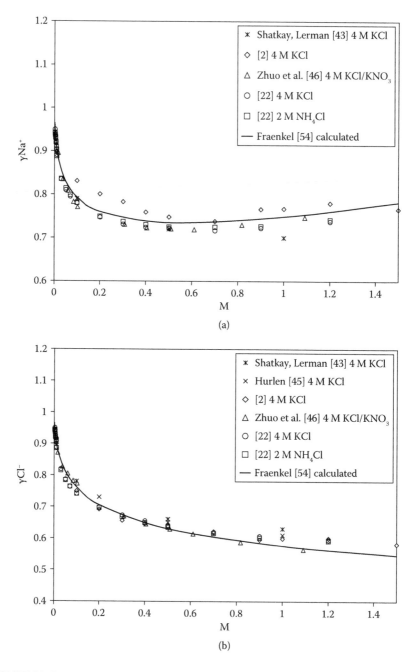

FIGURE 27.12 Comparison of experimental and calculated ionic activity coefficients in aqueous solutions of NaCl, CaCl$_2$, and K$_2$SO$_4$. (a) Individual activity coefficients of Na$^+$ in NaCl. (b) Individual activity coefficients of Cl$^-$ in NaCl. (Inner solutions of reference electrodes are listed in [54]. Calculated data for the systems were obtained from Dr. Dan Fraenkel through personal communication.)   *(Continued)*

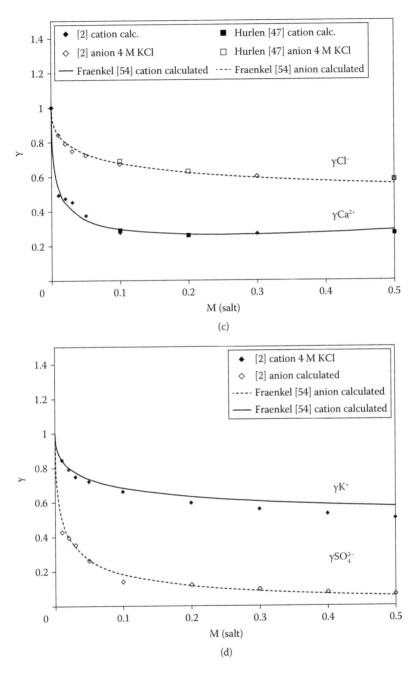

FIGURE 27.12 (Continued) Comparison of experimental and calculated ionic activity coefficients in aqueous solutions of NaCl, $CaCl_2$, and K_2SO_4. (c) $CaCl_2$. (d) K_2SO_4. (Inner solutions of reference electrodes are listed in [54]. Calculated data for the systems were obtained from Dr. Dan Fraenkel through personal communication.)

Liu and Eisenberg Poisson–Fermi Theory

Liu and Eisenberg [55] use a Poisson–Fermi (PF) model with a simple steric potential and a correlation length of ions. When the steric potential and the correlation length are omitted, the Gibbs–Fermi free energy of the model reduces to the classical Gibbs free energy of the Poisson–Boltzmann model. Figure 27.13 compares the results of the theory with experimental values for the activity of individual ions in an aqueous solution of NaCl and in an aqueous solution of $CaCl_2$.

These two electrolytes present quite different behavior. For the case of $CaCl_2$, the activity coefficients of the cation and the anion cross each other at a molar concentration slightly above unity. Both theories represent the experimental data well. The smaller-ion shell (SiS) theory of Fraenkel is simpler and fits the data better in the dilute region, while the PF theory of Liu and Eisenberg gives a better fit of the concentrated region.

As an interesting aside point, Fraenkel's manuscript also presents results reported by Abbas et al. for NaCl [57]. These values were generated by a Monte Carlo simulation using the unrestricted primitive model (MC-UPM), that is, the primitive model proposed by Friedman [58] but free from the restriction of equal-sized ions. Notably, the results of the Monte Carlo computer simulations disagree with both the theory and the experimental data.

In the modeling of electrolyte solutions, it is important to keep in mind that the DHT and its extensions are based on the McMillan–Mayer framework in which the solvent is just a background for the long-range interactions between ions. For modeling single-solvent dilute electrolyte solutions, the use of the Debye–Hückel expression is an acceptable approximation; however, for solvents other than water, the Debye–Hückel constant needs attention. From physics, this constant is given as

$$A_\gamma = \left[\frac{N_A^2}{8\pi} \left(\frac{e^2}{\varepsilon_0 RT} \right)^{3/2} \right] \left(\frac{2d_s}{D_s^3} \right)^{1/2} \tag{27.50}$$

The term in square brackets on the right-hand side of Equation 27.50 is independent of the nature of the solvent. In this term, N_A, R and T have their usual meaning as Avogadro's number, the gas constant, and the absolute temperature, while e and ε_0 are the charge of the electron and the vacuum permittivity. The density d_s and the dielectric constant D_s of the solvent change for solvents other than water. The thermodynamic treatment of solutions of electrolytes in mixed solvents needs to convert the equations from the McMillan–Mayer to the Lewis–Randall framework. This conversion has been discussed by Cabezas and O'Connell [59] and also by Haynes and Newman [60]. The extension of this treatment to mixed solvents is presented in [61].

As a closing remark, it must be said that the field of electrolyte solutions was stagnant for almost 100 years and received new attention only after the advent of biotechnology. There are many details that necessarily have been left out of the discussion here. An interesting new direction of development, which is beyond the scope of this book, is the relation of the activity of the individual ions with the transport number of the ions. This relation was first identified by Malatesta [62], developed further

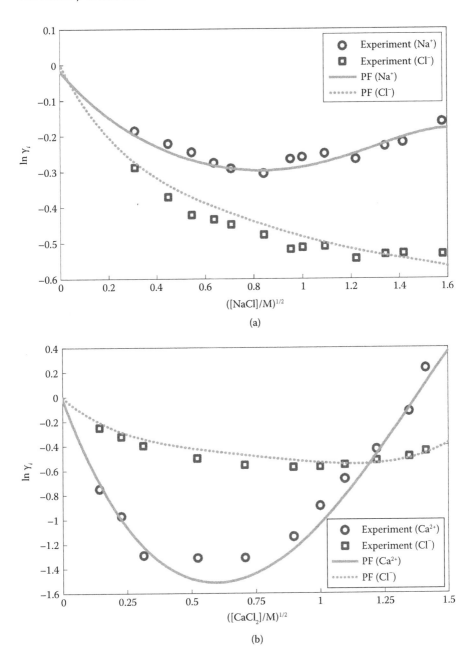

FIGURE 27.13 Comparison of experimental and calculated ionic activity coefficients in aqueous solutions of NaCl and CaCl$_2$. (a) NaCl. (b) CaCl$_2$. (Reprinted from *Chemical Physics Letters*, 637, Liu, J.-L., and Eisenberg, B., Poisson–Fermi model of single ion activities in aqueous solutions, 1–6, Copyright 2015, with permission from Elsevier.)

by Zarubin [7], and commented on by the authors of this text [8,24]. It is to be expected that new theories will shed light on its molecular basis.

CONCLUSIONS

Once we have made a case for supporting the measurability of ionic activities in concentrated aqueous solutions, we briefly address here both the evolution of thoughts on this matter and the potential use of the new thermodynamics of electrolyte solutions [26] that this process has generated.

Gilbert N. Lewis [63–65] introduced the concepts of activity and activity coefficients of individual ions and obtained values for these properties in dilute aqueous solutions of electrolytes from solubility and free energy of dilution data [66]. He also established the connection between the activity of individual ions and the electromotive force in an electrochemical cell [67]. In the first edition of their book *Thermodynamics*, G. N. Lewis and M. Randall [68] stated that "it would be of much theoretical interest if we could determine the actual activity of an ion in a solution of any concentration," and added, "This indeed might be accomplished if we had any general method of calculating the potential at a liquid junction." In the second edition of the same text, revised by K. S. Pitzer and L. Brewer [69], these new authors wrote, "In order to measure the activity of a single ion in solution, it is necessary to measure experimentally some process which transfers that single charged species into or out of the solution. The enormous magnitude of the space-charge energies prevents such measurements by conventional methods which require transfer of macroscopic amounts of materials." However, they added, "Nevertheless, single-ion properties are potentially measurable." In the third edition of the same text [41], this time edited by K. S. Pitzer alone, the activity coefficients of the individual ions appear only as analytical expressions derived from differentiation of the excess Gibbs energy based on mean ionic activity coefficients in what is called the Pitzer theory. Although in this latter text the possibility of measuring the activity of individual ions in concentrated solutions is ignored, the following statement in Pitzer's text [41] clearly justifies the new thermodynamics of electrolyte solutions [26] based on the actual measurement of these properties: "Indeed, for many applications to complex mixtures, it is simpler to use the single-ion expressions." These applications are well illustrated in another text by the same author [70].

REFERENCES

1. Wilczek-Vera, G., and Vera, J. H. 2005. On the measurement of individual ion activities. *Fluid Phase Equilib.* 236: 96–110.
2. Wilczek-Vera, G., Rodil, E., and Vera, J. H. 2004. On the activity of ions and the junction potential. Revised values for all data. *AIChE J.* 50: 445–462.
3. Malatesta, F. 2005. On a recent purported determination of individual ion activity coefficients. *AIChE J.* 52: 785–791.
4. Sandler, S. I. 2006. Preface to the letters to the editor. *AIChE J.* 52: 2306.
5. Vera, J. H. 2006. Letter to the editor. *AIChE J.* 52: 2306.
6. Lladosa, E., Arce, A., Wilczek-Vera, G., and Vera, J. H. 2010. Effect of the reference solution in the measurement of ion activity coefficients using cells with transference at T = 298.15 K. *J. Chem. Thermodyn.* 42: 244–250.

7. Zarubin, D. P. 2011. The nature of single-ion activity coefficients calculated from potentiometric measurements on cells with liquid junctions. *J. Chem. Thermodyn.* 43: 1135–1152.

8. Vera, J. H., and Wilczek-Vera, G. 2012. Comment on "The nature of single-ion activity coefficients calculated from potentiometric measurements on cells with liquid junctions" by Dmitri P. Zarubin, *J. Chem. Thermodyn.* 43 (2011) 1135–1152. *J. Chem. Thermodyn.* 47: 442–444.

9. Zarubin, D. P. 2012. The nature of single-ion activity coefficients calculated from potentiometric measurements on cell with liquid-junction. *J. Chem. Thermodyn.* 47: 445–448.

10. Wilczek-Vera, G., and Vera, J. H. 2012. Short answer to the reply from D.P. Zarubin to our comment on "The nature of single-ion activity coefficients calculated from potentiometric measurements on cell with liquid-junction." *J. Chem. Thermodyn.* 47: 449–450.

11. Khoshkbarchi, M. K., and Vera, J. H. 1996. Measurement and correlation of ion activity in aqueous single electrolyte solutions. *AIChE J.* 42(1): 249–258.

12. Khoshkbarchi, M. K., and Vera, J. H. 1996. Measurement and correlation of ion activity coefficients in aqueous solutions of mixed electrolyte with a common ion. *Fluid Phase Equilib.* 121(1–2): 253–265.

13. del Mar Marcos-Arroyo, M., Khoshkbarchi, M. K., and Vera, J. H. 1996. Activity coefficients of sodium, potassium, and nitrate ions in aqueous solutions of $NaNO_3$, KNO_3, and $NaNO_3 + KNO_3$ at 25°C. *J. Solution Chem.* 25(10): 983–1000.

14. Rabie, H. R., Wilczek-Vera, G., and Vera, J. H. 1999. Activities of individual ions. From infinite dilution to saturated solutions. *J. Solution Chem.* 28: 885–914.

15. Taghikhani, V., Modarress, H., and Vera, J. H. 1999. Individual anionic activity coefficients in aqueous electrolyte solutions of LiCl and LiBr. *Fluid Phase Equilib.* 166(1): 67–77.

16. Taghikhani, V., Modarress, H., and Vera, J. H. 2000. Measurement and correlation of the individual ionic activity coefficients of aqueous electrolyte solutions of KF, NaF and KBr. *Can. J. Chem. Eng.* 78(1): 175–181.

17. Rodil, E., and Vera, J. H. 2001. Individual activity coefficients of chloride ions in aqueous solutions of $MgCl_2$, $CaCl_2$ and $BaCl_2$ at 298.2 K. *Fluid Phase Equilib.* 187–188: 15–27.

18. Rodil, E., and Vera, J. H. 2001. Measurement and correlation of the activity coefficients of individual ions in aqueous electrolyte solutions of Na_2SO_4 and K_2SO_4. *Can. J. Chem. Eng.* 79(5): 771–776.

19. Rodil, E., Persson, K., Wilczek-Vera, G., and Vera, J. H. 2001. Determination of the activity of H+ ions within and beyond the pH meter range. *AIChE J.* 47(12): 2807–2817.

20. Rodil, E., and Vera, J. H. 2003. The activity of ions: Analysis of the theory and data for aqueous solutions of $MgBr_2$, $CaBr_2$ and $BaBr_2$ at 298.2 K. *Fluid Phase Equilib.* 205(1): 115–132. Erratum: *Fluid Phase Equilib.* 211(2): 289, 2003.

21. Wilczek-Vera, G., and Vera, J. H. 2003. Peculiarities of the thermodynamics of electrolyte solutions: A critical discussion. *Can. J. Chem. Eng.* 81: 70–79.

22. Wilczek-Vera, G., Rodil, E., and Vera, J. H. 2006. Towards accurate values of individual ion activities. Additional data for NaCl, NaBr and KCl, and new data for NH_4Cl. *Fluid Phase Equilib.* 241: 59–69.

23. Wilczek-Vera, G., Rodil, E., and Vera, J. H. 2006. A complete discussion of the rationale supporting the experimental determination of individual ionic activities. *Fluid Phase Equilib.* 244: 33–45.

24. Arce, A., Wilczek-Vera, G., and Vera, J. H. 2007. Activities of aqueous Na+ and Cl– ions from homoionic measurements with null junction potentials at different concentrations. *Chem. Eng. Sci.* 52: 3849–3857.

25. Rodil, E., Arce, A., Wilczek-Vera, G., and Vera, J. H. 2009. Measurement of ion activity coefficients in aqueous solutions of mixed electrolyte with a common ion: NaNO$_3$ + KNO$_3$, NaCl + KCl and NaBr + NaCl. *J. Chem. Eng. Data*: 345–350.
26. Wilczek-Vera, G., and Vera, J. H. 2009. On the predictive ability of the new thermodynamics of electrolyte solutions. *Ind. Eng. Chem. Res.* 48(13): 6436–6440.
27. Wilczek-Vera, G., and Vera, J. H. 2009. Answer to the comments by Francesco Malatesta on *J. Chem. Eng. Data*, 2009, 54, 345–350 by Rodil, E., Arce, A., Wilczek-Vera, G., Vera, J.H. *J. Chem. Eng. Data*: 2979.
28. Wilczek-Vera, G., Arce, A., and Vera, J. H. 2010. Answer to "Comment on individual ion activities of Na+ and Cl– by Arce, Wilczek-Vera and Vera" by F. Malatesta. *Chem. Eng. Sci.* 65: 2263–2264.
29. Wilczek-Vera, G., and Vera, J. H. 2011. On the measurement of the real values of individual ionic activities: A chemical engineering perspective. *Chem. Eng. Sci.* 66: 3782–3791.
30. Wilczek-Vera, G., and Vera, J. H. 2011. The activity of individual ions. A conceptual discussion of the relation between the theory and the experimentally measured values. *Fluid Phase Equilib.* 312: 79–84. Corrigendum to "The activity of individual ions. A conceptual discussion of the relation between the theory and the experimentally measured values" [*Fluid Phase Equilib.* 312 (2011) 79–84]. *Fluid Phase Equilib.* 314: 203, 2012.
31. Harned, H. S. 1926. Individual thermodynamic behaviors of ions in concentrated solutions including discussion of the thermodynamic method of computing liquid junction potentials. *J. Phys. Chem.* 30: 433–456.
32. Henderson, P. 1907. Zur thermodynamik der flüssigkeitsketten. *Z. Phys. Chem.* 59: 118–127.
33. Lewis, G. N., and Randall, M. 1921. The activity coefficient of strong electrolytes. *J. Am. Chem. Soc.* 43: 1112–1159.
34. Güntelberg, E. 1926. Interaction of ions. *Z. Phys. Chem.* 123: 199–247. Quoted from Guggenheim, E. A. 1950. *Thermodynamics: An Advanced Treatise for Chemists and Physicists*, 2nd ed. New York: Interscience Publishers.
35. Schneider, A. C., Pasel, C., Luckas, M., Schmidt, K. G., and Herbell, J.-D. 2003. Bestimmung von Ionenaktivitätskoeffizienten in wässrigen Lösungen mit Hilfe ionenselektiver Elektroden. *Chem. Ing. Tech.* 75: 244–249.
36. Atkins, P. W. 1986. *Physical Chemistry*, 3rd ed. New York: Freeman, p. 676.
37. (a) Bates, R. G. 1964. *Determination of pH, Theory and Practice*. New York: John Wiley. (b) 2nd ed., 1973.
38. Harper, H. W. 1985. Calculation of liquid junction potentials. *J. Phys. Chem.* 89(9): 1659–1664.
39. Guggenheim, E. A. 1929. The conceptions of electrical potential difference between two phases and the individual activity of ions. *J. Phys. Chem.* 33: 842–849.
40. Lin, C.-L., and Lee, L.-S. 2003. A two-ionic-parameter approach for ion activity coefficients of aqueous electrolyte solutions. *Fluid Phase Equilib.* 205: 69–88.
41. Pitzer, K. S. 1995. *Thermodynamics*, 3rd ed. New York: McGraw-Hill.
42. Wilczek-Vera, G., and Vera, J. H. 2016. How much do we know about the activity of individual ions? *J. Chem. Thermodyn.* 99: 65–69.
43. Shatkay, A., and Lerman, A. 1969. Individual activities of sodium and chloride in aqueous solutions of sodium chloride. *Anal. Chem.* 41: 514–517.
44. Shatkay, A. 1970. Further study of individual ion activities in pure aqueous sodium chloride solution. *Talanta* 17: 381–390.
45. Hurlen, T. 1979. Convenient single-ion activities. *Acta Chem. Scand. A* 33: 631–635.
46. Zhuo, K., Dong, W., Wang, W., and Wang, J. 2008. Activity coefficients of individual ions in aqueous solutions of sodium halides at 298.15 K. *Fluid Phase Equilib.* 274: 80–84.

47. Hurlen, T. 1979. Ion activities of alkaline-earth chlorides in aqueous solution. *Acta Chem. Scand. A* 33: 637–640.
48. Hurlen, T., and Breivik, T. R. 1981. Ion activities of alkali metal bromides in aqueous solution. *Acta Chem. Scand. A* 35: 415–418.
49. Hurlen, T. 1983. Single-ion activities of sodium sulfate in aqueous solution. *Acta Chem. Scand. A* 37: 739–742.
50. Szabo, Z. 1935. Über die Ionenaktivitäten der Salzsäure. *Z. Phys. Chem. A* 174: 22–32.
51. Schneider, A. C., Pasel, C., Luckas, M., Schmidt, K. G., and Herbell, J.-D. 2004. Determination of hydrogen single ion activity coefficients in aqueous HCl solutions at 25°C. *J. Solution Chem.* 33(3): 257–273.
52. Sakaida, H., and Kakiuchi, T. 2011. Determination of single-ion activities of H⁺ and Cl⁻ in aqueous hydrochloric acid solutions by use of an ionic liquid salt bridge. *J. Phys. Chem. B* 115(45): 13222–13226.
53. Wu, Y. Ch., Feng, D., and Koch, W. F. 1989. Evaluation of liquid junction potentials and determination of pH values of strong acids at moderate ionic strengths. *J. Solution Chem.* 18(7): 641–649.
54. Fraenkel, D. 2012. Single-ion activity: Experiment versus theory. *J. Phys. Chem. B* 116: 3603–3612.
55. Liu, J.-L., and Eisenberg, B. 2015. Poisson–Fermi model of single ion activities in aqueous solutions. *Chem. Phys. Lett.* 637: 1–6.
56. Valiskó, M., and Boda, D. 2015. Unraveling the behavior of the individual ionic activity coefficients on the basis of the balance of ion–ion and ion–water interactions *J. Phys. Chem. B* 119(4): 1546–1557.
57. Abbas, Z., Ahlberg, E., and Nordholm, S. 2009. Monte Carlo simulations of salt solutions: Exploring the validity of primitive models. *Phys. Chem. B.* 113: 5905–5916.
58. Friedman, H. L. 1960. Mayer's ionic solution theory applied to electrolyte mixtures. *J. Chem. Phys.* 32: 1134–1149.
59. Cabezas, H., and O'Connell, J. P. 1993. Some uses and misuses of dilute solution thermodynamic models. *Ind. Eng. Chem. Res.* 32: 2892–2904.
60. Haynes, C. A., and Newman, J. 1998. On converting from the McMillan-Mayer framework. I. Single-solvent system. *Fluid Phase Equilib.* 45: 255–268.
61. Curtis, R. A., Blanch, H. W., Prausnitz, J. M., and Newman, J. 2001. McMillan-Mayer solution thermodynamics for a protein in a mixed solvent. *Fluid Phase Equilib.* 192: 131–153.
62. Malatesta, F. 2010. Comment on the individual ion activities of Na+ and Cl– by Arce, Wilczek-Vera and Vera. *Chem. Eng. Sci.* 65: 675–679.
63. Lewis, G. N. 1907. Outlines of a new system of thermodynamic chemistry. *Proc. Am. Acad. Arts Sci.* 43: 259–296.
64. Lewis, G. N. 1907. Umriss eines neuen Systems der chemischen Thermodynamik. *Z. Phys. Chem.* 61: 129–165.
65. Lewis, G. N. 1909. The use and abuse of the ionic theory. *Z. Phys. Chem.* 70: 212–219.
66. Lewis, G. N. 1912. The activity of the ions and the degree of dissociation of strong electrolytes. *J. Am. Chem. Soc.* 34(12): 1631–1644.
67. Lewis, G. N. 1913. The free energy of chemical substances. Introduction. *J. Am. Chem. Soc.* 35: 1–30.
68. Lewis, G. N., and Randall, M. 1923. *Thermodynamics,*1st ed. New York: McGraw-Hill.
69. Lewis, G. N., and Randall, M. 1961. *Thermodynamics*, 2nd ed., rev. K. S. Pitzer and L. Brewer. New York: McGraw-Hill.
70. Pitzer, K. S. 1991. *Activity Coefficients in Electrolyte Solutions*, 2nd ed. Boca Raton, FL: CRC Press.

Section v

Appendices

ASOG-KT + PRSV = Design

Appendix A: Material Balances

MATERIAL BALANCES FOR SYSTEMS WITHOUT CHEMICAL REACTIONS

The basic idea of a material balance is that the mass of a closed system does not change unless there is interconversion between mass and energy. Thus, for processes in which the integrity of atoms is unaffected, the mass of the system changes only if mass enters or leaves a system. For this case, in a process occurring in an open system during a time interval dt, the differential change of mass of the system is given by

$$[\delta m]_{in} - [\delta m]_{out} = [dm]_{system} \tag{A.1}$$

The quantities $[\delta m]$ (in or out) represent the amount of material that enters and leaves the system during the time interval, dt, and $[dm]$ is the change in the mass of the system during that same time interval. Note that the differential symbol, d, is not used for the mass entering or leaving the system, as these are not exact differentials. Dividing all the terms in Equation A.1 by dt, we obtain

$$\dot{m}_{in} - \dot{m}_{out} = \frac{dm}{dt} \tag{A.2}$$

In Equation A.2, the symbol m with an upper dot is used for mass flow rate (lbm/h, kg/s, or else), and dm/dt is the rate of change of the mass, that is, the accumulation of mass in the system. This accumulation can be positive or negative. For the special case of "steady-state operation," the values of the properties of the system at any particular point do not change with time. Hence, for steady state there is no accumulation of material in the system and

$$\dot{m}_{in} = \dot{m}_{out} \tag{A.3}$$

In the case of a fluid of density ρ (kg/m³) flowing inside a duct or a pipe of cross-sectional area A (m²) at a linear velocity υ (m/s), the mass flow rate is given by

$$\dot{m} = \rho \upsilon A \tag{A.4}$$

Hence, combining Equations A.3 and A.4 for steady flow in a duct, we write

$$(\rho \upsilon A)_{in} = (\rho \upsilon A)_{out} \tag{A.5}$$

For the case of the flow of a liquid, that is, an incompressible fluid, the density is constant and the linear velocity is inversely proportional to the cross-sectional area of the flow.

MATERIAL BALANCES FOR SYSTEMS WITH CHEMICAL REACTIONS

When there are chemical reactions in a system at steady state, the mass entering the system per unit time is still equal to the mass leaving the system per unit time. However, the number of moles of different compounds entering the system per unit time can be different from the number of moles leaving the system per unit time. If there are no chemical reactions, the number of moles of each compound is conserved. If there *are* chemical reactions occurring in the system, neither the mass nor the number of moles of the individual compounds participating in the reactions is conserved. However, the total mass is conserved and the number of moles of each *chemical element* is conserved. Thus, in the case of chemical reactions, it is easier to perform material balances on each of the chemical elements rather than on the chemical compounds. If it is desired to perform a material balance on a *chemical compound i*, in terms of moles, Equation A.1 for steady state takes the form

$$[\text{mols of } i]_{\text{in}} + [\text{mols of } i]_{\text{produced}} - [\text{mols of } i]_{\text{consumed}} = [\text{mols of } i]_{\text{out}} \qquad (A.6)$$

At *unsteady* state, the difference between the right-hand side and the left-hand side of Equation A.6 gives the accumulation of moles of compound i in the system.

Appendix B: Working with the Virial EOS

VIRIAL EOS TRUNCATED AFTER THE THIRD TERM

Due to its direct connection with intermolecular forces, the virial equation of state (EOS) has been of permanent theoretical interest. An excellent presentation and compilation of information on this subject can be found in the text by Prausnitz et al. [1]. From a practical point of view, its use is limited to the calculation of fugacities in the gas phase at low pressures. The compressibility factor expansion in terms of density is

$$z = 1 + B\rho + C\rho^2 + \cdots \tag{25.6}$$

and in terms of pressure,

$$z = 1 + \frac{B}{RT}P + \frac{C - B^2}{(RT)^2}P^2 + \cdots \tag{25.7a}$$

The second and third virial coefficients of expansion in terms of density, B and C, are known with accuracy for many substances. Higher coefficients are seldom used in practice.

There are many correlations available for the calculation of values of the second and third virial coefficients of pure compounds. We reproduce here two simple ones that give satisfactory approximations.

For the second virial coefficient of nonpolar compounds, Tsonopoulos [2] recommended the form

$$B = \frac{RT_c}{P_c}\left(B^{(0)} + \omega B^{(1)}\right) \tag{B.1}$$

where T_c and P_c are the critical temperature and critical pressure of the compound, both in absolute scales. The two contributions to the second virial coefficient for nonpolar compounds are given by

$$B^{(0)} = 0.1445 - \frac{0.330}{T_r} - \frac{0.1385}{T_r^2} - \frac{0.0121}{T_r^3} - \frac{0.000607}{T_r^8} \tag{B.2}$$

$$B^{(1)} = 0.0637 + \frac{0.331}{T_r^2} - \frac{0.423}{T_r^3} - \frac{0.008}{T_r^8} \tag{B.3}$$

where $T_r = {T}/{T_c}$ is the reduced temperature and the acentric factor is defined as

$$\omega = -\log_{10} \frac{P^s_{at\, T_r = 0.7}}{P_c} - 1 \tag{B.4}$$

For the third virial coefficient of nonpolar compounds, Orbey and Vera [3] proposed the correlation

$$C = \left(\frac{RT_c}{P_c}\right)^2 \left(C^{(0)} + \omega C^{(1)}\right) \tag{B.5}$$

with

$$C^{(0)} = 0.01407 + \frac{0.02432}{T_r^{2.8}} - \frac{0.00313}{T_r^{10.5}} \tag{B.6}$$

$$C^{(1)} = -0.02676 + \frac{0.01770}{T_r^{2.8}} + \frac{0.040}{T_r^{3.0}} - \frac{0.003}{T_r^{6.0}} - \frac{0.00228}{T_r^{10.5}} \tag{B.7}$$

FUGACITY COEFFICIENT OF A PURE COMPOUND

From Equations 13.7a and 25.7a truncated after the third term, the fugacity coefficient of a pure nonpolar compound is given by

$$\ln \phi_i^0 = \frac{BP}{RT} + \frac{\left(C - B^2\right)P^2}{2(RT)^2} + \dots \tag{B.8}$$

For polar compounds, Abusleme and Vera [4] developed a group contribution method that reduces to Tsonopoulos's correlation for nonpolar compounds. For polar compounds, the correlation uses Tsonopoulos's correlation with a modified acentric factor that, following the work of Hayden and O'Connell [5], is calculated using the mean radius of gyration of the molecules that is obtained from a program available without cost from the supplemental material of the Hayden–O'Connell publication. In addition, the correlation uses an energetic contribution calculated in terms of group interaction parameters presented by Abusleme and Vera [4] in tabular form.

VIRIAL EOS FOR MIXTURES

As discussed in Chapter 25, the virial coefficients can be related to intermolecular potentials and the second virial coefficient of a mixture of nonpolar compounds is related to the respective pure compound virial coefficients by

$$B_m = \sum_j \sum_k y_j y_k B_{jk} \tag{B.9}$$

By a similar argument, it can be shown that

$$C_m = \sum_j \sum_k \sum_l y_j y_k y_l C_{jkl} \qquad \text{(B.10)}$$

Hence, for a binary mixture of nonpolar compounds,

$$B_m = y_1^2 B_{11} + 2 y_1 y_2 B_{12} + y_2^2 B_{22} \qquad \text{(B.9a)}$$

and

$$C_m = y_1^3 C_{111} + 3 y_1^2 y_2 C_{112} + 3 y_1 y_2^2 C_{122} + y_2^3 C_{222} \qquad \text{(B.10a)}$$

The problem here is to obtain the values of the cross virial coefficients B_{12}, C_{112}, and C_{122}. For this, empirical mixing rules need to be introduced and the treatment loses its elegance. This explains why the use of the virial EOS for mixtures is restricted to the pressure range up to which the second virial coefficient suffices to represent the volumetric behavior of the mixture.

FUGACITY COEFFICIENT OF A COMPOUND IN A MIXTURE

For the range of pressure, where the second virial term suffices, the fugacity coefficient of component i is given by

$$\ln \phi_i = \left(2 \sum_k y_k B_{ik} - B_m \right) \frac{P}{RT} \qquad \text{(B.11)}$$

In the optimum situation, if the second virial coefficient of the binaty mixture is known from experiment as a function of composition, the cross second virial coefficient can be evaluated by fitting the data with Equation B.9a. If this is not the case, crude mixing rules can be used. As suggested by Equations 25.19 and 25.20, the second virial coefficient is the result of two contributions: a hard-core contribution that is independent of temperature, $B_{ii,0}$, and a temperature-dependent contribution arising from the intermolecular potentials, $B_{ii,T}$,

$$B_{ii} = B_{ii,0} - B_{ii,T}$$

Judging by the success of the mixing rules used for cubic EOSs, Equations 25.21 and 25.22, although not tested, a logical approximation to obtain the cross second virial coefficient is

$$B_{jk,0} = \frac{B_{jj,0} + B_{kk,0}}{2}$$

$$B_{jk,T} = \sqrt{B_{jj,T} B_{kk,T}} \left(1 - \delta_{jk} \right)$$

$$B_{jk} = B_{jk,0} - B_{jk,T}$$

where δ_{jk} is a binary parameter. More complex mixing rules are available in the literature [1].

For the second virial coefficient of polar, associating, and quantum compounds, several correlations proposed in the literature are reviewed by Prausnitz et al. [1]. A compilation of values of virial coefficients is available in the literature [6].

REFERENCES

1. Prausnitz, J. M., Lichtenthaler, R. N., and Gomes de Azevedo, E. 1999. *Molecular Thermodynamics of Fluid Phase Equilibria*, 3rd ed. Upper Saddle River, NJ: Prentice Hall.
2. Tsonopoulos, C. 1974. An empirical correlation of second virial coefficients. *AIChE J.* 20: 263–272.
3. Orbey, H., and Vera, J. H. 1983. Correlation for the third virial coefficient using T_c, P_c and ω as parameters. *AIChE J.* 29: 107–113.
4. Abusleme, J. A., and Vera, J. H. 1989. A group contribution method for second virial coefficients. *AIChE J.* 35: 481–489.
5. Hayden, J. G., and O'Connell, J. P. 1975. A generalized method for predicting second virial coefficients. *Ind. Eng. Chem. Process Des. Dev.* 14: 209–216.
6. Diamond, J. H., and Smith, E. B. 1979. *The Virial Coefficients of Pure Gases and Mixtures*. Oxford: Clarendon.

Appendix C: Working with the PRSV EOS

PRSV EOS

The Peng–Robinson–Stryjek–Vera (PRSV) [1–3] is a cubic equation of state (EOS) specially designed to give reliable values of the vapor pressures and fugacities of nonpolar, polar, or associating pure compounds. Its extension to calculate fugacities of the components of a mixture is done using appropriate mixing rules. Basically, the PRSV EOS uses the framework of the Peng–Robinson (PR) EOS [4], namely,

$$P = \frac{RT}{(v-b)} - \frac{a_c \alpha\,(\omega, T_r)}{v^2 + 2bv - b^2} \tag{C.1}$$

The major modification differentiating the PRSV from the PR EOS is the form of the function $\alpha(\omega, T_r)$. Briefly stated, Soave [5] found that for different compounds, the value of $\alpha^{0.5}\,(\omega, T_r)$ of the Redlich–Kwong EOS [6] gave a straight line when plotted against $T_r^{0.5}$ and proposed the following form for a correlation satisfying the condition of $\alpha\,(\omega, T_r = 1) = 1$ at the critical point:

$$\alpha^{0.5} = 1 + \kappa\left(1 - T_r^{0.5}\right) \tag{C.2}$$

At reduced temperature 0.7, corresponding to the definition of the acentric factor ω,

$$\omega \equiv -\log_{10}\left(P^s at\ T_r = 0.7\right) + \log_{10} P_c - 1.000 \tag{C.3}$$

the value of the slope κ is given by

$$\kappa = \frac{\sqrt{\alpha_{T_r=0.7}} - 1}{1 - \sqrt{0.7}} \tag{C.4}$$

Reversing the logic of the calculations, the value of ω fixes the value of the saturation pressure at reduced temperature 0.7, which, in turn, fixes the value of $\alpha\,(\omega, T_r = 0.7)$, giving a single value of the slope κ. In the PRSV EOS, following Soave, the correlation of κ in terms of ω of the PR EOS was changed to satisfy the value of the slope κ at $T_r = 0.7$. Figure C.1 [7] compares the results obtained with the correlations used by the PR EOS, the PRSV EOS, and Soave's form for the Redlich–Kwong–Soave (RKS) EOS.

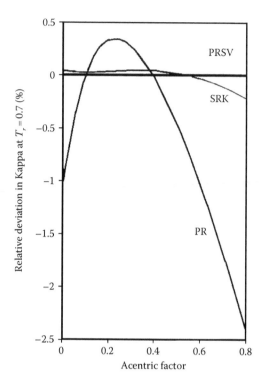

FIGURE C.1   Relative deviation in the value of κ at $T_r = 0.7$ by the Soave (1972) [5], PR (1976) [4], and PRSV (Stryjek and Vera, 1986) [1–3] EOSs. (Reprinted with permission from Zabaloy, M. S., and Vera, J. H., *Ind. Eng. Chem. Res.*, 37(5), 1591–1597, 1998. Copyright 2015 American Chemical Society.)

The PR EOS ω dependence is satisfactory only for medium-weight hydrocarbons, say from pentane to heptane, that are of interest in the petroleum industry. In addition to the change of the ω function of $\alpha(\omega, T_r)$, the temperature dependence was also changed to a form that corresponded to a function that had been previously proposed by Mathias [8]. With these changes, the PRSV EOS [1,2] extended the use of cubic EOSs to alcohols, acids, water, and all other kinds of functional compounds. PRSV2 [3] is an improved version of PRSV including two extra-pure compound parameters in the temperature dependence of the α (ω, T_r).

In this appendix, we first give the general solution of a cubic equation; then we give the cubic form of PRSV as a function of temperature and pressure. Thus, the values of the liquid and vapor molar volumes can be directly obtained, and the values of the fugacity coefficients can then be calculated. As a step for obtaining the equilibrium vapor pressure of a pure compound, we give a useful adaptation of the Antoine equation [12]. Tables with parameters for pure compounds for PRSV and PRSV2 are included. The extension of PRSV to calculate liquid densities, enthalpies, and heats of vaporization of pure compounds [13] follows, and the appendix is completed with a consideration of mixing rules and the expressions to calculate fugacity coefficients in mixtures.

EXACT SOLUTION OF A CUBIC EQUATION

The exact method to evaluate the solutions to a cubic equation, given in Perry's handbook [9] and presented in a simplified form elsewhere [10], is reproduced here with a clearer nomenclature. Consider the general form of a cubic equation:

$$x^3 + C_1 x^2 + C_2 x + C_3 = 0$$

where C_1, C_2, and C_3 are numeric coefficients. For simplicity, define

$$C_4 = C_1/3$$

$$D = \frac{1}{2}(C_2 C_4 - C_3) - C_4^3$$

$$E = C_4^2 - (C_2/3)$$

and

$$\Delta = D^2 - E^3$$

If $\Delta = 0$, say $|\Delta| < 10^{-3}$, there are three real roots with at least two equal roots:

$$x_1 = 2\sqrt[3]{D} - C_4$$

$$x_2 = x_3 = -\sqrt[3]{D} - C_4$$

If $\Delta > 0$, there is one real root and two complex conjugate roots. In this case, define

$$F = \sqrt[3]{D + \sqrt{\Delta}}$$

and

$$G = \sqrt[3]{D - \sqrt{\Delta}}$$

$$H = -\left[\frac{1}{2}(F + G) + C_4\right]$$

$$I = \frac{\sqrt{3}}{2}(F - G)$$

$$x_1 = F + G - C_4$$

$$x_2 = H + Ii$$

$$x_2 = H - Ii$$

If $\Delta < 0$, there are three real and unequal roots. In this case, define

$$\vartheta\,(radians) = arc\cos\left(\frac{D}{\sqrt{E^3}}\right)$$

$$x_1 = 2\sqrt{E}\,\cos(\vartheta/3) - C_4$$

$$x_2 = 2\sqrt{E}\,\cos(\vartheta/3 + (2/3)\pi) - C_4$$

$$x_2 = 2\sqrt{E}\,\cos(\vartheta/3 + (4/3)\pi) - C_4$$

VOLUME ROOTS FOR PRSV EOS AT GIVEN *T* AND *P*

The PRSV EOS cubic expansion is

$$v^3 + \left[b - \frac{RT}{P}\right]v^2 + \left[\frac{a_c\alpha(T_r)}{P} - 2\frac{RT}{P}b - 3b^2\right]v + \left[b^2 + \frac{RT}{P}b - \frac{a_c\alpha(T_r)}{P}\right]b = 0$$

So, to solve for the volume roots using the general solution of a cubic equation,

$$C_1 = b - \frac{RT}{P}$$

$$C_2 = \frac{a_c\alpha(T_r)}{P} - 2\frac{RT}{P}b - 3b^2$$

$$C_3 = \left[b^2 + \frac{RT}{P}b - \frac{a_c\alpha(T_r)}{P}\right]b$$

HOW TO CALCULATE THE SATURATION PRESSURE OF A PURE COMPOUND WITH PRSV EOS

This is the first question that a beginner asks when dealing with a cubic EOS, and although the answer is simple, the algebra is not. At saturation, the pressure of the liquid is equal to the pressure of the vapor phase, and also, the fugacity coefficient of the pure compound in the vapor phase (*sv*) is equal to its fugacity coefficient in the liquid phase (*sl*).

$$P^{sv}\left(v^{sv}, T\right) = P^{sl}\left(v^{sl}, T\right)$$

$$\phi^{sv}\left(v^{sv}, T\right) = \phi^{sl}\left(v^{sl}, T\right)$$

with

$$\ln\phi = (z - 1) - \ln(z - B_P) - \frac{A_P}{2B_P\sqrt{2}}\ln\left[\frac{z + B_P\left(1 + \sqrt{2}\right)}{z + B_P\left(1 - \sqrt{2}\right)}\right] \qquad (C.5)$$

$$z \equiv \frac{Pv}{RT} = \frac{v}{(v-b)} - \frac{va/RT}{v^2 + 2bv - b^2} \tag{C.6}$$

$$A_P \equiv \frac{Pa}{(RT)^2} \tag{C.7}$$

$$B_P \equiv \frac{Pb}{RT} \tag{C.8}$$

$$b = 0.07780 \frac{RT_c}{P_c} \tag{C.9}$$

$$a = a_c \alpha(\omega, T_r) = 0.45724 \frac{(RT_c)^2}{P_c} \alpha(\omega, T_r) \tag{C.10}$$

$$\alpha(\omega, T_r) = \left[1 + \kappa \left(1 - T_r^{0.5} \right) \right]^2 \tag{C.11}$$

For PRSV,

$$\kappa = \kappa_0 + \kappa_1 (1 + T_r^{0.5})(0.7 - T_r) \tag{C.12a}$$

and for PRSV2,

$$\kappa = \kappa_0 + [\kappa_1 + \kappa_2(\kappa_3 - T_r)(1 - T_r^{0.5})] \, (1 + T_r^{0.5})(0.7 - T_r) \tag{C.12b}$$

with

$$\kappa_0 = 0.37889 + 1.48971 \, \omega - 0.17132 \, \omega^2 + 0.01965 \, \omega^3 \tag{C.13}$$

Pure compound parameters κ_1, κ_2, and κ_3 are given in Table C.1A, together with T_c, P_c, and ω for various compounds of industrial interest. The pressure P_c is in atmospheres and the temperature T is in kelvin. The value of the parameter κ_1 is the same for PRSV and PRSV2 EOS. We observe here that when $\kappa_2 = 0$, PRSV2 EOS reduces to PRSV. The value of κ_3 is immaterial in this case.

For most cases of practical interest, PRSV gives enough accurate values of the vapor pressure and fugacity. Thus, in the extension to additional compounds [11] presented in Table C.1B and in the revised values of κ_1 for alkanes included in Table C.1C, only κ_1 is reported. The values of κ_1 for alkanes included in Table C.1C should not be combined with the values of κ_2 and κ_3 of Table C.1A. The values for T_c, P_c, and ω, however, are those of Table C.1A. The revised values for alkanes are recommended for use at temperatures around and above room temperature. The values included in Table C.1A, on the other hand, are for use at cryogenic temperatures and always together with the parameters κ_2 and κ_3.

TABLE C.1A
Pure Compound Parameters for the PRSV and the PRSV2 EOS

	T_c, K	P_c, kPa	ω	κ_1	κ_2	κ_3
		Inorganic				
Nitrogen	126.200	3400.	0.03726	0.01996	0.3162	0.535
Oxygen	154.77	5090.	0.02128	0.01512	–0.0090	0.490
Carbon dioxide	304.21	7382.43	0.22500	0.04285	0.0000	0.000
Ammonia	405.55	11289.52	0.25170	0.00100	–0.1265	0.510
Water	647.286	22089.75	0.34380	–0.06635	0.0199	0.443
Hydrogen chloride	324.60	8308.57	0.12606	0.01989	–0.0036	0.310
		Hydrocarbons				
Methane	190.555	4595.	0.01045	–0.00159	0.1521	0.517
Ethane	305.43	4879.76	0.09781	0.02669	0.1358	0.424
Propene	365.57	4664.55	0.14080	0.04400	0.2610	0.424
Propane	369.82	4249.53	0.15416	0.03136	0.2757	0.447
Butane	425.16	3796.61	0.20096	0.03443	0.6767	0.461
Pentane	469.70	3369.02	0.25143	0.03946	0.3940	0.457
Neopentane	433.75	3196.27	0.19633	0.04303	0.8697	0.615
Hexane	507.30	3012.36	0.30075	0.05104	0.8634	0.460
Heptane	540.10	2735.75	0.35022	0.04648	0.9331	0.496
Octane	568.76	2486.49	0.39822	0.04464	0.6214	0.509
Nonane	594.56	2287.90	0.44517	0.04104	0.6621	0.519
Decane	617.50	2103.49	0.49052	0.04510	0.8549	0.527
Undecane	638.73	1965.69	0.53631	0.02919	1.3288	0.568
Dodecane	658.2	1823.83	0.57508	0.05426	0.8744	0.505
Tridecane	675.8	1722.51	0.62264	0.04157	0.9387	0.528
Tetradecane	691.8	1621.18	0.66735	0.02686	0.9408	0.528
Pentadecane	706.8	1519.86	0.70694	0.01892	1.0908	0.559
Hexadecane	720.6	1418.54	0.74397	0.02665	0.0334	0.767
Heptadecane	733.4	1317.21	0.76976	0.04048	2.9805	0.571
Octadecane	745.2	1215.89	0.79278	0.08291	4.1441	0.577
Cyclohexane	553.64	4075.	0.20877	0.07023	0.6146	0.530
Bicyclohexyl	731.4	2563.50	0.39361	0.01805	3.0438	0.606
Benzene	562.16	4898.	0.20929	0.07019	0.7939	0.523
Toluene	591.80	4106.	0.26323	0.03849	0.5261	0.510
Ethylbenzene	617.20	3606.	0.30270	0.03994	0.5342	0.519
p-Xylene	616.23	3511.	0.32141	0.01277	0.5963	0.524
Indane	684.90	3950.	0.31000	0.01173	0.9246	0.548
n-Propylbenzene	638.32	3200.	0.34513	0.02715	0.7310	0.530
1,2,3-Trimethylbenzene	637.25	3127.	0.39970	–0.01384	0.4777	0.538
Naphthalene	748.35	4050.93	0.30295	0.03297	0.6634	0.510
1-Methyl-napthalene	766.[b]	3566.60	0.37666	–0.01842	–0.8140	0.577
2-Methyl-naphthalene	761.	3505.81	0.37119	–0.01639	–0.0750	0.597

(Continued)

TABLE C.1A *(Continued)*
Pure Compound Parameters for the PRSV and the PRSV2 EOS

	T_c, K	P_c, kPa	ω	κ_1	κ_2	κ_3
Biphenyl	769.15	3120.78	0.38095	0.11487	0.1077	0.407
Diphenylmethane	770.2	2857.34	0.43724	0.05955	0.3703	0.579
9,10-Dihydrophenanthrene	774.7	1314.17	0.54678	−0.01393	−5.8256	0.587
Ketones						
Acetone	508.1	4696.	0.30667	−0.00888	0.2871	0.537
Butanone	536.78	4207.	0.32191	0.00554	0.6847	0.613
2-Pentanone	561.08	3694.	0.34719	0.01681	0.7787	0.620
3-Pentanone	561.46	3729.	0.34377	0.03558	0.6180	0.610
Methylbutanone	555.	3790.	0.31314	0.04113	0.5138	0.615
2-Hexanone	587.	3320.	0.39385	0.00984	0.8448	0.550
3-Hexanone	582.82	3319.	0.37931	0.02321	0.8891	0.619
Dimethylbutanone	567.	3470.	0.32293	0.04005	0.7451	0.535
2-Heptanone	611.5	2990.	0.42536	0.02731	0.9236	0.561
5-Nonanone	640	2329.	0.51374	0.02002	0.9649	0.579
Alcohols						
Methanol	512.58	8095.79	0.56533	−0.16816	−1.3400	0.588
Ethanol	513.92	6148.	0.64439	−0.03374	−2.6846	0.592
1-Propanol	536.71	5169.55	0.62013	0.21419	−3.6816	0.640
2-Propanol	508.40	4764.25	0.66372	0.23264	−3.5578	0.652
1-Butanol	562.98	4412.66	0.59022	0.33431	−1.1748	0.642
2-Butanol	535.95	4248.52	0.58254	0.39045	0.0026	0.676
2-Methyl-1-propanol	547.73	4295.12	0.59005	0.37200	−1.2792	0.642
2-Methyl-2-propanol	506.15	3971.90	0.61365	0.43099	−0.0480	0.658
1-Pentanol	588.15	3909.[a]	0.57839	0.36781	0.2918	0.621
1-Hexanol	591.23[b]	3468.15[a]	0.77526	−0.00237	−3.3938	0.551
1-Octanol	684.8[b]	2860.00[a]	0.32420	0.82940	2.7372	0.543
1-Decanol	717.84[b]	2394.87[a]	0.38355	0.80898	1.4978	0.470
Ethers						
Dimethyl ether	400.1	5240.	0.18909	0.05717	−0.1211	0.481
Methyl ethyl ether	437.8	4410.	0.23479	0.16948	0.0515	0.768
Methyl *n*-propyl ether	476.25	3801.	0.27215	0.02300	0.9179	0.558
Methyl *i*-propyl ether	464.48	3762.	0.26600	0.04123	0.3833	0.562
Methyl *n*-butyl ether	512.78	3371.	0.31672	0.01622	0.6140	0.548
Methyl *t*-butyl ether	497.1	3430.	0.26746	0.05129	−0.2022	0.585
Ethyl *n*-propyl ether	500.23	3370.	0.33612	−0.01668	1.1679	0.553
Di-*n*-propyl ether	530.6	3028.	0.37070	−0.03162	1.4094	0.577
Di-*i*-propyl ether	500.32	2832.	0.33168	0.03751	0.8810	0.590
Methyl phenyl ether	645.6	4250.	0.34817	0.01610	1.0478	0.616

(Continued)

TABLE C.1A *(Continued)*
Pure Compound Parameters for the PRSV and the PRSV2 EOS

	T_c, K	P_c, kPa	ω	κ_1	κ_2	κ_3
		Various				
Nitromethane	588.	6312.49	0.34700	−0.10299	0.5905	0.463
Acetonitrile	545.5	4830.	0.33710	−0.13991	−0.3777	0.522
Acetic acid	592.71	5786.	0.45940	−0.19724	0.8136	0.541
Dimethylformamide	660.07[a]	5240.66[a]	0.26600	0.18999	−0.1857	0.470
2-Methoxyethanol	574.39[a]	5348.38[a]	0.65629	−0.42503	0.0020	0.407
1-Propylamine	497.	4742.	0.28037	0.14326	1.1002	0.614
2-Propylamine	476.	5066.20	0.28530	0.06001	2.2851	0.603
2-Methoxypropionitrile	636.11[a]	3602.55[a]	0.47656	−0.09508	0.0033	0.931
2-Methyl-2-propylamine	483.9	3840.	0.27417	0.13440	0.9308	0.635
Tetrahydrofuran	540.1	5190	0.22550	0.03961	0.5029	0.573
Pyridine	620.	5595.26[a]	0.23716	0.06946	0.4241	0.575
Furfural	652.48[a]	4345.45[a]	0.39983	−0.03471	4.5512	0.539
n-Methylpyrrolidone	719.33[a]	4057.72[a]	0.34478	0.11367	−0.7871	0.537
Hexafluorobenzene	516.7	3273.	0.39610	0.02752	0.8172	0.565
Nitrotoluene	743.	3207.	0.42200	−0.00901	0.5425	0.582
m-Cresol	705.15	4559.58	0.44492	0.24705	5.6910	0.597
Thianaphthene	752.	3880.71	0.29356	0.06043	−0.1210	0.592

Source: Stryjek, R., and Vera, J. H.: PRSV2: A cubic equation of state for accurate vapor-liquid equilibria calculations. *Can. J. Chem. Eng.* 1986. 64. 820–826. Copyright Wiley-VCH Verlag GmbH & Co. KGaA. Reproduced with permission.

[a] Estimated by group contribution method.
[b] Obtained from optimum fit.

The calculation of the saturation pressure, at a given temperature, follows the reverse path of the equations above. However, after five steps up, the calculation of the saturation volumes, necessary to calculate the fugacity coefficients, requires knowledge of the pressure that is precisely what we want to calculate. The solution to this impasse is to iterate values of pressure, evaluate the saturation volumes, and check whether the equality of fugacities is satisfied. The first step is to estimate the value of the saturation pressure with an equation that does not require knowing the saturation volumes. After the first estimate, a typical trial-and-error procedure is followed until the equality of fugacities is satisfied. For the first estimation of the saturation pressure, we follow the path proposed by Kolasinska and Vera [12] and use the Antoine equation written as

$$\log_{10} P^s = A_A - \frac{B_A}{C_A + T_r} \tag{C.14}$$

Applying this equation to the critical point, the normal boiling point, and the reduced temperature 0.7, where the saturation pressure is directly obtained from the value of the acentric factor ω using Equation C.3, Kolasinska and Vera [12] obtained

$$A_A = \frac{B_A}{C_A + \theta} \tag{C.15}$$

TABLE C.1B
Additional Pure Compound Parameters for the PRSV EOS

	T_c, K	P_c, kPa	ω	κ_1
Inorganic				
Sulfur dioxide	430.65	7883	0.25630	0.03962
Carbon monoxide	132.91	3496	0.04830	0.04279
Hydrogen sulfide	373.09	8943	0.10000	0.03160
Chlorine	416.95	7972	0.08700	0.03641
Nitrous oxide	309.58	7254	0.16033	0.06325
Hydrocarbons				
Acetylene	308.33	6139	0.18432	0.09919
Ethylene	282.35	5042	0.08652	0.04191
Isobutene	408.14	3655	0.18466	−0.00238
1-Butene	419.53	4023	0.19409	0.02222
2-Methylpropene	417.90	4000	0.19989	−0.00516
cis-2-Butene	435.58	4220	0.20308	0.04457
trans-2-Butene	428.63	4050	0.21246	0.00435
1,3-Butadiene	425.37	4330	0.19356	0.02327
Isopentane	460.43	3381	0.22802	0.02840
1-Pentene	464.78	3529	0.23329	0.03521
2,2,4-Trimethyl-pentane	543.96	2569	0.30365	0.03363
Cyclopentane	511.61	4502	0.19434	0.04602
Methylcyclohexane	572.12	3471	0.23637	0.04859
o-Xylene	630.33	3734	0.31130	0.02291
m-Xylene	617.05	3536	0.32570	0.01610
i-Propylbenzene	631.15	3209	0.32600	0.05509
Styrene	636.85	3678	0.26377	0.16109
Acids				
Propionic acid	612.65	5370	0.53595	−0.03093
n-Butyric acid	628.15	5269	0.67862	−0.14012
n-Valeric acid	651.15	4671	0.70048	−0.08171
Aldehydes				
Propanal	496.00	4762	0.30991	−0.03642
Esters				
Ethyl formate	508.45	4742	0.27596	0.1 1372
Methyl acetate	506.85	4691	0.32027	0.05791
Ethyl acetate	523.25	3830	0.36061	0.06464
n-Propyl acetate	549.35	3334	0.38978	0.06531
n-Butyl acetate	579.15	3090	0.40792	0.07587
Ethyl propionate	546.05	3362	0.38996	0.06362

(Continued)

TABLE C.1B *(Continued)*
Additional Pure Compound Parameters for the PRSV EOS

	T_c, K	P_c, kPa	ω	κ_1
Ethers				
Diethyl ether	466.70	3638	0.28115	0.05004
Ketones				
Cyclohexanone	629.15	3850	0.45102	−0.29473
Halogen Compounds				
Carbon tetrachloride	556.30	4557	0. 19200	0.05212
Carbon tetrafluoride (R-14)	227.50	3742	0.17977	0.02136
Bromotrifluoromethane (R-13B I)	340.20	3964	0.17196	0.04014
Chlorotrifluoromethane (R-13)	302.00	3870	0.17135	0.05054
Dichlorodifluoromethane (R-12)	385.00	4131	0.17875	0.04722
Trichlorofluoromethane (R-11)	471.20	4409	0.18749	0.03708
Chloroform	536.55	5472	0.21754	0.02899
Dichlorofluoromethane (R-21)	451.60	5168	0.20378	0.07915
Chlorodifluoromethane (R-22)	369.30	4989	0.21974	0.04513
Trifluoromethane (R-23)	299.07	4836	0.26537	0.00535
Difluoromethane (R-32)	351.55	5830	0.27704	−0.02874
Methyl chloride	416.30	6677	0.15600	0.01160
Methyl fluoride	317.80	5877	0.19096	−0.00374
Chloropentafluoroethane (R-115)	353.10	3157	0.25349	0.03220
1,2-Dichlorotetrafluoroethane (R-114)	419.00	3261	0.25098	0.06095
1,1,2-Trichlorotrifluoroethane (R-1 13)	487.50	3411	0.25153	0.05596
1,1-Trichloroethane	523.15	5066	0.24683	−0.00984
1,2-Trichloroethane	561.15	5370	0.28463	−0.04659
Ethyl chloride	460.35	5269	0.19000	0.03982
Chlorobenzene	632.35	4519	0.24942	0.03123
Nitrogen Compounds				
Methylamine	430.05	7446	0.28285	0.10475
Ethylamine	456.45	5612	0.30091	0.03536
Dimethylamine	437.75	5240	0.27869	0.20709
Trimethylamine	433.35	4075	0.19600	0.06414
Diethylamine	496.95	3708	0.29900	0.11210
Triethylamine	535.15	3034	0.32294	0.02951
Aniline	699.15	5308	0.38200	0.00316
o-Toluidine	694.15	3749	0.43500	0.03690
m-Toluidine	709.15	4154	0.40600	0.08293
Oxygen Compounds				
Ethylene oxide	469.00	7194	0.20387	0.01757
Propylene oxide	482.30	4924	0.26034	0.03693

(Continued)

TABLE C.1B *(Continued)*
Additional Pure Compound Parameters for the PRSV EOS

	T_c, K	P_c, kPa	ω	κ_1
Phenol	694.25	6130	0.44110	0.02100
o-Cresol	697.55	5005	0.43597	−0.00849
p-Cresol	704.55	5147	0.50819	−0.04098
Sulfur Compounds				
Thiophene	580.15	5694	0.18824	0.07654

Source: Proust, P., and Vera, J. H.: PRSV: The Stryjek-Vera modification of the Peng-Robinson equation of state. Parameters for other pure compounds of industrial interest. *Can. J. Chem. Eng.* 1989. 67. 170–173. Copyright Wiley-VCH Verlag GmbH & Co. KGaA. Reproduced with permission.

Note: An additional table of parameters for the PRSV for more than 800 compounds is given at the end of this appendix (Table C.4).

TABLE C.1C
Revised Pure Compound Parameters
for the PRSV EOS

Compound	κ_1
Methane	−0.00857
Ethane	0.01 181
Propane	0.00658
Butane	0.01668
Isobutane	0.01408
Pentane	0.01156
Isopentane	0.00925
Neopentane	0.04226
Hexane	0.01157
Heptane	0.02325
Octane	0.02569
2,2,4-Trimethylpentane	0.03425
Decane	0.01478

Source: Proust, P., et al.: Calculation of pure compound saturated enthalpies and saturated volumes with the PRSV equation of state. Revised κ_1 parameters for alkanes. *Can. J. Chem. Eng.* 1993. 71. 292–298. Copyright Wiley-VCH Verlag GmbH & Co. KGaA. Reproduced with permission.

$$B_A = \frac{(C_A + \theta)(C_A + 1)}{(1 - \theta)} \log_{10} P_c \qquad (C.16)$$

$$C_A = \frac{(D_A - 0.7)}{(1 - D_A)} \qquad (C.17)$$

with

$$D_A = \frac{(0.7 - \theta) \log_{10} P_c}{(1 - \theta)(\log_{10} P_c - \omega - 1)} \qquad (C.18)$$

and

$$\theta = \frac{T_b}{T_c} \qquad (C.19)$$

where T_b is the temperature of the normal boiling point in kelvin and the critical pressure P_c in Equation C.18 is in atmospheres.

HOW TO CALCULATE SATURATED LIQUID MOLAR VOLUMES WITH THE PRSV EOS

It has long been known that the cubic EOSs do not give reliable values of the molar volumes or enthalpies of pure compounds. In fact, the molar volumes obtained above do not correspond to the real values of the saturated molar volumes at the temperature of the calculation. They are just values that help in evaluating the fugacity coefficient of the pure compound and give a very accurate value of the vapor pressure. In order to obtain saturated liquid volumes, Proust et al. [13] proposed to use

$$v^{sl} = \left(v_{PRSV}^{sl} + D_v \right) \left[1 - \exp\left(A_v - \frac{B_v}{T_r^3} \right) \right] \qquad (C.20)$$

where D_v is a displacement parameter given in Table C.2 for some compounds of industrial interest. The temperature correction factor multiplying the displaced volume extends the use of the correlation to temperatures close to the critical point. The parameters A_v and B_v are given in Table C.3 for families of typical compounds. For compounds not listed in Table C.2, it suffices to know one single value of liquid density to calculate the value of the displacement D_v. As the temperature correction factor is given for classes of similar compounds, for estimates of liquid densities at other temperatures, it is only necessary to identify to what family the compound belongs.

HOW TO CALCULATE ENTHALPIES AND HEATS OF VAPORIZATION WITH THE PRSV EOS

From exact thermodynamics, the difference between the enthalpy of a pure compound fluid h and the enthalpy of the ideal gas h^{ig} at the same temperature is given by

$$h - h^{ig} = \int_v^\infty \left[P - T \left(\frac{\partial P}{\partial T} \right)_v \right] dv + Pv - RT$$

TABLE C.2
PRSV D_v Parameters for Pure Compounds

Compound	$D_v \cdot 10^3$ [m³/kmol]	Compound	$D_v \cdot 10^3$ [m³/kmol]	Compound	$D_v \cdot 10^3$ [m³/kmol]
Methane	4.050	Benzene	2.951	Ethanol	−3.957
Ethane	4.450	Toluene	0.764	1-Propanol	−1.319
Propane	4.708	o-Xylene	−0.261	2-Propanol	−3.196
n-Butane	4.974	Ethylbenzene	−0.664	Isobutanol	−0.877
n-Pentane	3.551	Methyl chloride	0.305	1-Butanol	−0.571
n-Hexane	0.823	Chloroform	5.889	1-Pentanol	1.110
n-Heptane	−0.745	1,2-Dichloroethane	1.435	1-Hexanol	5.416
n-Octane	−4.488	Nitrogen	4.056	Acetone	−9.661
n-Decane	−11.749	Oxygen	3.402	Butanone	−6.176
n-Dodecane	−20.204	Sulfur dioxide	0.526	Diethyl ether	1.253
n-Tetradecane	−55.068	Carbon monoxide	4.107	Ethyl acetate	−5.200
Acetylene	1.557	Carbon dioxide[a]	1.661	Propionic acid	−5.305
Ethylene	3.897	Chlorine	2.505	Ammonia	−3.164
Propylene	5.208	Hydrogen sulfide	2.977	Water	−3.566
Styrene	−0.750	Hydrogen chloride	−0.384	Acetic acid	−18.031
Cyclohexane	5.979	Methanol	−7.341	Acetonitrile	−33.772

Source: Proust, P., et al.: Calculation of pure compound saturated enthalpies and saturated volumes with the PRSV equation of state. Revised κ_1 parameters for alkanes. *Can. J. Chem. Eng.* 1993. 71. 292–298. Copyright Wiley-VCH Verlag GmbH & Co. KGaA. Reproduced with permission.
[a] At the triple-point temperature.

TABLE C.3
Parameters for the Volume Correction

Class	Typical Compounds	A_v	B_v
1	Nonpolar, slightly polar nonassociated (hydrocarbons, halogenated compounds, amines, N_2, O_2, SO_2, CO_2, Cl_2, H_2S, HCl)	0.808	2.575
2	Polar and associated compounds (alcohols, ketones, ethers, esters, organic acids other than acetic acid, ammonia)	0.983	2.392
3	Three special compounds: water, acetic acid, and acetonitrile	0.685	1.894

Source: Proust, P., et al.: Calculation of pure compound saturated enthalpies and saturated volumes with the PRSV equation of state. Revised κ_1 parameters for alkanes. *Can. J. Chem. Eng.* 1993. 71. 292–298. Copyright Wiley-VCH Verlag GmbH & Co. KGaA. Reproduced with permission.

For the PRSV EOS, then

$$h - h^{ig} = \frac{a_C \left[\alpha - (\alpha T_r)^{0.5} \right]}{2.828b} F(T_r, \omega) \ln\left(\frac{v - 0.414b}{v + 1.414b} \right) + Pv - RT \qquad (C.21)$$

with

$$F(T_r, \omega) = \kappa_1 (0.7 - T_r)\left(1 - T_r^{0.5}\right) - 2\kappa_1 T_r^{0.5}(1 - T_r) - \kappa \qquad (C.22)$$

The heat of vaporization is directly obtained as

$$\Delta h^s = \left(h - h^{ig}\right)^{sv} - \left(h - h^{ig}\right)^{sl} \qquad (C.23)$$

The PRSV EOS gives accurate values of the vapor pressure of pure compounds as a function of temperature. As the heat of vaporization is directly related by the Clausius–Clapeyron equation to the slope of the saturation pressure with respect to temperature, Equation C.23 gives reliable values of heats of vaporization [13], and Equation C.21 can also be used to calculate enthalpy changes.

HOW TO CALCULATE FUGACITY COEFFICIENTS IN MIXTURES WITH PRSV

To calculate the fugacity coefficient of a compound in a mixture, it is necessary to introduce mixing rules. For gas phase mixtures, commonly used mixing rules for cubic EOS have the form

$$b = \sum_k y_k b_k \qquad (C.24)$$

and

$$a = \sum_i \sum_j x_i x_j a_{ij} \qquad (C.25)$$

with

$$a_{ii} = a_{c,i}\, \alpha_i(\omega_i, T_{r,i}) \qquad (C.26)$$

and

$$a_{ij} = (1 - \delta_{ij})\sqrt{a_{ii} a_{jj}} \qquad (C.27)$$

where, in one way or another, the binary parameter δ_{ij} is adjusted to fit known experimental data. In many cases, the value of δ_{ij} is close to zero, so the value of $(1 - \delta_{ij})$ is close to unity. For mixtures of compounds belonging to the same homologous series, or of compounds of similar structure, the value of the adjustable parameter is closely independent of composition. For gas mixtures of very dissimilar compounds or for liquid mixtures, experience has shown that the binary parameter is a function of the composition of the mixture. For a component i in a multicomponent mixture, the fugacity coefficient expression obtained with the above mixing rules, where δ_{ij} is independent of composition, has the form

$$\ln \phi_i = \frac{b_i}{b}(z-1) - \ln(z - B_P) - \frac{A_P}{2 B_P \sqrt{2}} \left[\frac{2 \sum_k x_k a_{ik}}{a} - \frac{b_i}{b} \right] \ln \left[\frac{z + B_P(1 + \sqrt{2})}{z + B_P(1 - \sqrt{2})} \right]$$

(C.28)

In this expression, z, A_P, and B_P are given by Equations C.6 through C.8, respectively.

Very many mixing rules have been proposed. Here, we only comment in detail on the two-parameter extension of the mixing rule presented above, and also on what we think is a consistent method to combine the PRSV EOS with expressions for the excess Gibbs energy of liquid mixtures. In principle, the PRSV EOS can be used in at least three ways to calculate phase equilibria. First, it can be used to calculate fugacities in the vapor phase only, using an independent expression for the excess Gibbs energy to calculate activities in the liquid phase. This is the well-known γ–ϕ approach, and the simple mixing rule presented above suffices for the gas phase in most cases. An alternative to this, not used in practice, would be to use the PRSV EOS with different mixing rules (or different binary parameters) for the vapor and the liquid phase. Normally, PRSV is used with the same binary parameters and the same mixing rules for both the liquid and the vapor phase. For this third option, a two-binary-parameter mixing rule is necessary.

TWO-BINARY-PARAMETER MIXING RULE AND ITS EXTENSION TO MULTICOMPONENT MIXTURES

For strongly nonideal mixtures presenting an asymmetric pattern of the excess Gibbs energy, among other mixing rules, Stryjek and Vera [2] proposed to use Equation C.27 with a composition-dependent binary parameter of the form

$$\delta_{ij} = x_i k_{ij} + x_j k_{ji}$$

(C.29)

with

$$k_{ij} \neq k_{ji}$$

Almost simultaneously, Adachi and Sugie [14] and also Panagiotopoulos and Reid [15] presented slightly more complex, but perfectly equivalent, forms differing only in the details. In a study predicting ternary vapor–liquid equilibrium using binary parameters only, Sandoval et al. [16] found that both the Adachi–Sugie and the Panagiotopoulos–Reid mixing rules gave better results than the direct extension of the Stryjek form, which presented the Michelsen–Kistenmacher syndrome [17], discussed further by Zabaloy and Vera [18]. Thus, Sandoval et al. [16] proposed the following form to extend Stryjek two binary parameters to multicomponent mixtures:

$$\delta_{ij} = (1 - x_i - x_j)\overline{k}_{ij} + x_i k_{ij} + x_j k_{ji} \tag{C.30}$$

with

$$\overline{k}_{ij} = \frac{1}{2}\left(k_{ij} + k_{ji}\right) \tag{C.31}$$

For a binary mixture, Equation C.30 reduces to Equation C.29. Using Equation C.29 or C.30, the fugacity coefficient of compound i in a binary mixture is given by [2]

$$\ln \phi_i = \frac{b_i}{b}(z - 1) - \ln(z - B_P) - \frac{A_P}{2B_P\sqrt{2}}\left[\frac{\overline{a}_i}{a} + 1 - \frac{b_i}{b}\right]\ln\left[\frac{z + B_P(1 + \sqrt{2})}{z + B_P(1 - \sqrt{2})}\right] \tag{C.32}$$

with

$$\overline{a}_i = 2(x_i a_{ii} + x_j a_{ij}) - a + 2x_i x_j^2 (a_{ii} a_{jj})^{0.5}(k_{ji} - k_{ij}) \tag{C.33}$$

Using Equation C.30, the fugacity coefficient of compound i in a multicomponent mixture is given by Equation C.32, but with

$$\overline{a}_i = 2\sum_j x_j a_{ij} - a + 2\sum_{j \neq i} x_i x_j (a_{ii} a_{jj})^{0.5}\left[(x_i - x_j - 1)\Delta k_{ij}\right]$$

$$+ 2\sum_{j \neq i}\sum_{m > j} x_j x_m (a_{jj} a_{mm})^{0.5}\left[(x_j - x_m)\Delta k_{jm}\right] \tag{C.34}$$

with

$$\Delta k_{ij} = k_{ij} - \overline{k}_{ij} \tag{C.35}$$

For a binary system, the third term of the right-hand side of Equation C.34 cancels out, and it reduces to Equation C.33.

CONSISTENT METHOD TO COMBINE THE PRSV EOS
WITH EXCESS GIBBS ENERGY MODELS

Considering that parameters for group methods like ASOG or UNIFAC were available to calculate activities in the liquid phase, it was desirable to have a way to combine these group methods with the PRSV EOS. As early examples, Orbey et al. [19] used UNIFAC and Tochigi [20] used ASOG.

The activity coefficient of a component of a liquid mixture was defined in terms of volumetric properties by Equation 15.1:

$$\ln \gamma_{i,0}(T,P,x) \equiv \frac{1}{RT} \int_0^P \left[\bar{v}_i - v_i^0 \right] dP \tag{15.1}$$

This definition, combined with Equation 13.9, gave

$$\gamma_{i,0}(T,P,x) = \frac{\phi_i(T,P,x)}{\phi_i^0(T,P)} \tag{15.2a}$$

This is a result of exact thermodynamics. No assumptions or models are involved. Thus, we proposed [21] the use of Equation 15.2a as a consistent method to combine the PRSV EOS, or any other EOS, with excess Gibbs energy functions. Notably, this simple formulation gained acceptance and made it to the front cover of the *AIChE Journal* "Founders Tribute" dedicated to John M. Prausnitz [22].

Many other mixing rules are available in the literature. See, for example, the work of Zabaloy et al. for vapor–liquid equilibrium [23,24]. For the use of the PRSV EOS for liquid mixtures Ohta et al. [25] has presented studies of liquid–liquid equilibrium, of ternary vapor–liquid equilibrium [26], and even of excess enthalpies [27]. Just to name a few examples, Orbey et al. combined the PRSV EOS with UNIFAC [19] for vapor–liquid and liquid–liquid calculations using a mixing rule previously proposed by Wong and Sandler [28]. Tochigi used the PRSV EOS combined with ASOG to predict high-pressure vapor–liquid equilibrium for highly nonideal ternary mixtures [20].

ADDITIONAL TABLE OF PRSV PARAMETERS CONTRIBUTED
BY PROFESSOR CLAUDIO OLIVERA-FUENTES

Table C.4 was prepared by Professor Claudio Olivera-Fuentes based on the work of Figueira et al. [29] using data from the DIPPR compilation [30]. The presentation of this table departs from the traditional way of presenting the tables of parameters. Instead of giving the values of T_c, P_c, and ω for the calculation of b, a_c, and k_0, it directly gives the values of these three constants for each compound. In the preparation of this table, Olivera-Fuentes used high-precision numbers. For the numerical constants of Equations C.9 and C.11, he used 0.0777960739039 and 0.4572355289214 instead of 0.07780 and 0.45724, respectively.

For some calculations, if it is required to have the values of T_c or P_c, these values can be obtained from the information given in Table C.4. The value of T_c can be obtained by combining Equations C.9 and C.11:

$$T_c[K] = 0.020464 \frac{a_c}{b}$$

and then, using Equation C.9,

$$P_c[kPa] = 0.013238 \frac{a_c}{b^2}$$

where we have used

$$R = 8.31451(\text{kPa m}^3 \text{ K}^{-1} \text{ kmol}^{-1})$$

With the high-precision numbers used to prepare Table C.4, these expressions become

$$T_c(K) = 0.020463647 \frac{a_c}{b}$$

and

$$P_c(kPa) = 0.013236568 \frac{a_c}{b^2}$$

where for the gas constant the following CODATA 1998 value [31] was used

$$R = 8.314472(\text{JK}^{-1}\text{mol}^{-1})$$

We reproduce this information here for consistency with the fact that all values of the parameters are given to eight significant digits in Table C.4. In any case, if so desired, the values of T_c and P_c can also be generated using the group contribution methods given in Tables 2-336 and 2-337 of *Perry's Chemical Engineer's Handbook* [9].

The parameters κ_0 and κ_1 reported in Table C.4 were fitted to the saturation pressures of each individual compound over the entire range of vapor–liquid coexistence. Therefore, the values of κ_0 reported in this table are not constrained by Equation C.13. The value of the acentric factor, ω, if required, can be obtained from its definition (Equation C.3). These values can be correlated by the following equation, similar to Equation C.13, with a regression coefficient $R^2 = 0.99766$:

$$\kappa_0 = 0.37522952 + 1.4923448\omega - 0.18371717\omega^2 + 0.0259966\omega^3$$

The parameters reported in Table C.4 should not be used for PRSV2 unless new values for κ_2 and κ_3 are evaluated.

TABLE C.4
Olivera-Fuentes Additional Pure Compound Parameters for the PRSV EOS

Compound	a_c [kPa $(m^3/kmol)^2$]	b [m^3/kmol]	κ_0	κ_1
	1. ALCOHOLS			
1.1. Aromatic alcohols				
2-Phenylethanol	3777.4531	0.11301238	1.3976121	−0.35496631
2,3-Xylenol	3328.7344	0.094222346	1.1063922	−0.064106776
2,4-Xylenol	3637.8970	0.10519980	1.0964629	0.024473524
2,5-Xylenol	3229.1262	0.093458312	1.1678710	−0.12997002
2,6-di-*tert*-Butyl-*p*-cresol	7789.5233	0.22139174	1.3258175	0.047625599
2,6-Xylenol	3564.9169	0.10405991	1.0305397	−0.051116932
3,4-Xylenol	3325.9848	0.093241699	1.1890974	−0.044438156
3,5-Xylenol	4465.6561	0.12769316	1.0759485	0.19707395
Benzyl alcohol	3184.5472	0.096259158	1.3246100	−0.040642426
Dinonylphenol	20038.462	0.46282168	1.8762113	−0.097076076
m-Cresol	3455.6144	0.10018343	1.0027092	0.24875357
Nonylphenol	8751.2751	0.23656936	1.5955112	−0.039827477
o-Cresol	3040.5558	0.089199142	0.99643372	0.084189762
p-Cresol	3113.4831	0.090418249	1.0825546	0.10514202
p-Cumylphenol	8234.6529	0.20205160	1.2940233	0.004532896
p-Ethylphenol	3867.2954	0.11045986	1.1028754	0.053675336
p-*tert*-Amylphenol	5966.4208	0.16257620	1.1799403	0.016727111
p-*tert*-Butylphenol	5105.1254	0.14232900	1.1040679	0.10244591
p-*tert*-Octylphenol	8124.3042	0.21732404	1.2552972	0.028012406
Phenol	2514.6609	0.074121906	1.0089691	0.034254695
1.2. n-Alcohols				
1-Decanol	6147.8137	0.18232853	1.2376226	0.45744412
1-Dodecanol	8502.9057	0.24133212	1.2551663	0.53159853
1-Eicosanol	16030.680	0.41419973	1.6357748	0.57218529
1-Heptadecanol	13055.424	0.34696311	1.4722042	0.54683794
1-Heptanol	3967.7674	0.12849342	1.1984070	0.47200174
1-Hexadecanol	11245.301	0.30239141	1.4273216	0.50592828
1-Hexanol	3474.6722	0.11630730	1.1743997	0.27955353
1-Nonanol	5322.8952	0.16185119	1.2054542	0.51880190
1-Octadecanol	14150.758	0.37268483	1.5507037	0.45378492
1-Octanol	4716.4786	0.14791778	1.2055906	0.43568739
1-Pentanol	2809.2821	0.098077551	1.1867307	0.34811989
1-Tetradecanol	10020.020	0.27671545	1.3109096	0.57998135
1-Tridecanol	9344.6842	0.26159551	1.2379595	0.57803445
1-Undecanol	7511.9818	0.21835589	1.1795127	0.61542808
Ethanol	1377.1474	0.054588783	1.2557386	−0.11187817

(Continued)

TABLE C.4 *(Continued)*
Olivera-Fuentes Additional Pure Compound Parameters for the PRSV EOS

Compound	a_c [kPa $(m^3/kmol)^2$]	b [m^3/kmol]	κ_0	κ_1
Methanol	1023.2767	0.040852108	1.1675457	−0.15297483
n-Butanol	2283.3190	0.083003275	1.2257630	0.14564592
n-Propanol	1817.8969	0.069312667	1.275062853	−0.016346386
1.3. Other aliphatic alcohols				
1-Methylcyclohexanol	3007.7492	0.10207217	1.3331638	−0.015831479
2-Ethyl-1-butanol	3116.5374	0.10995814	1.3471562	0.26628077
2-Ethyl-1-hexanol	4749.4845	0.15180285	1.0469280	0.90530461
2-Heptanol	3674.2639	0.12787218	1.4091552	−0.13025589
2-Methyl-1-butanol	2604.0437	0.094315452	1.2559180	0.61186349
2-Methyl-1-pentanol	3151.5778	0.11081233	1.3868641	−0.15291931
2-Methyl-2-butanol	2418.9336	0.090801068	1.0664842	0.47681481
2-Octanol	4725.0957	0.15175813	1.0843306	0.56382178
2-Pentanol	2503.6324	0.092814220	1.3130143	0.20321816
2,2-Dimethyl-1-propanol	2441.5646	0.090842393	1.2379500	0.097659048
2,6-Dimethyl-4-heptanol	4520.1911	0.15339900	1.4739630	0.0073888279
3-Methyl-1-butanol	2727.9942	0.096340859	1.1936161	0.29342910
3-Methyl-2-butanol	2630.4450	0.093777871	0.86255118	0.85029331
3-Pentanol	2521.9929	0.094349492	1.3029060	0.052476805
4-Methyl-2-pentanol	3001.3776	0.10692746	1.1821588	0.068709508
Allyl alcohol	1670.1746	0.062705924	1.2142942	−0.15098591
cis-2-Methylcyclohexanol	3270.9450	0.10901541	1.2977436	−0.28406500
cis-3-Methylcyclohexanol	3242.9516	0.10738287	1.3414774	0.0041776831
cis-4-Methylcyclohexanol	3179.2901	0.10459786	1.3514890	0.067291833
Cyclohexanol	3289.5366	0.10767962	1.1012428	0.38722080
Isobutanol	2213.8892	0.082712738	1.1601736	0.56226259
Isodecanol	5745.6229	0.18257205	1.6106318	−0.045315963
Isopropanol	1713.1452	0.068968148	1.3130071	0.0096807165
Propargyl alcohol	1628.2522	0.057448242	1.1508768	−0.24272745
sec-Butanol	2108.4546	0.080496018	1.2081003	0.24966940
tert-Butanol	2048.3856	0.082808058	1.2339148	0.40386756
trans-2-Methylcyclohexanol	3184.1044	0.10577661	1.3177754	−0.26253362
trans-3-Methylcyclohexanol	3220.4036	0.10680908	1.3339598	0.011161879
trans-4-Methylcyclohexanol	3209.2084	0.10558217	1.3468914	0.033061990
1.4. Polyols				
1,2-Benzenediol	2462.7220	0.065963709	1.3543023	−0.56334402
1,2-Propylene glycol	2050.3545	0.067025129	1.7689859	−0.39321587
1,2,3-Benzenetriol	2471.7912	0.060942002	1.6441405	−0.64317938
1,3-Benzenediol	2741.0914	0.069250281	1.3254256	−0.013309769
1,3-Butanediol	2603.1194	0.082844972	1.826367	−0.23792323

(Continued)

TABLE C.4 *(Continued)*
Olivera-Fuentes Additional Pure Compound Parameters for the PRSV EOS

Compound	a_c [kPa $(m^3/kmol)^2$]	b [m^3/kmol]	κ_0	κ_1
1,3-Propylene glycol	2313.6377	0.071953595	1.8895127	−0.86473221
1,4-Butanediol	2880.2033	0.088365013	1.9389936	−0.70771683
1,5-Pentanediol	3255.7391	0.098995983	1.9973319	−0.66978444
1,6-Hexanediol	3915.4995	0.11959015	2.0301390	−0.40687037
2-Butyne-1,4-diol	2606.0034	0.076731417	1.8791104	−0.11439182
Bisphenol A	7527.4830	0.18143670	1.6647339	0.76571926
cis-2-Butene-1,4-diol	2780.3385	0.083932062	1.9054736	−0.64317668
Diethylene glycol	3857.7509	0.10602156	1.2381365	0.12212874
Dipropylene glycol	3769.3000	0.11794132	1.9383295	−0.98208594
Ethylene glycol	1758.6628	0.055796364	1.8500276	−0.75042577
Glycerol	4347.8057	0.12305942	2.0566400	−0.67389547
Hexylene glycol	3035.6154	0.10003182	1.9469537	−0.10562306
Neopentyl glycol	3080.6504	0.098042523	1.8917425	−0.17180118
p-Hydroquinone	2857.3397	0.071133324	1.3227115	−0.085643991
p-tert-Butylcatechol	5069.3021	0.13368094	1.3954447	−0.0077521208
Tetraethylene glycol	6369.0900	0.18051913	2.3543546	−0.20694433
trans-2-Butene-1,4-diol	2864.8833	0.086088047	1.9094683	−0.70850205
Triethylene glycol	4720.7113	0.13800424	2.1584085	−1.1394347
Trimethylolpropane	4063.2812	0.11727722	2.3485345	−1.0668890
2. ALDEHYDES				
1-Decanal	6349.7386	0.19777596	1.2539582	0.063879879
1-Dodecanal	7982.0787	0.23845612	1.4121641	−0.12156634
1-Heptanal	4101.8743	0.13920283	1.0503397	0.037703657
1-Hexanal	3409.4101	0.12049908	1.0015466	−0.027325714
1-Nonanal	5561.0697	0.17781214	1.1906436	0.022206238
1-Octanal	4775.9380	0.15738021	1.1328561	0.0096494503
1-Tridecanal	8891.4654	0.25993116	1.4311685	0.070801835
1-Undecanal	7151.1119	0.21776463	1.3385100	−0.10707945
2-Ethylhexanal	4503.9524	0.15184068	1.1152635	0.018839258
Acetaldehyde	1206.4464	0.053553781	0.7771264	0.12790222
Acrolein	1612.1025	0.065196635	0.82463387	−0.084975116
Benzaldehyde	3302.8971	0.097250822	0.82047774	0.081628669
Formaldehyde	798.01999	0.040025489	0.78503569	−0.092696316
Glyoxal	1317.0873	0.054449312	0.97019251	0.030978116
Isobutyraldehyde	1980.9833	0.079956890	0.89674647	−0.010838000
Methacrolein	2090.5272	0.080716623	0.73957949	0.039560879
n-Butyraldehyde	2185.8588	0.085201226	0.84392390	0.093135004
n-Propionaldehyde	1672.6626	0.069009631	0.81263770	−0.021573833
p-Tolualdehyde	4197.3902	0.12305718	1.0092089	0.036824060

(Continued)

TABLE C.4 *(Continued)*
Olivera-Fuentes Additional Pure Compound Parameters for the PRSV EOS

Compound	a_c [kPa $(m^3/kmol)^2$]	b [$m^3/kmol$]	κ_0	κ_1
trans-Crotonaldehyde	2440.6427	0.087468390	0.86815816	0.057018288
Valeraldehyde	2772.5712	0.10241321	0.91702877	0.066827024

3. ALKANES

3.1. Cycloalkanes

1,1-Dimethylcyclohexane	3758.8535	0.13011901	0.71596732	0.11576958
1,1-Dimethylcyclopentane	2735.1622	0.10232430	0.76291624	0.049335320
cis-1,2-Dimethylcyclohexane	3951.0907	0.13338897	0.71651742	0.14438800
cis-1,2-Dimethylcyclopentane	2929.4382	0.10607270	0.75962522	0.059067834
cis-1,3-Dimethylcyclohexane	3758.2827	0.13009925	0.72032252	0.13678178
cis-1,3-Dimethylcyclopentane	2781.1802	0.10329055	0.76969344	0.018263420
cis-1,4-Dimethylcyclohexane	3846.3868	0.13159091	0.71385102	0.15099432
cis-Decahydronaphthalene	4807.1897	0.14008207	0.78079605	0.13939813
Cyclobutane	1341.1568	0.059672036	0.64955475	−0.0064370216
Cycloheptane	3020.8312	0.10229559	0.72686916	0.067069769
Cyclohexane	2374.3440	0.087776381	0.68946108	0.038600844
Cyclopentane	1839.1438	0.073541484	0.65814630	0.065107850
Cyclopropane	903.29378	0.046454437	0.56538105	0.055790235
Ethylcyclohexane	3856.8583	0.12956642	0.72967925	0.15077781
Ethylcyclopentane	3018.0785	0.10844377	0.73965203	0.14489438
Methylcyclohexane	2971.3645	0.10626707	0.68774525	0.17819218
Methylcyclopentane	2373.2388	0.091152462	0.69114508	0.12456638
n-Butylcyclohexane	5462.2289	0.16758190	0.78520629	0.20665874
n-Propylcyclohexane	4596.8702	0.14717786	0.75597038	0.18037089
n-Propylcyclopentane	3830.2481	0.12998482	0.76633852	0.15479468
trans-1,2-Dimethylcyclohexane	3823.8978	0.13126041	0.72012993	0.12067994
trans-1,2-Dimethylcyclopentane	2806.3693	0.10382094	0.74573598	0.10601635
trans-1,3-Dimethylcyclohexane	3846.0061	0.13161089	0.71611508	0.15786375
trans-1,3-Dimethylcyclopentane	2809.9434	0.10398136	0.72436527	0.16154295
trans-1,4-Dimethylcyclohexane	3745.8127	0.12988730	0.72248944	0.11191860
trans-Decahydronaphthalene	5124.7819	0.15264061	0.73805425	0.20351006

3.2. Methylalkanes

2-Methylheptane	3993.2937	0.14601771	0.90596104	0.081592372
2-Methylhexane	3267.7930	0.12608361	0.82469338	0.13834835
2-Methyloctane	4754.2372	0.16581002	0.98124570	0.060507045
2-Methylpentane	2598.4021	0.10687997	0.76894140	0.073970634
3-Methylheptane	3992.4459	0.14494297	0.88013717	0.12761574
3-Methylhexane	3241.1098	0.12391392	0.82589318	0.094135029
3-Methyloctane	4700.7146	0.16299884	0.96395480	0.065004550
3-Methylpentane	2570.4589	0.10427803	0.70955648	0.23608391

(Continued)

TABLE C.4 *(Continued)*
Olivera-Fuentes Additional Pure Compound Parameters for the PRSV EOS

Compound	a_c [kPa $(m^3/kmol)^2$]	b [m^3/kmol]	κ_0	κ_1
4-Methylheptane	3942.3667	0.14361662	0.88747576	0.12022021
4-Methyloctane	4660.1233	0.16227877	0.96582751	0.050735709
Isobutane	1454.5432	0.072929041	0.61593026	0.13945189
Isopentane	1987.8537	0.088349448	0.69168484	0.083572804
3.3. n-Alkanes				
Ethane	605.16028	0.040546743	0.51946845	0.035633202
Methane	250.02652	0.026846754	0.39401837	−0.0034966480
n-Butane	1513.5590	0.072846645	0.66059669	0.055288924
n-Decane	5676.8304	0.18783839	1.0628541	0.073735336
n-Dodecane	7567.8366	0.23528644	1.1767082	0.077090006
n-Eicosane	17792.910	0.47469211	1.5734281	0.069258112
n-Heptadecane	12611.236	0.35189860	1.4166634	0.25194847
n-Heptane	3350.7582	0.12691803	0.86400112	0.13097652
n-Hexadecane	11642.561	0.33062623	1.3909216	0.073958472
n-Hexane	2682.7522	0.10819008	0.79879078	0.12382030
n-Nonadecane	16209.634	0.43880813	1.4873084	0.14057068
n-Nonane	4908.6517	0.16863748	1.0027846	0.015317067
n-Octadecane	14430.876	0.39624876	1.4670828	0.10038420
n-Octane	4124.5157	0.14837936	0.93566027	0.082977153
n-Pentadecane	10341.188	0.29940355	1.3381724	0.14329424
n-Pentane	2084.7613	0.09083747	0.72422164	0.12048929
n-Tetradecane	9345.3612	0.27619898	1.2831461	0.12978703
n-Tridecane	8373.1522	0.25354429	1.2364216	−0.0075047849
n-Undecane	6562.6013	0.21024290	1.1194446	0.10958350
Propane	1023.4796	0.056633293	0.58804342	0.069355822
3.4. Other alkanes				
2-Methyl-3-ethylpentane	3768.3284	0.13600307	0.83318233	0.11220805
2,2-Dimethylhexane	3786.4318	0.14093162	0.84343781	0.12468075
2,2-Dimethylpentane	3081.8792	0.12116520	0.79512405	0.046834454
2,2,3-Trimethylbutane	3031.1128	0.11677546	0.74027884	0.040322650
2,2,3-Trimethylpentane	3685.5185	0.13384055	0.79054432	0.10596082
2,2,3,3-Tetramethylpentane	4312.4064	0.14446683	0.78290552	0.086672350
2,2,3,4-Tetramethylpentane	4321.4836	0.14934276	0.79392605	0.15851460
2,2,4-Trimethylpentane	3672.6066	0.13816259	0.79364545	0.11486691
2,2,4,4-Tetramethylpentane	4369.0332	0.15648263	0.83557265	0.047220444
2,2,5-Trimethylhexane	4375.1128	0.15761071	0.89085356	0.069686789
2,3-Dimethylhexane	3827.9266	0.13903681	0.87503682	0.039989423
2,3-Dimethylpentane	3200.9110	0.12189879	0.78459267	0.086377149
2,3,3-Trimethylpentane	3697.5636	0.13193659	0.78576993	0.073843009

(Continued)

TABLE C.4 *(Continued)*
Olivera-Fuentes Additional Pure Compound Parameters for the PRSV EOS

Compound	a_c [kPa $(m^3/kmol)^2$]	b [m^3/kmol]	κ_0	κ_1
2,3,4-Trimethylpentane	3732.9837	0.13489398	0.82035644	0.074093599
2,4-Dimethylhexane	3804.1613	0.14064501	0.86885320	0.049421012
2,4-Dimethylpentane	3131.3865	0.12327976	0.79859061	0.10682246
2,5-Dimethylhexane	3864.9588	0.14380209	0.87698249	0.096299502
3-Ethylhexane	3894.4512	0.14095273	0.89569496	0.041821312
3-Ethylpentane	3183.2889	0.12048998	0.83313770	0.021415977
3,3-Diethylpentane	4398.6136	0.14754803	0.86611704	−0.0030473720
3,3-Dimethylhexane	3767.0190	0.13716539	0.81634050	0.12131257
3,3-Dimethylpentane	3108.4773	0.11858833	0.75180510	0.070230532
3,4-Dimethylhexane	3803.7548	0.13684721	0.86479082	0.037678248
Neopentane	1865.0510	0.087984107	0.66433810	0.017404661

4. ALKENES

4.1. 1-Alkenes

1-Butene	1419.2404	0.069217177	0.62766637	0.14979282
1-Decene	5544.5711	0.18387837	1.0360903	0.072006985
1-Dodecene	7191.9381	0.22400804	1.1665569	0.078771076
1-Eicosene	15373.637	0.40804238	1.5448676	0.18291081
1-Heptene	3232.8706	0.12312964	0.83526192	0.10907262
1-Hexadecene	11241.446	0.31861633	1.3850916	−0.048310908
1-Hexene	2557.6128	0.10383923	0.76404784	0.11398095
1-Nonene	4772.9252	0.16463794	0.95693462	0.12457744
1-Octadecene	12941.573	0.35405318	1.4329027	0.32868103
1-Octene	3980.9281	0.14377746	0.89544624	0.12206380
1-Pentene	1920.3225	0.084549252	0.66972773	0.19475786
1-Tetradecene	9140.7908	0.27030913	1.2574763	0.14094369
1-Tridecene	8174.5266	0.24782315	1.2338412	−0.031450505
1-Undecene	6324.8555	0.20286773	1.0888575	0.14740966
Ethylene	501.04665	0.036312657	0.50127555	0.05105929
Propylene	916.47576	0.051415825	0.57073782	0.078028419

4.2. Dialkenes

1,2-Butadiene	1385.0623	0.063836546	0.74551122	−0.080954304
1,2-Pentadiene	2078.3847	0.085062663	0.61503312	0.19863267
1,3-Butadiene	1320.6149	0.063531978	0.65732315	0.048193173
1,4-Pentadiene	1938.8670	0.082831503	0.52074824	0.24829997
cis-1,3-Pentadiene	2103.8342	0.086276795	0.60369696	0.20841027
Cyclopentadiene	1578.0881	0.063695144	0.68445465	−0.14529401
Isoprene	1922.8411	0.081298225	0.61172142	0.15576166
Propadiene	898.76649	0.046781229	0.57504748	0.14011472
trans-1,3-Pentadiene	2111.9592	0.086436776	0.55621700	0.24724884

(Continued)

TABLE C.4 *(Continued)*
Olivera-Fuentes Additional Pure Compound Parameters for the PRSV EOS

Compound	a_c [kPa $(m^3/kmol)^2$]	b [m^3/kmol]	κ_0	κ_1
4.3. Other alkenes				
2-Ethyl-1-butene	2620.8389	0.10474985	0.72099161	0.11524390
2-Ethyl-1-hexene	3392.6959	0.12095284	0.91329420	−0.040058389
2-Methyl-1-butene	1967.6008	0.086589868	0.71730110	0.091569556
2-Methyl-1-pentene	2571.8671	0.10380627	0.71862436	0.15315716
2-Methyl-2-butene	2054.8433	0.089277259	0.76355882	0.055974887
2-Methyl-2-pentene	2642.7483	0.10521453	0.73494641	0.13690857
2,3-Dimethyl-1-butene	2453.5958	0.10041904	0.69794888	0.13233294
2,3-Dimethyl-2-butene	2736.9231	0.10688440	0.72206709	0.15933617
2,4,4-Trimethyl-1-pentene	3676.8938	0.13606267	0.76874699	0.10744525
2,4,4-Trimethyl-2-pentene	3741.4945	0.13721258	0.76380443	0.18411163
3-Methyl-1-butene	1823.3116	0.082846562	0.68236740	0.098472275
4-Methyl-1-pentene	2410.5859	0.099454392	0.69621073	0.15331867
4-Methyl-*cis*-2-pentene	2442.8035	0.10017769	0.71881315	0.14001697
4-Methyl-*trans*-2-pentene	2464.9012	0.10068037	0.73379090	0.13221479
cis-2-Butene	1414.8288	0.066468979	0.66131489	0.098690607
cis-2-Heptene	3356.5799	0.12511451	0.81159071	0.032706082
cis-2-Hexene	2633.0507	0.10503279	0.78032989	0.034776042
cis-2-Pentene	1959.1054	0.084236002	0.70447356	0.13948331
cis-3-Heptene	3307.2070	0.12417893	0.77284990	0.18146432
Cyclohexene	2286.7465	0.083503164	0.66434675	0.14941548
Cyclopentene	1697.0139	0.068495253	0.64620803	0.090678880
Isobutene	1388.3781	0.067985834	0.64429294	0.10073019
trans-2-Butene	1422.9686	0.067935343	0.69042305	0.010787968
trans-2-Hexene	2631.4364	0.10496839	0.74775644	0.13817411
trans-2-Octene	4078.9903	0.14466381	0.87026053	0.024263103
trans-2-Pentene	1954.3480	0.084130441	0.71289567	0.093177115
trans-3-Octene	4038.0738	0.14396118	0.87534380	0.034028625
trans-4-Octene	4020.5183	0.14358546	0.86496050	0.074392244
5. ALKYLBENZENES				
5.1. n-Alkylbenzenes				
Benzene	2046.1605	0.074483967	0.68506759	0.062034848
Ethylbenzene	3345.5798	0.1109302	0.79976174	0.094566787
n-Butylbenzene	4741.5392	0.1468915	0.92591435	0.10507815
n-Decylbenzene	10090.526	0.2742217	1.3153224	0.0097323093
n-Dodecylbenzene	11975.879	0.3165218	1.4057441	0.23257127
n-Hexylbenzene	6468.9478	0.1896537	1.0528018	0.033166983
n-Nonylbenzene	9176.1936	0.2534121	1.2583675	−0.0035988938
n-Pentylbenzene	5590.1309	0.1682519	1.0125630	−0.073349094

(Continued)

TABLE C.4 *(Continued)*
Olivera-Fuentes Additional Pure Compound Parameters for the PRSV EOS

Compound	a_c [kPa $(m^3/kmol)^2$]	b [m^3/kmol]	κ_0	κ_1
n-Propylbenzene	4039.2871	0.1294817	0.85165106	0.10889984
n-Tridecylbenzene	12906.623	0.3373136	1.4775618	0.23303326
n-Undecylbenzene	11058.974	0.2962133	1.3509162	0.15980916
Toluene	2698.6040	0.093315668	0.74845373	0.084609266
5.2. Other alkylbenzenes				
1,2,3-Trimethylbenzene	4033.0074	0.12419310	0.91004185	−0.044546018
1,2,4-Trimethylbenzene	4111.1344	0.12960240	0.92766148	−0.053917890
1,2,4,5-Tetramethylbenzene	4902.3289	0.14858850	1.0027764	−0.047857355
3-Ethyl-*o*-xylene	5074.8679	0.15272104	0.89692161	0.11511670
Cumene	3923.7554	0.12721912	0.83005984	0.15741982
Isobutylbenzene	4393.7736	0.13829521	0.91934303	0.0074921716
m-Cymene	4657.2224	0.14505899	0.87029905	0.068777169
m-Diethylbenzene	4820.0907	0.14877320	0.87838812	0.13520073
m-Diisopropylbenzene	6039.7827	0.18069588	0.90739454	0.13445387
m-Ethyltoluene	4519.6880	0.14516095	0.84640454	0.11931100
m-Xylene	3410.7318	0.11311241	0.84226904	0.034174997
Mesitylene	4083.4797	0.13110783	0.94425092	0.011589387
o-Cymene	4729.3530	0.14619307	0.86500523	0.077574280
o-Diethylbenzene	4894.5542	0.14994076	0.87136793	0.11682717
o-Ethyltoluene	4411.2965	0.13863352	0.81792280	0.10163870
o-Xylene	3373.5706	0.10951593	0.82388628	0.044600673
p-Cymene	4756.6355	0.14902872	0.89337505	0.12991662
p-Diethylbenzene	4882.3264	0.15184845	0.95436285	0.026106151
p-Diisopropylbenzene	6128.1437	0.18200895	0.94206643	0.16088596
p-Ethyltoluene	4409.7974	0.14096780	0.84064119	0.12542512
p-Xylene	3517.1433	0.11679093	0.83040471	0.019601599
sec-Butylbenzene	4715.4738	0.14520690	0.75429246	0.28027395
tert-Butylbenzene	4624.3251	0.14337963	0.75753761	0.23843242
5.3. Other hydrocarbon rings				
1-Methylindene	4516.3235	0.13146579	0.86365775	0.050880053
1,2,3-Trimethylindene	6068.5583	0.17105349	0.99466506	−0.036145798
1,2,3,4-Tetrahydronaphthalene	4516.6342	0.12834383	0.84086957	0.098896237
2-Methylindene	4272.7396	0.12783017	0.85281631	0.048306179
α-Pinene	4579.7780	0.14828949	0.81486128	0.024979991
β-Pinene	4707.5270	0.14981831	0.82732938	0.13013014
Bicyclohexyl	6566.3249	0.18482937	0.97552062	−0.019335902
Indene	3884.9773	0.11572169	0.84523814	0.069539096
Vinylcyclohexene	3249.4977	0.11101264	0.79131312	0.14320684

(Continued)

TABLE C.4 *(Continued)*
Olivera-Fuentes Additional Pure Compound Parameters for the PRSV EOS

Compound	a_c [kPa $(m^3/kmol)^2$]	b [m^3/kmol]	κ_0	κ_1
5.4. Other monoaromatics				
α-Methylstyrene	4016.7100	0.12568278	0.85453375	0.026262519
Cyclohexylbenzene	6032.2753	0.16591714	0.92222112	0.22702774
m-Divinylbenzene	4852.0610	0.14348391	0.91453593	0.0027090368
p-Methylstyrene	4150.7783	0.12772942	0.83154458	0.10873647
Styrene	3337.4410	0.10539539	0.71026689	0.18538206
5.5. Polyaromatics				
1-Methylnaphthalene	5149.2253	0.13648506	0.88379322	0.031750442
1-*n*-Decylnaphthalene	14756.973	0.35155005	1.2737470	0.0054251448
1-*n*-Nonylnaphthalene	13563.1276	0.32691526	1.2417437	0.0091505710
1-Phenylnaphthalene	8651.0669	0.20851870	1.1289246	0.025658341
1,1-Diphenylethane	7088.5718	0.18717165	1.0310671	0.028585251
1,2-Diphenylethane	7261.7737	0.19051586	1.0646812	0.012439095
2-Methylnaphthalene	5643.8539	0.15176588	0.86601589	0.11942678
Acenaphthene	6565.9059	0.16729425	0.92321721	0.070964965
Anthracene	8307.9555	0.19474349	1.0831790	−0.29527534
Biphenyl	5159.2396	0.13376689	0.89622487	0.10941505
Chrysene	12684.376	0.26513647	1.2234185	0.013330917
Diphenylacetylene	7547.1535	0.18562775	0.92995161	0.040867808
Diphenylmethane	6342.9651	0.16901067	1.0286317	0.046811926
Fluoranthene	9972.4084	0.22549375	1.1900777	−0.31611267
Fluorene	5121.6237	0.12046793	0.87460108	0.031142057
m-Terphenyl	7659.6805	0.16948154	1.1650858	−0.027198465
Naphthalene	4451.5411	0.12172749	0.80320350	0.050534105
o-Terphenyl	6457.3422	0.14831446	1.0375328	0.028976008
p-Terphenyl	8152.9095	0.18018064	1.1240736	0.028212152
Phenanthrene	8245.7569	0.19411937	1.0724456	−0.12758150
Pyrene	10625.2949	0.23229945	1.0928389	−0.093348371
Tetraphenylethylene	18303.1639	0.37605370	1.3845765	−0.0082589990
trans-Stilbene	7843.4168	0.19573770	1.0623701	0.11942697
Triphenylethylene	12409.963	0.27968403	1.2187175	0.015086834
6. ALKYNES				
Acetylene	488.84754	0.03244552	0.65862824	−0.021336446
Dimethylacetylene	1480.9272	0.062081680	0.56837264	0.25148826
Ethylacetylene	1253.7020	0.057886543	0.72759068	0.076269391
Methylacetylene	910.58847	0.046308211	0.69012441	0.072257551

(Continued)

TABLE C.4 *(Continued)*
Olivera-Fuentes Additional Pure Compound Parameters for the PRSV EOS

Compound	a_c [kPa $(m^3/kmol)^2$]	b [m^3/kmol]	κ_0	κ_1
7. ANHYDRIDES				
Butyric anhydride	4847.7861	0.15524786	1.2202788	0.16604294
Maleic anhydride	2251.4721	0.063901985	1.1503399	−0.39468163
Phthalic anhydride	4187.2755	0.10832734	1.3603023	−0.84423779
Propionic anhydride	3911.8685	0.12953252	1.1961093	−0.11777679
Succinic anhydride	3111.4660	0.078510409	1.1244250	−0.27794003
Trimellitic anhydride	6136.2866	0.14109079	1.7796532	−0.094248163
8. AMINES				
8.1. Aliphatic amines				
1-Aminoheptane	4086.9654	0.13778290	1.0925863	−0.021862401
Di-*n*-Butylamine	3755.2410	0.12649535	1.0801879	0.18779859
Diethylamine	2117.7915	0.087268904	0.81297377	0.044729915
Diisopropylamine	2703.9594	0.10577876	0.79668306	0.66463218
Dimethylamine	1162.7048	0.054365773	0.79039080	0.16622220
Dodecylamine	8154.5462	0.23975827	1.4342054	−0.20599954
Ethylamine	1175.7853	0.052747681	0.78714949	0.11133117
Isobutylamine	1980.7633	0.078900671	0.89823606	0.017552039
Isopropylamine	1553.0821	0.067355568	0.77401877	0.18826950
Methylamine	788.49766	0.037520144	0.78220827	0.11418555
n-Butylamine	2130.7845	0.081977104	0.85323906	0.072252275
n-Decylamine	6371.1980	0.19664849	1.3073976	−0.15681712
n-Hexylamine	3368.1207	0.11822304	1.0387718	−0.060299274
n-Nonylamine	5601.9043	0.17690647	1.2304668	−0.099007710
n-Octylamine	4817.79434	0.15724026	1.1763967	−0.13125845
n-Pentylamine	2719.5150	0.10027242	0.96191467	−0.012840982
n-Propylamine	1681.5519	0.069243756	0.79661792	0.060363680
Piperazine	2324.8174	0.074567778	0.97110119	0.027596767
sec-Butylamine	2092.5394	0.083260719	0.79257734	0.035670577
tert-Butylamine	1927.6165	0.081516974	0.77732667	0.049637857
Tetradecylamine	9905.6831	0.28064018	1.5472622	−0.27075069
Tri-*n*-butylamine	7279.8222	0.23132253	1.3417221	−0.0045711459
Triethylamine	2980.3809	0.11396704	0.84242656	0.0032910651
Trimethylamine	1452.6990	0.068615164	0.66743979	0.099917910
Tripropylamine	4787.7964	0.16965502	1.3384916	0.006675618
8.2. Aromatic amines				
2-Methylpyridine	2783.7436	0.091731957	0.77739915	0.11066797
2,4,6-Trimethylpyridine	4055.3624	0.12708653	0.91638720	−0.0035983491
2,6-Diethylaniline	4641.6276	0.14009532	1.6634638	−0.054729063

(Continued)

TABLE C.4 *(Continued)*
Olivera-Fuentes Additional Pure Compound Parameters for the PRSV EOS

Compound	a_c [kPa $(m^3/kmol)^2$]	b [m^3/kmol]	κ_0	κ_1
2,6-Dimethylpyridine	3248.4175	0.10657230	0.87984174	0.087207730
3-Methylpyridine	3000.7246	0.095202744	0.76937642	0.11131135
4-Methylpyridine	2832.1886	0.089695750	0.81738286	0.010460029
Aniline	2896.9123	0.084808855	0.92371528	0.13135722
Benzylamine	3417.6743	0.10232345	0.96682747	0.031307198
Dibenzopyrrole	7828.1107	0.17818876	1.0882548	−0.14985598
Diphenylamine	6703.1904	0.16789684	1.1233050	−0.078441183
Hydrazobenzene	6414.7172	0.16574307	1.3209133	0.0089063040
Isoquinoline	4053.2850	0.10327460	0.79865696	0.057421375
m-Phenylenediamine	4141.4986	0.10285214	1.1444133	−0.19763062
m-Toluidine	3821.9070	0.11028718	0.97517602	−0.012796565
n-Methylaniline	2995.4824	0.087375803	1.0415383	−0.12713380
n,n-Diethylaniline	5465.4426	0.15932035	0.99539034	0.11854472
n,n-Dimethylaniline	4115.8063	0.12257063	0.95746593	−0.058398731
n,n-Diphenyl-*p*-phenylenediamine	11193.281	0.25282048	1.5695984	−0.040663287
o-Ethylaniline	4190.5540	0.12180969	1.0402785	0.025401387
o-Phenylenediamine	3719.0010	0.097444716	1.0783265	0.034740170
o-Toluidine	4047.2323	0.11931302	0.97708590	0.16955824
p-Aminodiphenyl	6424.1184	0.16090684	1.1461666	0.025291840
p-Aminodiphenylamine	7428.3351	0.17532967	1.3410122	0.00062096110
p-Phenylenediamine	3867.5254	0.099426726	1.1382081	0.026671923
p-Toluidine	3830.8213	0.11309612	1.0480644	−0.081468336
Phenylhydrazine	3727.2141	0.10022654	1.1055395	−0.088014016
Pyridine	2155.6861	0.071156060	0.72291745	0.091423562
Quinaldine	6383.1191	0.16898046	0.78000766	0.48201131
Quinoline	4067.4508	0.10641805	0.83193722	0.15010735
Toluenediamine	4768.3690	0.12136595	1.1698590	0.055713020
8.3. Other amines, imines				
1,3-Diphenyltriazene	7989.2224	0.19347767	1.2474316	0.014595097
Allylamine	1559.6614	0.063200714	0.85271825	0.040151102
Dicyclohexylamine	6786.7662	0.18844232	1.0876654	−0.055947529
Diethylene triamine	3428.9762	0.10380083	1.3317590	−0.39373502
Ethylenediamine	1765.3009	0.060918204	1.0586569	−0.092123129
Ethyleneimine	1330.5738	0.050704644	0.67843194	0.24470382
Hexamethylenediamine	4229.5496	0.13054602	1.2822895	0.010338717
Indole	4611.1375	0.11944391	0.89444768	0.27388875
n-Methylcyclohexylamine	3491.0215	0.11485375	0.93464115	0.040002553
n-Methylpyrrolidine	2277.3503	0.084732533	0.71522169	0.035795195
p-Aminoazobenzene	8386.9226	0.19569786	1.2626582	0.012590822

TABLE C.4 *(Continued)*
Olivera-Fuentes Additional Pure Compound Parameters for the PRSV EOS

Compound	a_c [kPa $(m^3/kmol)^2$]	b [m^3/kmol]	κ_0	κ_1
Piperidine	2399.6420	0.082662111	0.73246427	0.080821396
Propyleneimine	1632.8648	0.063165158	0.75326956	0.049017132
Pyrrole	2088.9409	0.066818833	0.78213460	0.16728841
Pyrrolidine	1821.0981	0.065546231	0.76308365	0.16497406
Tetraethylenepentamine	7486.0807	0.19792314	1.9899364	−0.11046364

9. CHLORIDES

9.1. Aliphatic chlorides

1-Chloropentane	3048.7582	0.10983928	0.84325952	0.0044550753
1,1-Dichloroethane	1703.3703	0.066648507	0.72398484	0.049619531
1,1-Dichloroethylene	1415.2297	0.060084568	0.74314992	0.034343292
1,1,1-Trichloroethane	2184.9493	0.082040423	0.69333475	0.070119383
1,1,1,2-Tetrachloroethane	3059.3126	0.10032803	0.72642782	0.13947838
1,1,2-Trichloroethane	2613.4536	0.088838523	0.74454571	0.076753698
1,1,2,2-Tetrachloroethane	3227.0775	0.10238415	0.73254081	0.31241555
1,2-Dichloroethane	1901.6704	0.069367400	0.77586388	0.0033469306
1,2-Dichloropropane	2418.9024	0.086537700	0.75370110	0.017773642
1,2,3-Trichloropropane	3472.2944	0.10898130	0.82387663	0.045753824
1,3-Dichloro-*trans*-2-butene	3185.0408	0.10546529	0.73225598	0.19282551
1,4-Dichloro-*trans*-2-butene	3482.6480	0.11032149	0.86196443	0.044292512
1,4-Dichlorobutane	3598.3685	0.11487635	0.84673984	0.040988880
1,5-Dichloropentane	4356.4042	0.13446142	0.93647090	0.027886580
2-Chloropropene	1533.7794	0.065662594	0.61202990	0.020379697
3-Chloropropene	1774.5865	0.070630188	0.61804941	0.0078959732
3,4-Dichloro-1-butene	2848.0346	0.098949364	0.81683421	0.039612571
Carbon tetrachloride	2153.2881	0.079202171	0.65810098	0.059990409
Chloroform	1637.4026	0.062466869	0.69937426	0.074815004
Chloroprene	2045.8775	0.079744980	0.67150756	0.021585083
cis-1,2-Dichloroethylene	1699.4314	0.065989685	0.74397947	0.026520219
Dichloromethane	1349.4050	0.054144603	0.65173483	0.16168954
Ethyl chloride	1227.3592	0.054559021	0.66728552	0.040568965
Hexachloro-1,3-butadiene	6111.0501	0.16876434	0.61886762	0.45300524
Hexachlorocyclopentadiene	5850.4407	0.16048439	0.90794151	0.15486681
Isopropyl chloride	1660.2822	0.069479404	0.60777142	0.39415110
Methyl chloride	818.5718	0.040242556	0.60388742	0.024419777
n-Butyl chloride	2385.2798	0.090896691	0.75552756	0.11589664
n-Propyl chloride	1746.1399	0.071017372	0.69688330	0.095832467
Pentachloroethane	3773.1485	0.11610884	0.71977811	0.16114569
Propargyl chloride	1745.0352	0.066006997	0.60484107	0.038957074
sec-Butyl chloride	2196.0424	0.086321622	0.78811355	−0.031077508

(Continued)

TABLE C.4 *(Continued)*
Olivera-Fuentes Additional Pure Compound Parameters for the PRSV EOS

Compound	a_c [kPa $(m^3/kmol)^2$]	b [m^3/kmol]	κ_0	κ_1
tert-Butyl chloride	2086.0404	0.084197228	0.66418016	0.051294027
Tetrachloroethylene	2705.3306	0.089291822	0.69294266	0.10126193
trans-1,2-Dichloroethylene	1572.2691	0.063335356	0.75886448	−0.061517268
Trichloroethylene	2100.1708	0.075266470	0.70405877	0.014657104
9.2. Aromatic chlorides				
1-Chloronaphthalene	5727.4865	0.14930607	0.88733613	0.16922787
1,2,4-Trichlorobenzene	4465.9027	0.12605332	0.89680159	−0.087923471
2,4-Dichlorotoluene	4451.3377	0.12920653	0.88509609	0.057206258
Benzotrichloride	5139.4284	0.14270210	0.76017815	0.041111857
Benzyl chloride	3869.8444	0.11543897	0.83201729	0.042321707
Benzyl dichloride	4630.4400	0.12962475	0.85518909	−0.11341992
Hexachlorobenzene	7482.7006	0.18560405	1.13768038	0.12606337
m-Dichlorobenzene	3633.0933	0.10870142	0.78808071	0.018297238
Monochlorobenzene	2790.5940	0.090307158	0.74654906	0.028144324
o-Chlorotoluene	3527.5152	0.11003937	0.81337350	0.053919954
o-Dichlorobenzene	3855.2138	0.11190317	0.69697201	0.18406882
p-Chlorotoluene	3540.8710	0.10978657	0.82261160	0.083760975
p-Dichlorobenzene	3641.3974	0.10882260	0.79071942	0.065305905

10. C, H, HALOGEN COMPOUNDS

10.1. C, H, Br compounds				
1-Bromobutane	2266.7368	0.080391164	0.77779123	0.11453937
1-Bromonaphthalene	5813.2500	0.14436929	0.91174173	0.041450827
1-Bromopropane	1735.0591	0.065267714	0.76734549	−0.034738973
1,1,2,2-Tetrabromoethane	4607.1128	0.11441545	0.59312880	0.52708317
1,2-Dibromoethane	2485.8993	0.078244353	0.66919076	0.054167724
2-Bromobutane	2198.5000	0.079346257	0.77614127	0.029406445
2-Bromopropane	1623.7905	0.062459917	0.72084723	−0.035210241
Bromobenzene	3140.8808	0.095909686	0.74150429	0.058264429
Bromoethane	1289.8074	0.052390160	0.71397619	−0.0091853261
Dibromomethane	1645.7476	0.055119472	0.67540318	0.064931635
Methyl bromide	861.99431	0.037772050	0.65299472	−0.070380509
p-Bromotoluene	3531.9732	0.10340065	0.83851674	−0.038020147
Tribromomethane	2515.5783	0.073962509	0.61055425	0.12964783
Vinyl bromide	984.83892	0.042607603	0.77462430	−0.32610635
10.2. C, H, F compounds				
1,1-Difluoroethane	1037.9149	0.054939276	0.76293871	−0.012124777
1,1-Difluoroethylene	650.1436	0.043937613	0.58597066	0.041912703
1,1,1-Trifluoroethane	1007.8093	0.059562322	0.74580479	0.030043818

(Continued)

TABLE C.4 *(Continued)*
Olivera-Fuentes Additional Pure Compound Parameters for the PRSV EOS

Compound	a_c [kPa $(m^3/kmol)^2$]	b [m^3/kmol]	κ_0	κ_1
Benzotrifluoride	2974.1371	0.10771981	0.78656636	0.088027038
Carbon tetrafluoride	437.53220	0.039356064	0.64999212	0.0018623429
Decafluorobutane	2026.6844	0.10734659	0.91558442	0.028158147
Difluoromethane	670.54306	0.039026611	0.77116880	0.029719974
Ethyl fluoride	885.65756	0.048290170	0.70791907	−0.062754594
Fluorobenzene	2182.2961	0.079733146	0.73969740	0.032694424
Hexafluorobenzene	2574.4857	0.10195531	0.94728273	0.0063861614
Hexafluoroethane	910.00178	0.063599574	0.73040640	0.10359936
Hexafluoropropylene	1505.7675	0.083732320	0.67738538	0.068237049
Methyl fluoride	527.48889	0.033976539	0.68716232	−0.057331558
Octafluoro-2-butene	2113.4831	0.11033054	0.79185326	0.047841882
Octafluorocyclobutane	1715.3618	0.090384319	0.89047275	0.039238509
Octafluoropropane	1415.5632	0.083951848	0.85368655	−0.11351326
Tetrafluoroethylene	752.42763	0.050244456	0.70873689	0.068877119
Trifluoromethane	584.66451	0.040029336	0.76000922	0.039509414
Vinyl fluoride	646.70703	0.040372130	0.64320722	−0.16843792

10.3. C, H, I, compounds
Diiodomethane	3236.6864	0.088667214	0.57387768	0.22755254
Ethyl iodide	1663.0406	0.060662882	0.67812989	−0.0051828372
Iodobenzene	3636.5087	0.10319106	0.73214161	0.082031988
Isopropyl iodide	2062.1064	0.073007295	0.70481823	−0.010371280
Methyl iodide	1195.7425	0.046343282	0.66115212	−0.086030288
n-Propyl iodide	2211.2593	0.076307638	0.75238670	−0.089021888

10.4. C, H, multihalogen compounds
1-Chloro-1,1-difluoroethane	1288.9019	0.064299448	0.73259104	−0.16845853
1,1,2-Trichlorotrifluoroethane	2166.8393	0.091003456	0.75843566	0.052313342
1,1,2,2-Tetrachlorodifluoroethane	2794.2343	0.10377536	0.80235301	0.055508491
1,2-Dibromotetrafluoroethane	2232.3268	0.093648110	0.74610909	0.018972871
1,2-Dichlorotetrafluoroethane	1702.9055	0.083198417	0.73976650	0.10709837
2-Chloro-1,1-difluoroethylene	1137.1747	0.058096971	0.70377218	0.037080295
Bromochlorodifluoromethane	1347.5338	0.064708333	0.63855297	0.072571636
Bromochloromethane	1437.6351	0.052817338	0.70489493	0.042253562
Bromotrichloromethane	2339.6394	0.079005866	0.66119156	0.025938629
Bromotrifluoroethylene	1317.3808	0.062403739	0.63846315	0.043065145
Bromotrifluoromethane	922.0936	0.055473757	0.62535640	0.059071939
Chlorodifluoromethane	849.57495	0.047076637	0.68175459	0.15350398
Chloropentafluoroethane	1253.4148	0.072630435	0.74167008	0.035455245
Chlorotrifluoroethylene	1116.3040	0.060249640	0.69118084	0.21475426
Chlorotrifluoromethane	743.00038	0.050352687	0.64206246	−0.0090716394

(Continued)

TABLE C.4 *(Continued)*
Olivera-Fuentes Additional Pure Compound Parameters for the PRSV EOS

Compound	a_c [kPa (m³/kmol)²]	b [m³/kmol]	κ_0	κ_1
Dibromodifluoromethane	1354.7936	0.058000036	0.67319593	−0.061534056
Dichlorodifluoromethane	1139.7861	0.060590156	0.62456046	0.099439002
Dichlorofluoromethane	1242.2222	0.056292124	0.69134332	0.0016072868
Halothane	2187.1338	0.085905438	0.51854669	0.30441548
p-Chlorobenzotrifluoride	3790.2998	0.12905717	0.91587269	0.034490433
Trichlorofluoromethane	1586.8071	0.068913114	0.65043218	0.061939584

11. C, H, N, O, S COMPOUNDS

11.1. C, H, S compounds

Compound	a_c	b	κ_0	κ_1
Benzothiophene	4323.9582	0.11735272	0.81046971	0.043204381
Diethyl disulfide	3364.9001	0.10725565	0.86052535	0.030003758
Diethyl sulfide	2475.9482	0.090939479	0.80098249	0.024135906
Dimethyl disulfide	2166.3786	0.073155127	0.74361724	0.044663647
Dimethyl sulfide	1478.4696	0.060144084	0.63724449	0.072243952
Ethyl mercaptan	1430.1143	0.058630381	0.64148602	0.082089195
Methyl mercaptan	964.65227	0.042005115	0.58478758	0.18776185
n-Butyl mercaptan	2577.9072	0.092712447	0.76062339	0.11275636
n-Dodecyl mercaptan	8989.7828	0.25409357	1.2916485	−0.095469034
n-Hexyl mercaptan	3963.5830	0.13019159	0.90553495	0.056847731
n-Pentyl mercaptan	3257.1938	0.11146165	0.83956866	0.054683578
n-Propylmercaptan	1961.4340	0.074884501	0.70202346	0.11149194
Phenyl mercaptan	3173.6298	0.094258404	0.75942512	0.20497960
tert-Butyl mercaptan	2186.2935	0.084414223	0.65982446	0.094358058
tert-Octyl mercaptan	4806.3005	0.15686513	0.82755909	0.034696588
Tetrahydrothiophene	2443.7560	0.079133095	0.65957564	0.12411374
Thiophene	1856.5827	0.065577721	0.65994924	0.10371980

11.2. C, H, NO₂ compounds

Compound	a_c	b	κ_0	κ_1
1-Nitropropane	2659.2957	0.089948576	0.95573244	−0.076449231
2-Nitropropane	2508.5548	0.086421180	0.91676358	−0.096056203
2,4-Dinitrotoluene	6160.8164	0.15488056	1.3672416	−0.71090600
2,6-Dinitrotoluene	5493.2526	0.14598959	1.3969156	−0.0083118060
3,4-Dinitrotoluene	6589.9091	0.16015864	1.3941784	−0.0073235598
m-Dinitrobenzene	5333.9956	0.13559379	1.3225690	0.0090607399
m-Nitrotoluene	4482.5195	0.12497098	1.0758846	−0.33371368
Nitrobenzene	3712.8684	0.10567292	1.0053108	−0.10450033
Nitroethane	2155.1689	0.074372034	0.90647300	−0.07151607
o-Dinitrobenzene	5695.8517	0.14026221	1.3286568	0.010298334
o-Nitrotoluene	4329.9744	0.12306537	1.0604948	−0.14173868
p-Dinitrobenzene	5278.8638	0.13452653	1.3317130	−0.0071327734
p-Nitrotoluene	4504.0825	0.12523092	1.1364876	−0.40697642

(Continued)

TABLE C.4 *(Continued)*
Olivera-Fuentes Additional Pure Compound Parameters for the PRSV EOS

Compound	a_c [kPa $(m^3/kmol)^2$]	b $[m^3/kmol]$	κ_0	κ_1
11.3. Other monofunctional C, H, O, N				
Cyclohexyl isocyanate	3686.2999	0.11917084	1.1162716	0.033052091
Methyl isocyanate	1552.3296	0.062903613	0.63739245	0.045790383
n-Butyl isocyanate	2966.5446	0.10687733	0.97414355	0.031335588
Phenyl isocyanate	3268.7897	0.10322741	1.0008073	−0.17814072
Toluene diisocyanate	5548.9214	0.15407214	1.1624697	0.063558379
11.4. Polyfunctional C, H, N, (O), halide				
1-Chloro-2,4-dinitrobenzene	6000.2031	0.15088543	1.3874004	−0.0044978098
1,2-Dichloro-4-nitrobenzene	5046.2891	0.13623414	1.1375561	0.026672155
3-Nitrobenzotrifluoride	4993.5664	0.15320327	1.1235395	0.021771208
3,4-Dichloroaniline	4909.3024	0.12557779	1.0452990	0.034619238
4-Chloro-3-nitrobenzotrifluoride	5449.6309	0.16256461	1.2090581	−0.035625130
m-Chloroaniline	3884.7390	0.10585343	0.96125778	0.090019597
m-Chloronitrobenzene	4341.6222	0.11973777	1.0762473	−0.19902248
o-Chloroaniline	3588.4696	0.10170800	0.97552703	−0.041666560
o-Chloronitrobenzene	4521.3205	0.12222286	1.0675917	−0.20666206
p-Chloronitrobenzene	4483.4251	0.12216675	1.0696122	0.047834585
11.5. Polyfunctional C, H, O, halides				
2-Chloroethanol	1827.4156	0.063924082	1.3018544	−0.41065883
Acetyl chloride	1426.4704	0.057462178	0.79945754	0.081824941
α-Epichlorohydrin	2398.7112	0.080469476	0.76209761	0.15019934
Benzoyl chloride	3783.6635	0.11108688	0.98127798	−0.13268202
bis(Chloromethyl)ether	2314.2756	0.081793643	0.84787400	0.047500026
Carbonyl fluoride	483.93642	0.033343785	0.80198736	0.1268412
Chloroacetaldehyde	1812.0647	0.066813427	0.85666285	0.047273446
Chloroacetic acid	2605.7168	0.077729545	1.1429189	0.048235544
Chloroacetyl chloride	2084.3284	0.073413014	0.89027363	0.078541361
Dichloroacetaldehyde	1953.6478	0.072033798	0.87869362	0.043361502
Dichloroacetyl chloride	2300.4937	0.081306548	0.91023863	0.028648327
Ethyl chloroformate	1812.0480	0.072972766	1.5166836	−0.026740606
Hexafluoroacetone	1417.7884	0.081237389	0.90401330	0.11574155
Isophthaloyl chloride	5656.0813	0.15070840	1.2674364	0.017693856
m-Chlorobenzoyl chloride	4538.5227	0.12828001	1.0212183	−0.056623118
m-Chlorophenol	3190.2349	0.089552593	1.0582744	−0.28753934
Methyl chloroacetate	2530.3028	0.086298704	0.99718789	0.055190425
Methyl chloroformate	1626.4885	0.063397879	0.94298599	0.036694094
o-Chlorobenzoic acid	4918.5154	0.12708430	1.3013590	0.0039672154
o-Chlorophenol	2884.9190	0.087460687	0.99792426	−0.36328295

(Continued)

TABLE C.4 *(Continued)*
Olivera-Fuentes Additional Pure Compound Parameters for the PRSV EOS

Compound	a_c [kPa $(m^3/kmol)^2$]	b [$m^3/kmol$]	κ_0	κ_1
p-Chlorophenol	3281.4821	0.090990639	1.0573930	−0.23697911
Trichloroacetaldehyde	2289.4309	0.082920541	0.85985452	0.044738838
Trichloroacetic acid	3128.1841	0.093043686	1.1525970	0.055019479
Trichloroacetyl chloride	2686.1534	0.093166941	0.87910466	0.025427066
Trifluoroacetic acid	2345.5204	0.097705651	1.1147483	−0.078213188
11.6. Polyfunctional C, H, O, S				
3-Methyl sulfolane	4980.4996	0.12474809	0.98037871	0.031024710
Carbonyl sulfide	707.33894	0.038212076	0.52624200	0.064791637
Di-*n*-Butyl sulfone	7317.0418	0.19521951	1.3316847	0.0021959431
Di-*n*-Propyl sulfone	5944.9999	0.15944480	1.1883972	0.032165327
Sulfolane	3197.9073	0.077079913	0.94237060	0.024432281
11.7. Polyfunctional C, H, O				
2-(2-Ethoxyethoxy)ethanol	4019.3490	0.13014326	1.5588855	−0.22407809
2-(2-Methoxyethoxy)ethanol	3550.1123	0.11531467	1.5116417	−0.15029013
2-Butoxyethanol	3508.3923	0.11965750	1.4530104	−0.20645545
2-Ethoxyethanol	2414.3905	0.086831697	1.3431685	−0.10232741
2-Methoxyethanol	1999.6757	0.072554358	1.3482114	−0.21410771
Citric acid	5640.8378	0.14042836	2.6824919	−0.39615614
Diethylene glycol ethyl ether acetate	5667.1527	0.17571305	1.3613476	−0.32027769
Diethylene glycol monobutyl ether	5284.4653	0.16535081	1.5145987	0.23841343
Diglycolic acid	4804.1953	0.11989190	1.8172800	−0.11044920
Ethyl acetoacetate	4027.9094	0.12818929	1.1700672	−0.16139209
Ethyl lactate	2832.5652	0.098579276	1.4581123	−0.77861293
Furfural	2451.8155	0.076366951	1.0120984	−0.12197605
Furfuryl alcohol	2358.4243	0.076363865	1.3921231	−0.26299568
Glycolic acid	1649.6256	0.054800902	1.7884171	−0.11073781
Hydroxycaproic acid	4973.9872	0.13428221	1.9126828	−0.11592734
Lactic acid	2027.0679	0.067339613	1.7567190	−0.085424482
Methoxyacetic acid	3023.2941	0.089533463	1.2393917	0.11848404
Methyl acetoacetate	3514.2462	0.11201603	1.1079688	0.017428755
Methyl salicylate	3785.7656	0.11051437	1.1360756	0.34215209
p-Methoxyphenol	3650.5252	0.098552849	1.1408396	0.024608856
Salicylaldehyde	2925.4984	0.088038775	1.2433335	−0.48527491
Tartaric acid	4177.0801	0.10323465	2.8481211	−0.53600569
Tetrahydrofurfuryl alcohol	2770.5168	0.088724377	1.3231888	−0.32111670
Vanillin	4730.4017	0.12458336	1.4236148	−0.026639145

(Continued)

TABLE C.4 *(Continued)*
Olivera-Fuentes Additional Pure Compound Parameters for the PRSV EOS

Compound	a_c [kPa $(m^3/kmol)^2$]	b [m^3/kmol]	κ_0	κ_1
11.8. Polyfunctional C, H, O, N				
1-Amino-2-propanol	2112.0365	0.070390829	1.4504471	−0.35288772
2-Pyrrolidone	3212.4634	0.083003430	0.99878459	0.033642366
3-Amino-1-propanol	2413.2822	0.076093305	1.5088299	−0.026854360
3-Methoxypropionitrile	3519.1522	0.11287569	1.0446425	−0.10345078
8-Hydroxyquinoline	4503.5855	0.11695404	1.1158712	0.031010468
Acetanilide	5743.6886	0.14246887	1.1737524	0.0025324270
Acetone cyanohydrin	3112.5055	0.098443916	1.3773837	−0.73392663
bis(Cyanoethyl)ether	7079.6320	0.18502566	1.4204766	0.019745540
Dehydroabietylamine	13791.814	0.32703455	1.4021430	−0.0091061545
Epsilon-caprolactam	4307.0243	0.10935164	1.0557071	0.020892266
Formanilide	4767.7933	0.12397260	1.1457701	0.025676928
L-Glutamic acid	6032.8110	0.13933783	1.9516167	−0.16858266
Lactonitrile	2588.7466	0.082387553	1.4697344	−0.021051493
Lysine	6041.5089	0.15058624	1.7349738	−0.10914624
Methyl cyanoacetate	3916.3440	0.11665601	1.1521583	0.022162572
Methyl diethanolamine	3740.7893	0.11290589	2.0669461	−0.13483684
Monoethanolamine	1928.6426	0.061860598	1.4477034	−0.21966904
Morpholine	2261.1517	0.074872833	0.87874008	0.034459321
n-Aminoethyl ethanolamine	3442.4519	0.10092424	1.7747197	−0.075539317
n-Methyl-2-pyrrolidone	3431.5481	0.096991696	0.89861870	0.035036314
n-Methylformamide	2924.9995	0.083018248	0.96680736	0.036516683
n,n-Dimethylacetamide	3395.4820	0.10559870	0.90333759	0.039040866
n,n-Dimethylformamide	2998.7957	0.094847444	0.85487690	0.13014625
Nitroglycerine	5086.4225	0.15306876	1.9149758	−0.095169695
o-Nitroanisole	5144.8392	0.13463194	1.1683604	0.019277578
Oxazole	1536.0541	0.056738752	0.72283726	0.037000604
p-Dimethylaminobenzaldehyde	7100.7215	0.17464743	1.1214359	−0.27101896
p-Phenetidine	5034.9430	0.13664893	1.1579228	0.020735895
Pentaerythritol tetranitrate	6416.5541	0.19423979	2.2471341	−0.30655400
Triethanolamine	8123.9451	0.21123957	1.7944963	−0.13684108
12. ELEMENTS				
Argon	146.99772	0.019939742	0.37414086	0.0029062082
Bromine	1049.6354	0.036770298	0.56728738	−0.0011322542
Chlorine	705.80912	0.034624065	0.48735538	0.075796202
Deuterium	27.958434	0.014918684	0.15918813	−0.059285425
Fluorine	125.89156	0.017851849	0.46282373	0.0015521138
Helium-3	2.9981803	0.018535862	−0.36700701	−0.15714048
Helium-4	3.7413227	0.014723290	−0.22409080	−0.13537160

TABLE C.4 *(Continued)*
Olivera-Fuentes Additional Pure Compound Parameters for the PRSV EOS

Compound	a_c [kPa $(m^3/kmol)^2$]	b [m³/kmol]	κ_0	κ_1
Hydrogen	26.492357	0.016339067	0.051526807	−0.11738005
Mercury	595.90532	0.007028470	0.11805156	−0.022005780
Neon	23.487290	0.010825126	0.31024345	0.024655119
Nitrogen	148.62678	0.024119318	0.43743964	0.010707724
Oxygen	150.43964	0.019915536	0.40832681	0.020801738
13. EPOXIDES AND PEROXIDES				
1,2-Epoxybutane	1995.8558	0.077647318	0.73762517	0.014905147
1,2-Propylene oxide	1491.5234	0.063290839	0.75270077	0.068254857
1,3-Propylene oxide	1488.8405	0.058590590	0.67249097	0.038812753
1,4-Dioxane	2122.1611	0.073981527	0.77243968	0.041781418
2,5-Dihydrofuran	1687.2894	0.063704974	0.71522067	0.049168186
Ethylene oxide	967.1134	0.042184093	0.68750738	0.012995286
Furan	1378.1845	0.057538878	0.67315670	0.091589917
Tetrahydrofuran	1772.6274	0.067156201	0.71024113	0.041312022
Trioxane	1989.2831	0.067397330	0.85927304	0.047841607
14. ESTERS				
14.1. Aromatic esters				
Benzyl acetate	4830.4317	0.14141380	1.0303396	0.056424738
Benzyl benzoate	8145.2037	0.20326899	1.2504283	0.014857767
Dibutyl phthalate	11139.9201	0.29188655	1.5499898	0.21565362
Diethyl phthalate	7789.2252	0.21056269	1.4005204	0.052544349
Dimethyl phthalate	6672.4211	0.17825336	1.2860778	−0.12449116
Dimethyl terephthalate	6781.3149	0.17975445	1.2611592	−0.33446437
Dioctyl phthalate	17395.991	0.44166926	1.9019842	0.025851568
Ethyl benzoate	4784.1144	0.14025849	1.0637752	−0.14755105
Methyl benzoate	4228.9156	0.12487596	0.97468747	0.031451080
14.2. Other saturated aliphatic esters				
Dibutyl sebacate	14358.1528	0.38257835	1.8655572	−0.20101677
Diethyl malonate	4780.3554	0.14980629	1.2247330	−0.049599112
Diethyl oxalate	4270.5601	0.13528055	1.1778395	0.018126655
Diethyl succinate	5419.0899	0.16802173	1.3935650	0.027684410
Dihexyl adipate	14112.247	0.37651636	1.8213368	−0.033515161
Ethyl isovalerate	3847.5576	0.13391455	0.98024756	−0.021859326
Ethyl propionate	2824.2551	0.10585084	0.92997999	0.11176239
Ethylene glycol diacetate	4364.5717	0.13677650	1.0881057	0.81965576
Gamma-butyrolactone	2904.8818	0.080439074	0.92601492	−0.16443755
Glyceryl triacetate	6797.8774	0.19759853	1.5202189	−0.024120604
Isobutyl isobutyrate	4408.2310	0.14984798	0.95510575	0.12383530

(Continued)

TABLE C.4 *(Continued)*
Olivera-Fuentes Additional Pure Compound Parameters for the PRSV EOS

Compound	a_c [kPa $(m^3/kmol)^2$]	b [m^3/kmol]	κ_0	κ_1
Isopentyl isovalerate	5823.8983	0.18709293	1.1973192	−0.36308778
Methyl dodecanoate	9198.3741	0.26437118	1.3234423	0.11429634
Methyl *n*-butyrate	2816.0113	0.10392400	0.90576228	0.12581828
Methyl propionate	2247.5025	0.086679415	0.87566154	0.078826526
n-Butyl *n*-butyrate	4722.0861	0.15686867	1.0717582	0.018934418
n-Butyl propionate	4111.2973	0.14163659	1.0394833	0.045970120
n-Butyl stearate	16578.8345	0.44406206	1.7596807	−0.045961461
n-Propyl *n*-butyrate	4003.9954	0.13793998	1.01913712	−0.037815918
n-Propyl propionate	3391.5485	0.12007517	0.92670194	0.050682799
Vinyl propionate	2561.6704	0.096009376	0.86025894	0.0016081910
15. ETHERS				
1,2-Dimethoxyethane	2352.5045	0.089789838	0.87437233	−0.019487712
Acetal	3105.5388	0.11746885	1.0093134	0.078745965
Anethole	5691.2641	0.16108440	1.0669792	0.029967840
Anisole	3116.5919	0.099395053	0.90724769	0.0067410215
Benzyl ethyl ether	4449.6633	0.13754734	0.99593703	−0.031577730
Butyl vinyl ether	2900.9670	0.11075441	0.92694131	0.044499615
Di-*n*-Butyl ether	4372.4217	0.15400292	1.0431002	0.040259132
Di-*n*-Hexyl ether	7532.9898	0.23427423	1.3624551	−0.0028267190
Di-*n*-Propyl ether	2941.7473	0.11345435	0.87539258	0.11721086
Dibenzyl ether	7456.5144	0.19638028	1.2452426	0.080565570
Diethyl ether	1909.4888	0.083726389	0.76760851	0.14440991
Diisopropyl ether	2755.0548	0.11274567	0.87264874	−0.064554698
Dimethyl ether	965.17439	0.049365129	0.66710318	0.0098302824
Diphenyl ether	5876.8170	0.15761613	1.0457127	−0.16939040
Divinyl ether	1594.8502	0.070489095	0.80020021	0.036192868
Ethyl propyl ether	2345.0083	0.095930716	0.80765157	0.24986080
Ethyl vinyl ether	1753.2416	0.075508194	0.76946362	0.040335995
Methyl ethyl ether	1356.6407	0.063412100	0.67488632	0.68013086
Methyl tert-butyl ether	2281.8834	0.093936143	0.77022242	0.047964831
Methyl vinyl ether	1317.0667	0.061675031	0.72605707	0.060751760
Methylal	1848.3387	0.078701104	0.80297845	0.038067161
n-Butyl ethyl ether	2983.6923	0.11498536	0.94121183	0.031523589
Phenetole	3866.2089	0.12225409	0.97494659	0.031529239
Tetraethylene glycol dimethyl ether	7945.1758	0.23062025	1.6931668	−0.42429078
16. FORMATES AND ACETATES				
2-Ethylhexyl acetate	5952.5789	0.19062828	1.2614381	0.00039260867
Allyl acetate	2651.8717	0.097078651	0.93693961	−0.12445373

(Continued)

TABLE C.4 (Continued)
Olivera-Fuentes Additional Pure Compound Parameters for the PRSV EOS

Compound	a_c [kPa $(m^3/kmol)^2$]	b [m^3/kmol]	κ_0	κ_1
Ethyl acetate	2248.1324	0.087913218	0.90207766	0.055381792
Ethyl formate	1735.3270	0.069848779	0.78589032	0.055433003
Ethylidene diacetate	3921.2828	0.12636810	1.0587419	0.29894956
Isobutyl formate	2493.9604	0.092564659	0.95118738	−0.12899941
Isopentyl acetate	4010.2986	0.13700390	0.97422068	0.036069623
Isopropyl acetate	2555.3570	0.097196883	0.89521687	−0.0035489757
Methyl acetate	1745.5688	0.070482839	0.83818916	0.061918731
Methyl formate	1284.9215	0.053969993	0.73105842	0.097081783
n-Butyl acetate	3409.4015	0.12046756	0.96676661	0.051625813
n-Butyl formate	2811.2928	0.10291468	0.94609348	−0.067112430
n-Pentyl acetate	4039.8853	0.13824546	1.0748246	0.024651643
n-Propyl acetate	2854.8407	0.10633501	0.92558516	0.12423781
n-Propyl formate	2292.6621	0.087204885	0.82953845	0.055333415
tert-Butyl acetate	2961.7165	0.11120646	0.87398446	0.048305412
Vinyl acetate	2050.8044	0.080089576	0.87571466	−0.030224144
Vinyl formate	1592.6367	0.065444087	0.78366914	0.15321973
17. INORGANIC GASES				
Arsine	671.01844	0.036813631	0.39095266	0.042992568
Carbon disulfide	1202.9106	0.044594089	0.54625774	−0.019869276
Carbon monoxide	159.83481	0.024607306	0.45098475	0.039868702
Diborane	702.39723	0.049598375	0.52781942	0.095435474
Disilane	1148.7332	0.054414981	0.52717691	−0.035910443
Hydrazine	915.40685	0.028680338	0.83283219	0.0051948669
Hydrogen sulfide	490.08867	0.026849253	0.52175647	0.045558612
Nitric oxide	157.44303	0.017884311	1.1966190	0.058053276
Nitrogen dioxide	582.04962	0.027625787	1.5312032	−0.79316762
Nitrogen trifluoride	381.23074	0.033360579	0.58053421	−0.0091212901
Nitrosyl chloride	672.55400	0.031233195	0.82236572	−0.22636125
Nitrous oxide	416.20319	0.027512469	0.59455708	0.15322754
Ozone	386.09791	0.030271920	0.66687090	0.079545135
Perchloryl fluoride	793.48025	0.044075732	0.62541153	0.064745462
Phosgene	1145.4199	0.051515315	0.65746421	0.090821885
Phosphine	509.85398	0.032127704	0.43265394	0.0035315482
Silane	475.11290	0.036049472	0.48420872	0.016652831
Sulfur dioxide	746.20839	0.035450134	0.73328102	0.093364814
Sulfur trioxide	929.65995	0.038757733	1.0007079	0.67478535
Tetrafluorohydrazine	815.90611	0.053972571	0.70497870	−0.30616308

(Continued)

TABLE C.4 (Continued)
Olivera-Fuentes Additional Pure Compound Parameters for the PRSV EOS

Compound	a_c [kPa $(m^3/kmol)^2$]	b [m^3/kmol]	κ_0	κ_1
18. KETONES				
2-Heptanone	4048.6982	0.13547728	0.96698716	0.088062742
2-Hexanone	3289.0558	0.11465135	0.94780030	0.040867249
2-Pentanone	2693.2460	0.098227767	0.88849954	−0.089041441
5-Methyl-2-hexanone	3838.2389	0.13068946	0.97898941	0.089556900
Acetone	1733.5551	0.069804917	0.82114231	−0.017586622
Acetophenone	4040.9197	0.11796285	0.98587807	−0.065552953
Acetylacetone	2893.4151	0.098355192	1.0812389	0.032638569
Benzophenone	6994.0164	0.17539593	1.1501039	−0.17301392
Camphor	5299.9172	0.15296986	0.84536409	−0.010016609
Cyclohexanone	3147.5914	0.10237813	1.0193956	−0.20223933
Cyclopentanone	2128.7215	0.069586909	0.90264751	−0.11635843
Diethyl ketone	2655.8330	0.09688569	0.88104494	−0.0072863217
Isophorone	4851.6616	0.13885691	0.94921118	−0.16908669
Ketene	751.30436	0.041552506	0.57363314	0.069469146
Mesityl oxide	3338.5125	0.11386357	0.84877108	0.10548778
Methyl ethyl ketone	2199.9697	0.084069848	0.83933590	0.0084544073
Methyl isobutyl ketone	3149.0830	0.11277865	0.88985029	0.18470690
Methyl isopropyl ketone	2516.4044	0.093119007	0.78436109	0.36916243
19. n-ALIPHATIC ACIDS				
19.1. Aromatic carboxilic acids				
Isophthalic acid	8138.41965	0.16538406	1.794398509	−0.134193535
o-Toluic acid	4618.38633	0.12584424	1.292968353	0.010838362
p-Toluic acid	4892.58155	0.12952143	1.297555235	−0.008631397
Phthalic acid	5104.23613	0.13056411	1.792846681	−0.12699126
Pyromellitic acid	8015.89277	0.18368914	2.656669216	−0.505780886
Terephthalic acid	9931.31966	0.18259750	1.790947037	−0.139377479
19.2. Inorganic acids				
Hydrogen bromide	488.69005	0.027537879	0.48121536	0.13349968
Hydrogen chloride	398.67837	0.025129873	0.57793434	−0.046943151
Hydrogen fluoride	1036.1812	0.045980802	0.92137589	−0.50671496
Hydrogen iodide	688.62073	0.033246883	0.42983469	0.082471874
Perchloric acid	3263.4499	0.10583532	0.45721839	0.12540479
19.3. Inorganic bases				
Ammonia	460.26755	0.023218914	0.74545487	−0.012061178

(Continued)

TABLE C.4 *(Continued)*
Olivera-Fuentes Additional Pure Compound Parameters for the PRSV EOS

Compound	a_c [kPa $(m^3/kmol)^2$]	b [m^3/kmol]	κ_0	κ_1
19.4. Inorganic halides				
Antimony trichloride	4134.4449	0.10655645	0.62641676	0.11139889
Boron trichloride	1650.6602	0.074739523	0.60002024	0.17733657
Boron trifluoride	428.43586	0.033604294	1.0108459	−0.044808936
Dichlorosilane	1437.7143	0.065525341	0.62969367	0.082773087
Silicon tetrachloride	2236.8459	0.090284071	0.72261540	0.057156476
Sulfuryl chloride	2037.6484	0.076509574	0.64107352	0.18165662
Titanium tetrachloride	2745.5820	0.088063669	0.77923727	−0.10523728
Trichlorosilane	1742.2479	0.074431620	0.68817984	−0.012170739
19.5. n-Aliphatic acids				
Acetic acid	1955.2632	0.067506565	1.0216356	−0.17966872
Formic acid	1439.2961	0.050781460	1.0492666	−0.52864419
n-Butyric acid	2847.2519	0.092778914	1.2180017	−0.014442586
n-Decanoic acid	7095.2627	0.20363948	1.5495994	0.69737178
n-Dodecanoic acid	8752.2381	0.24400914	1.6823928	0.22405382
n-Heptanoic acid	4883.9928	0.14697692	1.3705469	0.27719333
n-Hexadecanoic acid	12623.382	0.33288717	1.7637494	0.76607256
n-Nonanoic acid	6364.8572	0.18527481	1.4958342	0.22841814
n-Octanoic acid	5660.3961	0.16738779	1.4345819	0.25973482
n-Tetradecanoic acid	10684.839	0.28922060	1.7474202	0.065771850
Pentadecanoic acid	11668.979	0.31173613	1.7288166	0.68778397
Propionic acid	2549.0796	0.086363354	1.1344665	−0.034737088
Stearic acid	14948.838	0.382863260	1.8133525	−0.27017975
Valeric acid	3518.8555	0.11061231	1.2443862	0.075962748
19.6. Other aliphatic acids				
2-Methylbutyric acid	3362.6133	0.10701607	1.2081557	0.0059066559
Acrylic acid	2112.0131	0.070275594	1.1356440	−0.14560751
Azelaic acid	8079.4619	0.20386591	1.9242514	−0.13829119
cis-Crotonic acid	2815.9062	0.089062923	1.1818168	0.022801532
Glutaric acid	5128.7720	0.13005375	1.6598487	0.051204460
Isobutyric acid	2893.7510	0.097212014	1.2465735	−0.13187224
Isovaleric acid	3267.2164	0.10545609	1.2813614	−0.0035128126
Linoleic acid	13431.545	0.35465598	1.9232537	−0.098878848
Maleic acid	3779.9353	0.10006631	1.7182121	−0.086617539
Methacrylic acid	2792.4506	0.088870488	1.0472514	0.045461466
Sebacic acid	8946.8357	0.22464404	1.9634932	−0.25428501
Succinic acid	4334.6773	0.11005373	1.7134879	−0.10089981
trans-Crotonic acid	2957.0770	0.090859729	1.1895622	0.10397067

(Continued)

TABLE C.4 *(Continued)*
Olivera-Fuentes Additional Pure Compound Parameters for the PRSV EOS

Compound	a_c [kPa $(m^3/kmol)^2$]	b [m^3/kmol]	κ_0	κ_1
19.7. Other inorganics				
Deuterium oxide	604.05272	0.019197567	0.90091414	−0.047223017
Hydrogen peroxide	777.20957	0.021782568	0.89040053	−0.063312378
Water	603.34119	0.019078950	0.86975396	−0.062787920
19.8. Other salts				
Potassium chloride	21144.339	0.12469461	0.17378161	0.31508072
20. NITRILES				
1.4-Dicyano-2-butene	6109.2715	0.16558672	1.3044724	0.0063192095
Acetonitrile	1938.6867	0.07272704	0.86591088	−0.17479070
Acrylonitrile	2019.4879	0.07724502	0.86564622	−0.13055942
Adiponitrile	6815.2753	0.17857284	1.3121771	0.0024393034
Benzonitrile	3674.3322	0.10751446	0.88943360	−0.0054126211
cis-Dicyano-1-butene	5120.6849	0.15164673	1.3112027	0.0029824494
Cyanogen	851.50508	0.04354592	0.77998536	0.0064955920
Cyanogen chloride	1074.6603	0.04897878	0.82991437	0.0085544278
Glutaronitrile	6154.1275	0.16104334	1.1923210	−0.22240893
Hexanenitrile	4192.8430	0.13793242	1.0677150	−0.23000489
Hydrogen cyanide	1231.4180	0.05518297	0.96227219	−0.34796116
Isobutyronitrile	2684.8287	0.09724139	0.86939331	0.037120775
Methacrylonitrile	2498.1932	0.09227824	0.81216017	−0.14037982
Methylglutaronitrile	6047.5416	0.16678539	1.2697683	0.0026382615
n-Butyronitrile	2829.6282	0.09944957	0.90127759	−0.011749574
Propionitrile	2402.7240	0.08711640	0.83358725	−0.017768606
Succinonitrile	5293.1032	0.14067038	1.1640566	0.025905582
trans-Crotonitrile	2782.5987	0.09717085	0.94462581	−0.11770785
trans-Dicyano-1-butene	5088.1113	0.15111947	1.3016477	0.0050616206
Valeronitrile	3521.5237	0.11950782	0.94047244	0.03220135
Vinylacetonitrile	2779.6098	0.097398892	0.92320370	−0.073845497
21. ORGANIC/INORGANIC COMPOUNDS				
Diethyl carbonate	3103.4112	0.11025540	1.0413448	0.052649200
Dimethyl sulfate	3519.6866	0.095020613	0.51660430	0.40377576
Ethylene carbonate	2914.0387	0.075483365	0.97328705	0.13253627
Tri-*n*-butyl borate	8781.0766	0.24179890	0.66491664	0.021304369

REFERENCES

1. Stryjek, R., and Vera, J. H. 1986. An improved cubic equation of state. In *Equations of State: Theories and Applications*, ed. K. C. Chao and R. L. Robinson ACS Symposium Series 300. Washington, DC: American Chemical Society, pp. 560–570.
2. Stryjek, R., and Vera, J. H. 1986. PRSV: An improved Peng-Robinson equation of state for pure compounds and mixtures. *Can. J. Chem. Eng.* 64(2): 323–333.

3. Stryjek, R., and Vera, J. H. 1986. PRSV2: A cubic equation of state for accurate vapor-liquid equilibria calculations. *Can. J. Chem. Eng.* 64: 820–826.

4. Peng, D. Y., and Robinson, D. B. 1976. A new two-constant equation of state. *Ind. Eng. Chem. Fundam.* 15: 50–64.

5. Soave, G. 1972. Equilibrium constants from a modified Redlich-Kwong equation of state. *Chem. Eng. Sci.* 27: 1197–1203.

6. Redlich, O., and Kwong, J. N. S. 1949. On the thermodynamics of solutions. V. An equation of state. Fugacities of gaseous solutions. *Chem. Rev.* 44(1): 233–244.

7. Zabaloy, M. S., and Vera, J. H. 1998. The Peng-Robinson sequel—An analysis of the particulars of the second and third generations. *Ind. Eng. Chem. Res.* 37(5): 1591–1597.

8. Mathias, P. M. 1983. A versatile phase equilibrium equation of state. *Ind. Eng. Chem. Process Des. Dev.* 22: 385–391.

9. Green, D. W., and Perry, R. H. 2007. *Perry's Chemical Engineers' Handbook*, 8th ed. Blacklick, OH: McGraw-Hill Professional Publishing.

10. Wilczek-Vera, G., and Vera, J. H. 2015. Understanding cubic equations of state: A search for the hidden clues of their success. *AIChE J.* 61(9): 2824–2831.

11. Proust, P., and Vera, J. H. 1989. PRSV: The Stryjek-Vera modification of the Peng-Robinson equation of state. Parameters for other pure compounds of industrial interest. *Can. J. Chem. Eng.* 67: 170–173.

12. Kolasinska, G., and Vera, J. H. 1986. On the prediction of heats of vaporization of pure compounds. *Chem. Eng. Commun.* 43: 185–194.

13. Proust, P., Meyer, E., and Vera, J. H. 1993. Calculation of pure compound saturated enthalpies and saturated volumes with the PRSV equation of state. Revised k_1 parameters for alkanes. *Can. J. Chem. Eng.* 71: 292–298.

14. Adachi, Y., and Sugie, H. 1986. A new mixing rule—Modified conventional mixing rule. *Fluid Phase Equilib.* 28: 103–118.

15. Panagiotopoulos, A. Z., and Reid, R. C. 1986. A new mixing rule for cubic equations of state for highly polar, asymmetric system. *ACS Symp. Ser.* 300: 571–582.

16. Sandoval, R., Wilczek-Vera, G., and Vera, J. H. 1989. Prediction of ternary vapor-liquid equilibria with the PRSV equation of state. *Fluid Phase Equilib.* 52: 119–126.

17. Michelsen, M. L., and Kistenmacher, H. 1990. On composition-dependent interaction coefficients. *Fluid Phase Equilib.* 58: 229–230.

18. Zabaloy, M. S., and Vera, J. H. 1996. Identification of variant and invariant properties in the thermodynamics of mixtures: Tests for models and computer codes. *Fluid Phase Equilib.* 119: 27–49.

19. Orbey, H., Sandler, S. I., and Wong, D. S. H. 1993. Accurate equation of state predictions at high temperatures and pressures using the existing UNIFAC model. *Fluid Phase Equilib.* 85: 41–54.

20. Tochigi, K. 1995. Prediction of high-pressure vapor-liquid equilibria using ASOG. *Fluid Phase Equilib.* 104: 253–260.

21. Wilczek-Vera, G., and Vera, J. H. 1989. A consistent method to combine the PRSV EOS with excess Gibbs energy models. *Fluid Phase Equilib.* 51: 197–208.

22. Founders tribute dedicated to John M. Prausnitz. 2015. *AIChE J.* 61(9): 2673–3144.

23. Zabaloy, M. S., Brignole, E. A., and Vera, J. H. 1999. A conceptually new mixing rule for cubic and non-cubic equations of state. *Fluid Phase Equilib.* 158–160:245–257.

24. Zabaloy, M. S., Brignole, E. A., and Vera, J. H. 2002. A flexible mixing rule satisfying the ideal-solution limit for equations of state. *Ind. Eng. Chem. Res.* 41(5): 922–930.

25. Ohta, T., Todoriki, H., and Yamada, T. 2004. Representation of liquid–liquid equilibria at low and high pressures using the EOS-G E mixing rules. *Fluid Phase Equilib.* 225: 23–27.

26. Ohta, T. 1991. Prediction of multicomponent vapor-liquid equilibria using the PRSV and PRSV2 equations of state with the Huron-Vidal mixing rules. *Thermochim. Acta* 185: 283–293.

27. Ohta, T. 1992. Prediction of ternary excess enthalpies using the PRSV and PRSV2 equations of state. *Thermochim. Acta* 202: 51–63.
28. Wong, D. S. H., and Sandler, S. I. 1992. A theoretically correct mixing rule for cubic equations of state. *AIChE J.* 38: 671–680.
29. Figueira, F. L., Lugo, L., and Olivera-Fuentes, C. 2007. Generalized parameters of the Stryjek-Vera and Gibbs-Laughton cohesion functions for use with cubic EOS of the van der Waals type. *Fluid Phase Equilib.* 259: 105–119.
30. Daubert, T. F., and Danner, R. P. 1988. *Physical and Thermodynamic Properties of Pure Chemicals—Data Compilation.* Washington, DC: Hemisphere.

Appendix D: Working with ASOG-KT

As discussed in Chapter 26, the analytical solution of groups (ASOG) is based on studies carried out in the early sixties at Shell Research Co., Emeryville, California, by G. M. Wilson, C. H. Deal, and E. L. Derr, among others. It basically recognizes the importance in the behavior of mixtures of functional groups characteristic of the homologous series. Although the basic elements of the method were clearly set forward by Wilson and Deal in 1962 [1], its use only became possible with the publication of parameters for the group interactions by K. Kojima and K. Tochigi [2]. For this reason, the ASOG method used with the Kojima–Tochigi parameters is known as ASOG-KT. Variations of ASOG, such as PRASOG [3], are not included here.

The use of ASOG-KT is simple, but a word of warning is required. It is necessary to remember that not all interactions between groups are physical in nature. The most common cases of *nonphysical interaction* are those of the hydroxyl (–OH) group that reacts with the carboxyl group (–COOH) and the amino (–NH$_2$) group that reacts with the ketone (–CO–) group.

BASIC EQUATIONS

According to ASOG, the activity coefficient of component i of a liquid phase mixture is given by the sum of a combinatorial term and a term representing the group interactions.

$$\ln \gamma_i = (\ln \gamma_i)_{\text{combinatorial}} + (\ln \gamma_i)_{\text{interactions}} \tag{D.1}$$

For the combinatorial term, ASOG uses

$$(\ln \gamma_i)_{\text{combinatorial}} = \left(1 - \frac{r_i}{\sum_k x_k r_k} + \ln \frac{r_i}{\sum_k x_k r_k} \right) \tag{D.2}$$

where x_k is the mole fraction of compound k in the liquid mixture and r_i is the number of atoms different from hydrogen atoms in molecule i. The contribution of the interactions between groups is given by

$$(\ln \gamma_i)_{\text{interactions}} = \sum_k \nu_{ki} \left[\ln \Gamma_k - \ln \Gamma_k^{(i)} \right] \tag{D.3}$$

where v_{ki} is the total number of atoms different from hydrogen in all the groups of type k of molecule i. Only for three groups is this counting done differently:

1. The value of v_{ki} is considered to be 1.6 for water.
2. The group saturated methine (CH) is considered to be 0.8 methylene (CH_2) groups.
3. The group quaternary carbon (C) is considered to be 0.5 methylene (CH_2) groups.

The methyl (CH_3) and the methylene (CH_2) groups are considered to be the same group.

The group activity coefficients Γ_k are given by the Wilson equation, namely,

$$\ln \Gamma_k = 1 - \ln \sum_l X_l A_{kl} - \sum_l \frac{X_l A_{lk}}{\sum_m X_m A_{lm}} \tag{D.4}$$

and

$$\ln \Gamma_k^{(i)} = 1 - \ln \sum_l X_l^{(i)} A_{kl} - \sum_l \frac{X_l^{(i)} A_{lk}}{\sum_m X_m^{(i)} A_{lm}} \tag{D.5}$$

In the liquid mixture, the (mole) fraction of group k is given by

$$X_k = \frac{\sum_j x_j v_{kj}}{\sum_l \sum_m x_m v_{lm}} \tag{D.6}$$

and in pure compound i,

$$X_k^{(i)} = \frac{v_{ki}}{\sum_l v_{li}} \tag{D.7}$$

In both cases,

$$A_{kl} = \exp\left(m_{kl} + \frac{n_{kl}}{T} \right) \tag{D.8}$$

HOW TO READ PARAMETERS IN TABLE D.1

Parameters m_{kl} and n_{kl} (K) for 341 group pairs formed by 43 groups were reported by K. Tochigi et al. [4], and an extension and revision was presented by Tochigi and Gmehling [5]. These values are reproduced by permission here in the rearranged tables. Due to the format used in the original tables of parameters, and also used here in Table D.1, it is necessary to pay special attention to the order of reading the values of the parameters for groups k and l. For the m_{kl} and the n_{kl} parameters, the *leading* group k is found down in the column of the right-hand side of the table and the *secondary* group l is found in the heading (upper) line of the table. For interaction of the groups OH and Cl, for example, the parameters are

$$m_{OH/Cl} = -1.7798 \text{ and } n_{OH/Cl} = 0.1; \; m_{Cl/OH} = -0.6453 \text{ and } n_{Cl/OH} = 0.2$$

We have kept the group number assigned to each of the groups in the original publication of ASOG-KT [4], which included the first 31 groups, and also in the extension that followed [5]. We have reorganized the table of parameters and presented it in a compact form by dividing it into separate sections. In Tables D.2 through D.7, we give the parameters for those groups that are not in included in Table D.1. This was done with the purpose of avoiding an excessive number of blank spaces. In addition, in Table D.2 we have indicated clearly what the groups k and l are and listed separately the values of m_{kl} and n_{kl} from the values of the m_{lk} and n_{lk}. The nomenclature used for the new groups included in the extension of the method is reproduced (with permission) at the end of the rearranged tables of parameters. The nomenclature for the first 31 groups is quite simple, as it follows the one used for the homologous series in organic chemistry: CH_2 includes the groups CH_3, CH_2, CH, and C (with the different way of counting the number of carbon atoms indicated above); CO, COO, CHO, COOH, CN, and CON stand for the groups ketone, ester, aldehyde, carboxylic acid, nitrile, and amide, respectively. Formic acid is a group by itself: HCOOH. The group OH in an alcohol is differentiated from the group GOH in a polyalcohol. The groups NH_2, NH, and N denote primary, secondary, and tertiary amine groups, while NO_2 is the group in a nitro compound. The prefix *Ar* indicates a group that is part of an aromatic ring, like ArCH, or that is linked to an aromatic ring, like in ArCl, ArF, $ArNH_2$, $ArNO_2$, or ArOH for aromatic chloride, fluoride, amine, nitro compound, or a phenol, respectively. The prefix *Cy* indicates a group in a cyclo compound, while CCl_4 and CS_2, carbon tetrachloride and carbon disulfide, are groups by themselves. The groups CCl_2 and CCl_3 are carbon dichloride and carbon trichloride.

HEXANE–METHANOL SYSTEM

As our mind works better with numbers than with symbols, the easiest way to use ASOG is to assign numbers to the groups, write all equations, and only in the final step go back to replace the group numbers by their symbols.

Let us consider the prediction of vapor–liquid equilibrium at 60°C for the system hexane (1)–methanol (2). To do this, we need only two groups, (CH_2) and (OH). Hexane has six (CH_2) groups as the methyl (CH_3) and the methylene (CH_2) groups are not distinguished by the ASOG method. Methanol has one (CH_2) and one (OH) group. Let us denote (CH_2) as group 1 and (OH) as group 2. Example D.1 shows in detail how to use Equations D.1 through D.8 for this system.

The results of these calculations are presented in Figure D.1.

Example D.1: Step-by-Step Calculation for the Hexane (1)–Methanol (2) System

Part I. Calculations for Compounds in the Mixture

Hexane ($i = 1$) at Mole Fraction x_1

$r_1 = 6, k = 1(CH_2), v_{11} = 6, v_{21} = 0$

Methanol ($i = 2$) at Mole Fraction x_2

$r_2 = 2, k = 1(CH_2), v_{12} = 1, k = 2(OH), v_{22} = 1$

$$\sum_k x_k r_k = 6x_1 + 2x_2; \text{ then, using Equation D.2}$$

$$(\ln \gamma_1)_{combinatorial} = \left(1 - \frac{6}{6x_1 + 2x_2} + \ln \frac{6}{6x_1 + 2x_2}\right)$$

$$(\ln \gamma_2)_{combinatorial} = \left(1 - \frac{2}{6x_1 + 2x_2} + \ln \frac{2}{6x_1 + 2x_2}\right)$$

Using Equation D.3

$$(\ln \gamma_1)_{interactions} = 6(\ln \Gamma_1 - \ln \Gamma_1^{(1)})$$

$$(\ln \gamma_2)_{interactions} = (\ln \Gamma_1 - \ln \Gamma_1^{(2)}) + (\ln \Gamma_2 - \ln \Gamma_2^{(2)})$$

where $\Gamma_1^{(1)}$ is Γ_1 calculated for $x_1 = 1$

where $\Gamma_k^{(2)}$ is Γ_k calculated for $x_2 = 1$

Part II. Calculations of X_k and A_{kl} for the Groups Constituting the system, Equations D.4 and D.6

(CH_2) Group, $k = 1$

$$X_1 = \frac{6x_1 + x_2}{6x_1 + 2x_2}$$

$$\ln \Gamma_1 = 1 - \ln(X_2 A_{12} + X_1) - \left(\frac{X_1}{X_1 + X_2 A_{12}} + \frac{X_2 A_{21}}{X_1 A_{21} + X_2}\right)$$

(OH) Group, $k = 2$

$$X_2 = \frac{x_2}{6x_1 + 2x_2}$$

$$\ln \Gamma_2 = 1 - \ln(X_1 A_{21} + X_2) - \left(\frac{X_1 A_{12}}{X_1 + X_2 A_{12}} + \frac{X_2}{X_1 A_{21} + X_2}\right)$$

$$A_{kl} = \exp\left(m_{kl} + \frac{n_{kl}}{T}\right) \quad (D.8)$$

A_{12}		A_{21}	
m_{12}	n_{12} [K]	m_{21}	n_{21} [K]
−41.2503	7686.4	4.7125	−3060

Note: Values of parameters are taken from Table D.1.

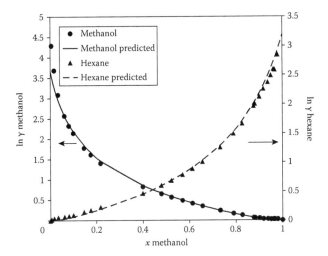

FIGURE D.1 Comparison of values of the activity coefficients predicted by the ASOG method with the experimental data [6] for the hexane (1)–methanol (2) system at 60°C. (Experimental data from Koennecke, H. G., *J. Prakt. Chem.*, 311, 974–982, 1969.)

TABLES OF PARAMETERS

Tables D.1 through D.7 contain the ASOG parameters for the temperature range 298.15 K–423.15 K. Explanation of symbols: NMP, C_5H_9NO; furfural, $C_5H_4O_2$; DMSO, $(CH_3)_2SO$; ACRY, $CH_2=CH_2$ group; GOH, OH group in glycols.

TABLE D.1

ASOG Group Interaction Parameters (298.15 K–423.15 K)

k	1. CH_2 m	n	2. C=C m	n	3. ArCH m	n	4. CyCH m	n	5. H_2O m	n	6. OH m	n	9. CO m	n
1. CH_2	0.0	0.0	0.7767	0.0	-0.7457	146.0	0.1530	2.1	-0.2727	277.3	-41.2503	7686.4	2.6172	-865.1
2. C=C	-0.4816	-58.9	0.0744	1.2487	-0.0622	-140.0	-1.0732	263.4	0.8390	-331.0	-4.1886	566.7	-1.093	367.8
3. ArCH	0.7297	-176.8	-9.5958	-0.4867	0.0	0.0	-0.3288	156.3	-0.0649	-252.4	2.2682	-1111.5	0.9273	-185.8
4. CyCH	-0.1842	0.3	-94.4	0.0	0.5301	-251.0	0.0	0.0	n.a.	n.a.	-11.9939	-2231.6	0.8476	-281
5. H_2O	0.5045	-2382.3	88.8	-347.0	-2.1939	150.3	n.a.	n.a.	0.0	0.0	1.4318	-280.2	0.0585	-278.8
6. OH	4.7125	-3060	498.6	-751.8	-0.5859	-939.1	n.a.	n.a.	-5.8341	1582.5	0.0	0.0	-0.7262	2.9
9. CO	-1.7588	169.6	2.8184	-1212.8	-0.4021	-216.8	0.0319	-350.3	0.3198	-91.2	0.3283	1.3	0	0
10. O	0.7666	-444.0	-0.8602	476.9	-2.4476	562.6	0.1546	0.2	-3.2419	1037.9	-1.2619	380.7	0.265	0.2
11. CHO	-1.1266	0.2	-4.9564	1355.3	-0.5546	-0.1	-1.2628	0.1	-5.0228	1562.0	0.9824	-0.1	0.3871	72.9
12. COO	-0.3699	162.6	-0.1323	114.2	-0.1541	97.5	-0.0991	2.4	-2.5548	659.9	-0.0296	2.6	-0.1212	180
16. CN	1.2569	-990.6	-1.4116	-48.1	-0.1163	-379.9	-4.509	478.7	-0.2016	-85.6	0.6616	-230.8	-0.2673	0
19. N	0.1993	2.8	0.1398	72.4	2.324	-733	-0.4799	0.2	1.9691	-304.6	2.6807	-115.6	0.8531	0.6
21. NO_2	-1.4089	228.5	-1.0057	141.4	-0.1225	-161.6	-0.86	0.1	0.7062	-341.5	2.8755	-916.7	0.0617	1.9
23. Cl	-1.2497	0.3	2.2753	-690.8	-0.7402	-0.2	-0.797	0	n.a.	n.a.	-0.6453	0.2	-0.7939	0
24. CCl_2	0.2849	-151.1	-1.0081	379	-0.5189	0	-0.0744	0.1	0.043	-335.3	-0.6986	0.2	0.1845	0.1
25. CCl_3	-0.1134	41.1	0.4263	-1.2	0.2511	1	-0.2819	3.9	-3.2238	670.1	-2.2978	605.7	0.3823	3.1
26. CCl_4	0.6926	-358.5	-3.5607	1161.7	0.8304	-374.5	0.0193	-78.1	3.7046	-1414.1	-9.7985	2539	0.8583	-252.5
27. ArCl	3.1729	-520.4	-7.268	53.7	2.4031	-195.3	0.8643	243.5	n.a.	n.a.	1.4497	0.1	1.5244	0
28. ArF	-0.5051	-0.1	n.a.	n.a.	-0.7661	-111.1	-0.3937	0	n.a.	n.a.	-0.9964	0.5	n.a.	n.a.
29. Br	-7.0135	1842.7	0.5317	-419	-3.7012	1050.8	-0.6985	0.2	n.a.	n.a.	0.7523	-589.5	0.1539	-280.2
30. I	-2.0131	2.8	n.a.	n.a.	-1.8259	0.1	n.a.	n.a.	n.a.	n.a.	-0.8586	-421.2	-0.8678	-0.1
31. CS_2	-0.0033	6	-0.1252	2.3	-1.3705	279.3	0.3599	-178.5	-2.2182	15.4	0.9985	-4056.7	-0.1141	-88.2

(Continued)

TABLE D.1 (Continued)

ASOG Group Interaction Parameters (298.15 K–423.15 K)

l / k	10. O m	10. O n	11. CHO m	11. CHO n	12. COO m	12. COO n	16. CN m	16. CN n	19. N m	19. N n	21. NO₂ m	21. NO₂ n	23. Cl m	23. Cl n
1. CH₂	-1.3836	606.4	0.1147	0.1	-15.2623	515	-0.0786	10.4	-14.6389	5.9	-0.3228	-70.8	0.2909	0.3
2. C=C	-0.0489	-407.5	3.358	-1057.2	-2.4963	-31.6	0.772	-186.5	-4.9244	-14.9	0.7995	-355.5	-1.7799	218.2
3. ArCH	-0.4061	370.9	0.1448	0.1	-0.5812	-249.3	-0.2142	176.4	-3.2954	728.4	-0.407	131.4	0.391	0.2
4. CyCH	-0.4055	0.1	0.1354	0.1	-2.0465	10.5	1.4239	-278.4	-1.3776	0.1	-0.4029	0.2	0.0409	0.1
5. H₂O	-0.3108	369.2	7.98	-2720.9	-2.4686	565.7	-0.3984	-107.2	-3.8611	1497.3	-2.1995	162.5	n.a.	n.a.
6. OH	0.4251	-474.9	-1.7642	0.1	0.0583	-455.3	-1.9168	166.7	-2.1231	231.6	0.4399	-821.2	-1.7798	0.1
9. CO	-0.266	-0.1	-1.4943	176.3	-2.5152	489.5	0.3168	0	-3.4855	-0.6	0.0734	0.8	0.5934	-0.1
10. O	0	0	-0.4918	0	-7.8816	-0.3	n.a.	n.a.	n.a.	n.a.	-0.3103	0	-0.9684	0.3
11. CHO	0.4562	0.1	0	0	-1.1887	0	n.a.	n.a.	n.a.	n.a.	n.a.	n.a.	n.a.	0.1527
12. COO	1.005	0.7	0.5393	0.1	0	0	0.0960	0.0	-4.3706	1400	-0.1814	0	0.8513	0
16. CN	n.a.	n.a.	n.a.	n.a.	-0.5146	0	0	0	0	0	-0.7931	-0.6	0.7529	0
19. N	n.a.	n.a.	n.a.	n.a.	n.a.	n.a.	-0.3426	-301.5	0	0	n.a.	n.a.	-2.357	368.5
21. NO₂	0.5306	0.1	n.a.	n.a.	0.0238	0.1	0.5731	0.2	1.4946	-471.8	0	0	2.3999	-659.6
23. Cl	0.5451	0.3	-0.2703	1.1	-2.3239	0.1	-0.7719	0.1	-1.1286	0.8	-0.9281	111.5	0	0
24. CCl₂	0.2922	-0.1	n.a.	n.a.	0.9944	0.2	-1.6043	0.4	0.0056	-0.2	n.a.	n.a.	0.5065	0.2
25. CCl3	0.7693	-0.1	n.a.	n.a.	-16.8658	-3637.1	0.0081	0.1	-3.0713	503.1	n.a.	n.a.	0.4207	0
26. CCl₄	0.4234	-9	n.a.	n.a.	-2.8851	-27.1	n.a.	n.a.	1.6708	-0.2	-1.0435	229.9	0.0817	0.1
27. ArCl	-11.339	137	n.a.	n.a.	1.9759	2.1	1.0743	0.1	-0.5503	11.6	2.5829	-236.2	1.9183	0.2
28. ArF	-4.3765	-0.2	n.a.	n.a.	n.a.	n.a.	n.a.	n.a.	n.a.	n.a.	n.a.	n.a.	n.a.	n.a.
29. Br	n.a.	n.a.	n.a.	n.a.	0.6142	-280.1	n.a.	n.a.	n.a.	n.a.	0.1008	2.3	0.7114	0
30. I	0.2383	0	n.a.	n.a.	-1.6441	0.0	n.a.	n.a.	n.a.	n.a.	-1.4827	-0.1	n.a.	n.a.
31. CS₂	0.0461	0.1	n.a.	n.a.	-1.5267	0.1	-0.5953	0.2	n.a.	n.a.	-2.0453	344.4	0.3352	0

(Continued)

TABLE D.1 (Continued)
ASOG Group Interaction Parameters (298.15 K–423.15 K)

l / k	24. CCl$_2$ m	n	25. CCl$_3$ m	n	26. CCl$_4$ m	n	27. ArCl m	n	28. ArF m	n	29. Br m	n	30. I m	n	31. CS$_2$ m	n
1. CH$_2$	0.2805	−96.9	0.2352	−119.8	−0.3917	227.9	−0.0904	−414.4	0.3653	0.1	−10.7665	−1628.9	0.3598	−1.1	−0.3104	11.8
2. C=C	2.2117	−769.4	−0.5531	8.2	4.1529	−1370.5	−0.0676	−86.3	n.a.	n.a.	−5.4517	453.5	n.a.	n.a.	−0.0706	1.1
3. ArCH	0.439	0	−0.2599	1.9	−0.5769	270.4	0.1546	−469.3	0.3856	98.6	−9.6405	502.6	0.5627	0.1	−0.2253	128.7
4. CyCH	−0.0065	−0.1	0.139	1.9	0.0103	55.07	4.1582	−1746.2	0.1225	0.1	−2.4054	0.3	n.a.	n.a.	−0.2922	114
5. H$_2$O	−4.0265	577.5	−3.4302	−811.2	−11.816	215.5	n.a.	n.a.	n.a.	n.a.	n.a.	n.a.	0.5339	−836.5	−1.8482	12.8
6. OH	−2.4298	0.2	2.804	−1898.7	5.9993	−3241	−3.5499	0.1	−7.3618	0.5	1.67	−1217.7	−0.0874	0.1	5.3401	−2977
9. CO	−0.185	−0.1	−1.118	−101.3	0.6643	−536.3	−1.1822	0.6	n.a.	n.a.	−4.9986	1582.6	−1.3801	0	−0.0999	−247.9
10. O	0.688	0.1	0.2244	0.1	−2.2383	631	0.7735	−121.3	0.5774	−0.6	n.a.	n.a.	n.a.	n.a.	−0.3254	0
11. CHO	36.2	n.a.	n.a.	n.a.	n.a.	n.a.	n.a.	n.a.	n.a.	n.a.	n.a.	n.a.	n.a.	n.a.	n.a.	n.a.
12. COO	0	0.3375	0.1	−0.4238	−0.5166	253.3	−4.8012	−0.4	n.a.	n.a.	−5.4226	1582.6	−0.0622	0	−0.2434	0.1
16. CN	0.4827	−1.3	−0.4294	0	n.a.	n.a.	−1.8328	0.1	n.a.	n.a.	n.a.	n.a.	n.a.	n.a.	−1.2178	0.1
19. N	1.2659	−0.4	1 6569	−0.2	−0.7196	346.8	1.3450	−0.1	−2.1126	10.8	−0.792	2.1	n.a.	n.a.	n.a.	n.a.
21. NO$_2$	n.a.	n.a.	n.a.	n.a.	−0.436	−79.1	−4.7073	15.9	n.a.	n.a.	−6.0004	1.1	0.3076	0.2	−1.5109	244.8
23. Cl	−0.7878	−0.3	−0.7114	0.1	−0.6544	0.1	−5.5731	0.4	n.a.	n.a.	n.a.	n.a.	n.a.	n.a.	−0.9223	0
24. CCl$_2$	0	0	0.1852	57.6	−0.4902	92.2	−62.946	723.4	n.a.	n.a.	−0.2827	4.2	−0.073	0.1	n.a.	n.a.
25. CCl$_3$	−0.535	24.2	0	0	−0.0063	−43.9	n.a.	n.a.	n.a.	n.a.	−2.5136	0.2	−0.4744	130.2	−0.2393	138.5
26. CCl$_4$	0.073	2.1	0.0195	24.1	0	0	−1.4077	225.8	n.a.	n.a.	1.2242	0.8	0.308	0.1	0.7329	−255.8
27. ArCl	3.8955	−854.7	n.a.	n.a.	2.7659	−478	0	0	0.119	0.1	n.a.	n.a.	n.a.	n.a.	n.a.	n.a.
28. ArF	n.a.	n.a.	n.a.	n.a.	−0.2123	0	n.a.	n.a.	0	0	n.a.	n.a.	n.a.	n.a.	n.a.	n.a.
29. Br	n.a.	n.a.	−0.9142	●10.0	−0.3827	0.2	0.1938	−0.5	n.a.	n.a.	0	0	0.3107	74.6	n.a.	n.a.
30. I	−0.2628	0	−2.7033	694.7	−1.2747	0.3	n.a.	n.a.	n.a.	n.a.	0.9094	−441.6	0	0	n.a.	n.a.
31. CS$_2$	n.a.	n.a.	−1.3831	269.1	−0.7212	218.8	n.a.	n.a.	n.a.	n.a.	n.a.	n.a.	n.a.	n.a.	0	0

Source: Reproduced and modified from Tochigi, K., et al., *J. Chem. Eng. J.*, 23, 453–463, 1990; Tochigi, K., and Gmehling, J., *J. Chem. Eng. J.*, 44, 304–306, 2011. With permission of the Society of Chemical Engineers, Japan. Copyright 2015 Society of Chemical Engineers, Japan.

TABLE D.2
ASOG Group Interaction Parameters for Groups 7, 8, and 13–15 (298.15 K–423.15 K)

l		7 ArOH		8 GOH		13 COOH		14 HCOOH		15 CON	
k		m	n	m	n	m	n	m	n	m	n
1. CH$_2$	k,l	-9.5152	0.7	-7.1548	64.0	9.7236	-3797.5	0.2365	-95.3	1.0067	-378
	l,k	-3.8090	0.5	-17.925	102.0	-10.9719	4022.0	-0.0721	-264.8	-1.3137	-103.2
2. C=C	k,l	11.708	-4898.9	n.a.	n.a	n.a.	n.a.	n.a.	n.a.	0.9654	-323.4
	l,k	-6.4189	1634.6							-1.4194	116.2
3. ArCH	k,l	-5.8576	1.0	-7.6896	-7.5	1.4405	-429.9	0.1022	12.0	-1.1344	162.6
	l,k	-2.6414	0.6	-0.9602	5.6	-0.2256	-213.7	-0.3000	-232.2	0.5928	-252.7
4. CyCH	k,l	-14.4465	317.8	n.a.	n.a.	n.a.	n.a.	n.a.	n.a.	0.5728	-236.6
	l,k	1.2180	-1955.2							-0.1198	-397.7
5. H$_2$O	k,l	0.7186	-320.2	-7.9975	2397.4	-0.4492	7.4	-0.6400	423.3	-08550	513.7
	l,k	0.8632	-357.2	2.2157	-450.2	-2.1113	797.7	1.5229	-921.5	-1.2225	159.3
6. OH	k,l	-0.2115	-0.2	0.8185	-1318.4	3.8786	-1712.0	n.a.	n.a.	0.1081	-23.7
	l,k	0.9971	0.1	0.8906	15.5	1.7000	-664.5			0.3059	-58.8
9. CO	k,l	-12.348	363.0	n.a.	n.a.	1.0434	-626.0	n.a.	n.a.	-0.6028	-133.4
	l,k	-4.3851	2209.5			1.8864	-543.0			-0.7286	388.4
10. O	k,l	n.a.	n.a.	-4.2085	2478.6	3.9356	41.4	n.a.	n.a.	n.a.	n.a.
	l,k			-5.3133	1673.0	-4.8241	1433.0				

(Continued)

TABLE D.2 (Continued)

ASOG Group Interaction Parameters for Groups 7, 8, and 13–15 (298.15 K–423.15 K)

l	k	7 ArOH		8 GOH		13 COOH		14 HCOOH		15 CON	
		m	n	m	n	m	n	m	n	m	n
20. $ArNH_3$	k, l	0.4097	0.0	n.a.	n.a.	n.a.	n.a.	n.a.	n.a.	n.a.	n.a.
	l, k	0.1214	0.0								
23. Cl	k, l	−24.647	188.4	n.a.	n.a.	n.a.	n.a.	0.0373	−274.0	n.a.	n.a.
	l, k	−46.055	201.3					−4.0112	915.5		
25. CCl_3	k, l	n.a.	n.a.	n.a	n.a.	−0.3039	3.7	−0.1506	3.3	n.a.	n.a.
	l, k					−0.6070	168.3	−0.5650	2.1		
26. CCl_4	k, l	−36.680	1301.4	n.a.	n.a.	5.2636	−2014.2	n.a.	n.a.	0.2965	4.9
	l, k	−0.9416	−1006.5			−5.0606	1679.3			−1.9351	5.0

Source: Reproduced and modified from Tochigi, K., et al., *J. Chem. Eng. J.*, 23, 453–463, 1990; Tochigi, K., and Gmehling, J., *J. Chem. Eng. J.*, 44, 304–306, 2011. With permission of the Society of Chemical Engineers, Japan. Copyright 2015 Society of Chemical Engineers, Japan.

TABLE D.3
ASOG Group Interaction Parameters for Groups 17, 18, 20, and 22
(298.15 K–423.15 K)

l		17 NH$_2$		18 NH		20 Ar NH$_2$		22 Ar NO$_2$	
k		*m*	*n*	*m*	*n*	*m*	*n*	*m*	*n*
1. CH$_2$	k, l	0.6435	−159.8	−12.3803	−785.1	−9.2417	1.2	−3.0243	854.1
	l, k	1.1005	−346.7	0.2778	−274.9	−12.764	1.8	4.4726	−1571.0
2. C=C	k, l	−0.0924	5.0	−1.8432	7.2	−4.3664	10.5	−0.7495	7.2
	l, k	−1.3899	4.6	−0.5681	12.0	−9.8620	55.7	0.031	−4.4
3. ArCH	k, l	0.9525	−268.4	−2.3433	454.7	−2.3077	−1799.9	3.7681	−985.1
	l, k	−0.6231	−183.8	−0.5387	34.7	0.3858	−832.8	−4.857	1152.7
4. CyCH	k, l	−0.1458	165.3	−2.4289	0.3	n.a.	n.a.	−70.4174	619.0
	l, k	−2.1740	−212.2	−0.7841	0.3			−4.4362	1354.8
5. H$_2$O	k, l	4.2174	−1082.2	−0.5958	489.1	n.a.	n.a.	n.a.	n.a.
	l, k	4.4468	−1847.2	−2.6892	919.5				
6. OH	k, l	0.7008	0.0	−3.5886	1484.0	−0.9734	0.1	−7.9764	0.2
	l, k	0.3741	−0.1	5.9417	−1834.8	−0.2167	0.1	0.4607	0.6
9. CO	k, l	n.a.	n.a.	n.a.	n.a.	−5.0439	−0.7	n.a.	n.a.
	l, k					0.0578	1.6		
12. COO	k, l	n.a.	n.a.	0.6579	0.0	−5.1419	0.8	n.a.	n.a.
	l, k			−1.2709	0.0	−0.4415	1.2		
17. NH$_2$	k, l	0.0	0.0	−0.8988	0.0	n.a.	n.a.	n.a.	n.a.
	l, k			0.6143	0.1				
18. NH	k, l	0.6143	0.1	0.0	0.0	n.a.	n.a.	n.a.	n.a.
	l, k	−0.8988	0.0						
19. N	k, l	n.a.	n.a.	−12.4922	231.6	n.a.	n.a.	n.a.	n.a.
	l, k			2.6204	−769				
20. ArNH$_3$	k, l	n.a.	n.a.	n.a.	n.a.	0.0	0.0	−1.5563	2.1
	l, k							0.0849	−0.4
21. NO$_2$	k, l	n.a.	n.a.	n.a.	n.a.	n.a.	n.a.	0.5023	0.0
	l, k							−1.3655	0.3
22. ArNO$_2$	k, l	n.a.	n.a.	n.a.	n.a.	0.0849	−0.4	0.0	0.0
	l, k					−1.5563	2.1		
23. Cl	k, l	n.a.	n.a.	n.a.	n.a.	−9.3156	1.1	n.a.	n.a.
	l, k					−2.6502	0.8.		
26. CCl$_4$	k, l	1.4075	−305.3	−6.9598	1261.6	−16.841	129.9	0.0442	5.0
	l, k	−1.6955	−34.3	1.5993	−511.7	−37.843	220.3	−0.7072	1.6
27. ArCl	k, l	0.0936	4.7	−34.169	13.5	n.a.	n.a.	n.a.	n.a.
	l, k	−0.8748	3.2	−1.6453	548.5				

Source: Reproduced and modified from Tochigi, K., et al., *J. Chem. Eng. J.*, 23, 453–463, 1990; Tochigi, K., and Gmehling, J., *J. Chem. Eng. J*, 44, 304–306, 2011. With permission of the Society of Chemical Engineers, Japan. Copyright 2015 Society of Chemical Engineers, Japan.

TABLE D.4
ASOG Group Interaction Parameters for Groups 32–35 (298.15 K–423.15 K)

l		32 Pyridine		33 Furfural		34 ACRY		35 Cl(C=C)	
k		m	n	m	n	m	n	m	n
1. CH_2	k, l	0.3012	−137.6	−0.7638	114.8	0.5607	−4.6	1.1392	−171.2
	l, k	0.0106	−39.0	0.6961	−241.2	−1.4541	9.1	−1.9268	227.5
2. C=C	k, l	−0.0156	2.4	−0.1147	−47.1	n.a.	n.a.	0.9875	−20.9
	l, k	−0.1709	2.2	−0.0993	16.7			−2.4412	31.8
3. ArCH	k, l	0.2311	−5.9	−0.6636	78.0	0.0965	89.6	0.1170	182.9
	l, k	−0.2436	−12.3	0.5358	−91.7	−0.5230	−41.6	−0.2341	−250.5
4. CyCH	k, l	0.6196	−210.1	n.a.	n.a.	n.a.	n.a.	0.9884	−129.7
	l, k	−0.4535	83.6					−1.9390	327.9
5. H_2O	k, l	−0.6486	5.4	−0.6920	−262.3	−0.8319	1.8	n.a.	n.a.
	l, k	−0.5628	3.0	−0.1105	−80.2	−0.8244	7.7		
6. OH	k, l	−3.5249	862.8	−2.8028	241.5	−2.1018	9.8	−9.3298	56.5
	l, k	1.2271	−704.9	2.5170	−817.9	−0.5398	6.8	−1.7019	35.5
9. CO	k, l	−0.6859	8.7	−0.9171	144.5	n.a.	n.a.	0.7515	−140.8
	l, k	0.4533	−43.4	2.6700	−794.0			−0.4612	−73.3
12. COO	k, l	n.a.	n.a.	−0.4431	6.5	n.a.	n.a.	0.473	1.4
	l, k			0.0487	−6.4			−2.5822	4.0
16. CN	k, l	−0.5626	1.5	n.a.	n.a.	−0.2184	3.4	1.0854	−130.9
	l, k	0.0608	4.2			−0.1420	1.9	−11.477	156.3
23. Cl	k, l	n.a.	n.a.	n.a.	n.a.	n.a.	n.a.	−0.9831	611.2
	l, k							−81.178	611.17
24. CCl_2	k, l	0.0115	1.6	n.a.	n.a.	n.a.	n.a.	n.a.	n.a.
	l, k	0.0393	3.2						
25. CCl_3	k, l	−0.4599	3.3	n.a.	n.a.	n.a.	n.a.	0.5687	5.1
	l, k	0.5132	4.8					−1.0044	−4.8
26. CCl_4	k, l	0.2895	−0.6	0.207	4.4	0.2042	34.9	−0.0045	166.4
	l, k	−0.4234	−15.4	−0.6327	0.2	−0.7635	−11.1	−0.8124	48.1
31. CS_2	k, l	n.a.	n.a.	n.a.	n.a.	n.a.	n.a.	0.2582	2.4
	l, k							−0.5219	2.8

Source: Reproduced and modified from Tochigi, K., et al., *J. Chem. Eng. J.,* 23, 453–463, 1990; Tochigi, K., and Gmehling, J., *J. Chem. Eng. J.,* 44, 304–306, 2011. With permission of the Society of Chemical Engineers, Japan. Copyright 2015 Society of Chemical Engineers, Japan.

TABLE D.5
ASOG Group Interaction Parameters for Groups 36–39 (298.15 K–423.15 K)

l		36 DMSO		37 NMP		38 C≡C		39 SH	
k		*m*	*n*	*m*	*n*	*m*	*n*	*m*	*n*
1. CH₂	k, l	1.1476	−340.7	−0.4505	62.5	−1.3791	478.8	0.6493	−90.7
	l, k	−0.8839	−93.8	0.2854	−127.4	1.2210	−563.2	−1.0967	−203.0
2. C=C	k, l	0.3592	−95.8	−1.5635	365.1	−0.1650	3.8	n.a.	n.a.
	l, k	−0.4881	−41.6	2.1023	−639.4	−0.0150	1.5		
3. ArCH	k, l	0.4226	−59.9	1.0714	−230.2	0.2346	−46.8	0.2005	7.4
	l, k	−0.4264	−41.9	−1.5972	368,8	−0.5775	114.2	−0.9086	−2.7
4. CyCH	k, l	1.1713	−253.5	0.6173	−182.3	n.a.	n.a.	−2.6058	316.6
	l, k	−2.3215	−37.7	−0.3014	−34.0			2.2088	−821.8
5. H₂O	k, l	−0.0058	−181.9	−0.4469	9.7	n.a.	n.a.	n.a.	n.a.
	l, k	−0.3146	321.6	−1.4178	421.2				
6. OH	k, l	−1.2893	−185.6	0.5490	−407.2	0.6644	−797.6	−2.3877	12.3
	l, k	0.1689	275.2	−0.2521	56.3	1.1483	−895.2	−1.5491	11.9
7. ArOH	k, l	n.a.	n.a.	−0.4991	0.0	n.a.	n.a.	n.a.	n.a.
	l, k			1.221	0.0				
9. CO	k, l	0.3498	−17.2	n.a.	n.a.	−0.2678	1.9	0.0602	3.6
	l, k	−0.5942	31.2			0.0865	2.0	−0.5705	1.9
10. O	k, l	n.a.	n.a.	n.a.	n.a.	n.a.	n.a.	0.0751	1.1
	l, k							0.0055	1.8
11. CHO	k, l	n.a.	n.a.	1.5366	−1094.2	n.a.	n.a.	n.a.	n.a.
	l, k			−5.4464	1447.3				
12. COO	k, l	0.1131	3.4	n.a.	n.a.	n.a.	n.a.	n.a.	n.a.
	l, k	−0.6784	3.3						
16. CN	k, l	n.a.	n.a.	n.a.	n.a.	−0.783	280.9	0.0864	−0.3
	l, k					−0.3102	−109.4	−0.7913	3.0
17. NH₂	k, l	n.a.	n.a.	n.a.	n.a.	n.a.	n.a.	−0.3014	1.6
	l, k							0.1122	2.6
19. N	k, l	0.7100	1.6	n.a.	n.a.	n.a.	n.a.	n.a.	n.a.
	l, k	−0.3215	−1.5						
21. NO₂	k, l	n.a.	n.a.	n.a.	n.a.	0.1228	0.6	n.a.	n.a.
	l, k					−0.4273	−3.0		
24. CCl₂	k, l	0.4910	24.2	n.a.	n.a.	n.a.	n.a.	n.a.	n.a.
	l, k	−0.5936	5.8						
25. CCl₃	k, l	0.3529	154.0	−1.0846	−182.2	n.a.	n.a.	n.a.	n.a.
	l, k	−1.2394	119.5	0.5672	175.9				
26. CCl₄	k, l	0.2529	21.0	n.a.	n.a.	n.a.	n.a.	n.a.	n.a
	l, k	−0.8684	−16.5						
27. ArCl	k, l	n.a.	n.a.	1.711	16.5	n.a.	n.a.	n.a.	n.a.
	l, k			−0.8995	−18.9				
37. NMP	k, l	n.a.	n.a.	0.0	0.0	0.7622	−17.1	n.a.	n.a.
	l, k					−0.6922	−39.7		

(Continued)

TABLE D.5 *(Continued)*
ASOG Group Interaction Parameters for Groups 36–39 (298.15 K–423.15 K)

l		36 DMSO		37 NMP		38 C≡C		39 SH	
k		m	n	m	n	m	n	m	n
40. DMF	k, l	n.a.	n.a.	n.a.	n.a.	0.0021	3.7	2.8472	−805.0
	l, k					0.1352	1.2	0.3152	−219.6

Source: Reproduced and modified from Tochigi, K., et al., *J. Chem. Eng. J.*, 23, 453–463, 1990; Tochigi, K., and Gmehling, J., *J. Chem. Eng. J.*, 44, 304–306, 2011. With permission of the Society of Chemical Engineers, Japan. Copyright 2015 Society of Chemical Engineers, Japan.

TABLE D.6
ASOG Group Interaction Parameters for Groups 40–43 (298.15 K–423.15 K)

l		40 DMF		41 EDOH		42 DEG		43 Sulfolane	
k		m	n	m	n	m	n	m	n
1. CH₂	k, l	−0.1442	112.0	0.4982	−132.8	0.9532	−186.0	0.9906	−326.9
	l, k	1.5415	−794.2	−4.3182	121.3	−7.5046	63.1	−1.4478	247.4
2. C=C	k, l	−1.2730	452.7	n.a.	n.a.	−0.0696	−3.0	n.a.	n.a.
	l, k	2.2580	−862.2			−0.6872	13.0		
3. ArCH	k, l	1.629	−394.0	0.7659	−100.1	−0.3610	−5.7	−0.3259	−11.4
	l, k	−2.8986	724.5	−10.598	208.8	0.0648	8.8	−0.4483	−30.0
4. CyCH	k, l	−1.6170	511.4	n.a.	n.a.	n.a.	n.a.	n.a.	n.a.
	l, k	2.8862	−1057.6						
5. H₂O	k, l	0.2704	−235.2	−0.5995	141.1	n.a.	n.a.	n.a.	n.a.
	l, k	−1.4479	455.7	−0.3272	51.8				
6. OH	k, l	−0.9392	254.1	n.a.	n.a.	n.a.	n.a.	n.a.	n.a.
	l, k	−0.6461	−251.6						
9. CO	k, l	−0.2038	1.0	n.a.	n.a.	n.a.	n.a.	n.a.	n.a.
	l, k	0.0236	17.4						
21. NO₂	k, l	n.a.	n.a.	0.0240	3.3	n.a.	n.a.	n.a.	n.a.
	l, k			−1.0633	3.3				
23. Cl	k, l								
	l, k								
26. CCl₄	k, l	−0.2233	114.9	n.a.	n.a.	n.a.	n.a.	n.a.	n.a.
	l, k	−1.9153	163.4						
29. Br	k, l	1.7273	−332.3	n.a.	n.a.	n.a.	n.a.	n.a.	n.a.
	l, k								
38. C≡C	k, l	0.1352	1.2	n.a.	n.a.	n.a.	n.a.	n.a.	n.a.
	l, k	0.0021	3.7						
39. SH	k, l	0.3152	−219.6	n.a.	n.a.	n.a.	n.a.	n.a.	n.a.
	l, k	2.8472	−805.0						
40. DMF	k, l	0.0	0.0	−0.0851	1.8	n.a.	n.a.	n.a.	n.a.
	l, k			0.0321	2.1				

Source: Reproduced and modified from Tochigi, K., et al., *J. Chem. Eng. J.*, 23, 453–463, 1990; Tochigi, K., and Gmehling, J., *J. Chem. Eng. J.*, 44, 304–306, 2011. With permission of the Society of Chemical Engineers, Japan. Copyright 2015 Society of Chemical Engineers, Japan.

TABLE D.7
Values of v_{ki} and r_i for More Complex Groups

Group	General Name	Name	Formula	v_{ki}	r_i
32. pyridine	Pyridines	Pyridine	C_5H_5N	Pyridine: 6	6
		3-Methylpyridine	C_6H_7N	CH_2: 1, Pyridine: 6	7
33. furfural		Furfural	$C_5H_4O_2$	Fufural: 7	7
34. ACRY		Acrylonitrile	$CH_2 = CHCN$	ACRY: 4	4
35. Cl(C=C)	Chlorides (C=C)	Trichloroethylene	$Cl_2C = CHCl$	C=C: 2, Cl(C=C): 3	5
36. DMSO		Dimethylsulfoxide	$(CH_3)_2SO$	DMSO: 4	4
37. NMP		N-Methylpyrrolidone	C_5H_9NO	NMP: 7	7
38. C:C	Alkynes	1-Hexyne	$CH_3CH_2CH_2CH_2C\equiv CH$	CH_2: 4, C≡C: 2	6
		2-Hexyne	$CH_3CH_2CH_2C\equiv CCH_3$	CH_2: 4, C≡C: 2	6
39. SH	Thiols	Methanethiol	CH_3SH	CH_2: 1, SH: 1	2
		Ethanethiol	CH_3CH_2SH	CH_2: 2, SH: 1	3
40. DMF		Dimethylformamide	$(CH_3)_2NCHO$	DMF: 5	5
		Diethylformamide	$(CH_3CH_2)_2NCHO$	CH_2: 2, DMF: 5	7
41. EDOH		1,2-Ethanediol	$(CH_2OH)_2$	EDOH : 4	4
42. DEG		Diethyleneglycol	$HOCH_2CH_2OCH_2CH_2OH$	DEG: 7	7
43. Sulfolane		Sulfolane	$C_4H_8O_2S$	Sulfolane: 7	7

Source: Reproduced and modified from Tochigi, K., et al., *J. Chem. Eng. J.,* 23, 453–463, 1990; Tochigi, K., and Gmehling, J., *J. Chem. Eng. J.,* 44, 304–306, 2011. With permission of the Society of Chemical Engineers, Japan. Copyright 2015 Society of Chemical Engineers, Japan.

REFERENCES

1. Wilson, G. M., and Deal, C. H. 1962. Activity coefficients and molecular structure. *Ind. Eng. Chem.* 1: 20–23.
2. Kojima, K., and Tochigi, K. 1979. *Prediction of Vapor-Liquid Equilibria by the ASOG Method.* Tokyo: Kodansha-Elsevier.
3. Tochigi, K. 1998. Prediction of vapor–liquid equilibria in non-polymer and polymer solutions using an ASOG-based equation of state (PRASOG). *Fluid Phase Equilib.* 144: 59–68.
4. Tochigi, K., Tiegs, D., Gmehling, J., and Kojima, K. 1990. Determination of new ASOG parameters. *J. Chem. Eng. J.* 23: 453–463.
5. Tochigi, K., and Gmehling, J. 2011. Determination of ASOG parameters—Extension and revision. *J. Chem. Eng. J.* 44: 304–306.
6. Koennecke, H. G. 1969. Ergebnisse von Dampfflüssigkeits-Phasengleichgewichtsunter suchungen am System n-Hexan-Methylcyclohexan-Methanol bei 60°C. *J. Prakt. Chem.* 311: 974–982.

Index